Slope Stability Analysis by the Limit Equilibrium Method

Companion Title

LEAME Software and User's Manual: Analyzing Slope Stability by the Limit Equilibrium Method, **by Yang H. Huang, Sc.D., P.E.** (ASCE Press, 2014). Offers a PC-based software program for performing two- and three-dimensional slope stability analyses, accompanied by supporting documentation and worked examples. (ISBN 978-0-7844-7799-1; online at http://dx.doi.org/10.1061/9780784477991)

Other Titles of Interest

Geotechnical Testing, Observation, and Documentation, **Second Edition, by Tim Davis.** (ASCE Press, 2008). Presents an in-depth field manual for soil technicians and geotechnical engineers for use during the investigation, grading, and construction phases of geotechnical projects. (ISBN 978-0-7844-0949-7)

Slope Stability and Earth Retaining Walls, **edited by Hanlong Liu, Ph.D.; Robert Y. Liang, Ph.D., P.E.; Ke Yang, Ph.D., P.E.; and Jamal Nusairat, Ph.D., P.E.** (ASCE Geotechnical Special Publication No. 216, 2011). Addresses contemporary geotechnical subjects including slope stability evaluations; slope and landslide movement monitoring techniques; and recent developments of earth retaining structures. (Online only; http://dx.doi.org/10.1061/9780784476277)

Slope Stability, Retaining Walls, and Foundations: Selected Papers, **edited by Louis Ge, Ph.D., P.E.; Jinyuan Liu, Ph.D.; James-C. Ni, Ph.D., P.E.; and Zhaoyi He, Ph.D.** (ASCE Geotechnical Special Publication No. 197, 2009). Showcases recent developments and advancements in geotechnical, focusing soil stabilization, dynamic behavior of soils and foundations, earth retaining walls, and slope stability. (ISBN 978-0-7844-1049-3)

Slopes and Retaining Structures under Seismic and Static Conditions, **edited by Mohamed A. Gabr, Ph.D., P.E.; John J. Bowders, Ph.D., P.E.; David Elton, Ph.D., P.E.; and Jorge G. Zornberg, Ph.D., P.E.** (ASCE Geotechnical Special Publication No. 140, 2005). Examines the design of slopes and retaining structures that pose challenges to geotechnical engineers. (Online only; http://dx.doi.org/10.1061/9780784407875)

Slope Stability Analysis by the Limit Equilibrium Method

Yang H. Huang, Sc.D., P.E.

ASCE PRESS

Cataloging-in-Publication Data on file with the Library of Congress.

Huang, Yang H. (Yang Hsien), 1927–

Slope stability analysis by the limit equilibrium method : fundamentals and methods / Yang H. Huang, Sc.D., P.E.

pages cm

Includes bibliographical references and index.

ISBN 978-0-7844-1288-6 (print : alk. paper) – ISBN 978-0-7844-7800-4 (ebook) 1. Slopes (Soil mechanics) I. Title.

TA710.H785 2014

624.1′5136–dc23

2013032422

Published by American Society of Civil Engineers
1801 Alexander Bell Drive
Reston, Virginia, 20191-4382
www.asce.org/bookstore | ascelibrary.org

Errata: Errata, if any, can be found at http://dx.doi.org/10.1061/9780784412886.

ISBN 978-0-7844-1288-6 (paper)
ISBN 978-0-7844-7800-4 (PDF)
Manufactured in the United States of America.

Instructors: A PDF solution manual providing worked answers for the problems presented in this book is available at no charge. Contact ASCE by email (pubsful@asce.org) or telephone (1-800-548-2723, domestic U.S.; 1-703-296-6300, international). Ask for *Slope Stability Analysis by the Limit Equilibrium Method: Solution Manual*, ISBN 978-0-7844-7802-8, Stock No. 47802.

Contents

Preface

During the past 40 years, I have been engaged in a study on the stability analysis of earth slopes. The study was initiated in 1973 when I received a research grant from the Institute for Mining and Minerals Research, University of Kentucky. When the research project was completed in 1977, the U.S. Congress passed the Surface Mining Control and Reclamation Act, which requires stability analysis for refuse dams, hollow fills, and spoil banks created by surface mining, thus putting the research into practical use. The results of the research were published in several journals and reports and also were presented in a number of short courses. Both the simplified and the computerized methods of stability analysis, as developed from this research, have been widely used by practicing engineers throughout Kentucky for the application of mining permits. The large number of out-of-state participants in the short courses indicates that the methods developed have widespread applications.

In 1983, my book *Stability Analysis of Earth Slopes* was published by the Van Nostrand Reinhold Company. The book was well received by the engineering profession and was reprinted quite a few times. It was recommended by the Professional Civil Engineering Book Club as a feature selection and was translated into Chinese and Russian by foreign publishers. Two computer programs, one called SWASE (Sliding Wedge Analysis of Sidehill Embankments) for analyzing plane or noncircular failure surfaces, and the other called REAME (Rotational Equilibrium Analysis of Multilayered Embankments) for cylindrical failure surfaces, written in both Fortran and Basic languages, were listed in the book. In 1994, the SWASE program was incorporated into the REAME program, and a separate program for three-dimensional analysis, named REAME3D, was developed. In 1996, the first Windows version of REAME for both two- and three-dimensional analyses was released and used widely by the mining industries. The program has been continuously improved, and a new version has been released every four years, culminating in the latest REAME2012. Because the name REAME is a misnomer and the computer software associated with this book is very similar to REAME2012 with only some minor changes, the name REAME has been changed to LEAME (Limit Equilibrium Analysis of Multilayered Earthworks) in this book to reflect the capabilities of the software better.

Further evolutions of the book and software have prompted their separation into two separate products. *Slope Stability Analysis by the Limit Equilibrium Method: Fundamentals and Methods* presents the basic principles at work in the analysis of slope stability and common methods for analyzing slope stability in two and three dimensions. A companion product, *LEAME Software and User's Manual: Analyzing Slope Stability by the Limit Equilibrium Method* provides both the software program and the supporting documentation for its use. The software can be obtained at http://dx.doi.org/10.1061/9780784477991.

Although some of the materials presented in this book, such as the five chapters in Part 1 and Chapter 7 in Part 2, are essentially the same as the 1983 book, this revised and updated volume is dramatically different in the following aspects:

1. Many new sections have been added, such as the back-calculation of shear strength, undrained shear strength varying linearly with depth, granular materials with curved strength envelope, unsteady-state seepage, and external and internal forces.
2. Some new stability charts have been added and some others have been deleted, because they are too cumbersome for hand calculations. With the availability of the LEAME software, no one likes to resort to stability charts for preliminary analysis unless they are very simple to use.
3. Only the limit equilibrium method is covered here, and the section on finite element method is eliminated. Also, only the methods incorporated in LEAME are presented in detail, while the sections on Janbu's method and Morgenstern's and Price's method are eliminated. The section on the probabilistic method has been expanded greatly, and a new chapter on reliability is presented.
4. The three-dimensional analysis, which was not even mentioned in the previous book, is presented here in a full chapter. It covers the theoretical aspect by showing how the failure mass is divided into columns and deriving the equations used for LEAME.
5. Spreadsheets have been added to solve many of the examples, and the results are compared with the LEAME computer program. It is amazing that many problems involving iterations or using trial-and-error can be solved easily by spreadsheets. Although spreadsheets can be used to check the correctness of a computer program, they cannot serve as a substitute, because they involve only a single failure surface; to determine the minimum factor of safety, hundreds of failure surfaces need to be analyzed.
6. Homework problems and more examples have been added so the book can serve as a college text for senior and graduate courses in soil mechanics and geotechnical engineering.

This volume is divided into two parts. Part 1 presents the fundamentals of slope stability and consists of five chapters. Chapter 1 describes slope movements and discusses some of the more well-known methods for stability analysis. Chapter 2 explains the mechanics of slope failures and defines the factor of safety for both cylindrical and plane failures. Chapter 3 discusses both the laboratory and the field methods for determining the shear strength of soils used for stability analysis. Chapter 4 illustrates some methods for estimating the location of the phreatic surface and determining the pore pressure ratio. Chapter 5 outlines remedial measures for correcting slides.

Part 2 presents methods of stability analysis and also consists of five chapters. Chapter 6 derives some simple formulas for determining the factor of safety for plane failures. Chapter 7 presents a number of stability charts for determining the factor of safety for cylindrical failures, as well as the well-known friction circle and logarithmic spiral methods. Chapter 8 discusses methods of slices for two-dimensional analysis and derives the equations used in developing LEAME. Chapter 9 discusses the three-dimensional analysis with both ellipsoidal and planar ends and derives the equations used in LEAME. Chapter 10 discusses various methods to determine the reliability of slope design, including the use of Taylor's expansion for closed-form solutions and the mean-value first order second moment (MFOSM) method for computer solutions.

The Appendix provides an introduction to the LEAME software to encourage readers to obtain the software. The LEAME computer software is completely different from the REAME program listed in the 1983 book. It is an excellent and well-tested software program to determine the factors of safety for both two- and three-dimensional slopes and contains many new features not available elsewhere. It can be downloaded and used right away to solve practical problems in slope stability.

Finally, I want to thank ASCE Press for giving me the opportunity to publish these books. It is my sincere hope that the books, especially the LEAME computer software, will be helpful to civil engineers in their engineering practice and to college professors in teaching courses in slope stability.

Yang H. Huang, Sc.D., P.E.
Professor Emeritus of Civil Engineering
University of Kentucky

Part 1

Fundamentals of Slope Stability

Chapter 1

Introduction

Problems associated with failures of natural and artificial slopes often pose formidable challenges in geotechnical engineering. In general, an exposed inclined ground surface that is unrestrained may be prone to mass movement due to gravitational forces. The resulting shear stresses, induced along a potential or known failure surface, could exceed the shear strength of the soil and cause slope failure. The ratio of available shear strength to induced shear stress on a potential failure surface most commonly is referred to as the factor of safety. The intent of any limit equilibrium stability analysis is to determine this factor, by which the soil strength is divided or reduced, to bring the slope to the threshold of instability. The types of slope movements and the use of various limit equilibrium methods to determine the factor of safety for the sliding types of mass movements are reviewed in this chapter.

1.1 Slope Movements

The stability analysis of slopes plays a very important role in civil engineering. Stability analysis is used in the construction of transportation facilities such as highways, railroads, airports, and canals; the development of natural resources such as surface mining, refuse disposal, and earth dams; as well as many other human activities involving building construction and excavation. Failures of slopes in these applications may be caused by movements within the human-created cut or fill, in the natural slope, or a combination of both. These movement phenomena usually are studied from two different points of view. Geologists consider the movement phenomena as a natural process and study the crux of their origin, their courses, and the resulting surface characteristics. Engineers, however, investigate the safety of slopes based on the principles of soil mechanics and develop methods for a reliable assessment of the stability of slopes, as well as the controlling and corrective measures needed. The most viable stability studies can be achieved only by the combination of both approaches. The quantitative assessment of the stability of slopes by the methods of soil mechanics must be based on knowledge of the geological structure of the area, the detailed composition and orientation of strata, and the geomorphological history of the land surface. On the other hand, geologists may obtain a clearer picture of the origin and character of the movement process by checking their considerations against the results of engineering analyses based on soil mechanics. For example, it is well known that one of the most favorable settings for landslides is where permeable or soluble beds overlie or are interbedded with relatively impervious beds. This geological phenomenon was explained by Henkel (1967) using the principles of soil mechanics.

Slopes failures involve such a variety of processes and causative factors that they afford unlimited possibilities of classification. For instance, they can be divided according to the form of failures, the type of materials moved, the age, the stage of development, or the cause of movements.

One of the most comprehensive references on landslides or slope failures is a special report published by the Transportation Research Board (Turner and Schuster 1996). According to this report, the type of slope movements is divided into five main groups: falls, topples, slides, spreads, and flows (Varnes 1978). A sixth group, complex slope movements, includes the combination of two or more of these five types. The type of materials is divided into two classes: rock and soil. Soil is further divided into debris and earth. Table 1-1 shows the classification of slope movements.

Recognizing the types of slope movements is important because they dictate the method of stability analysis and the remedial measures to be employed. Varnes (1978) described the types of movements as follows:

- In falls, a mass of any size is detached from a steep slope or cliff, along a surface on which little or no shear displacement takes place, and descends mostly through the air by free fall, leaping, bouncing, or rolling. Movements are very rapid and may or may not be preceded by minor movements leading to progressive separation of the mass from its source.
- In topples, one or more units of mass rotate forward about some pivot point, below or low in the unit, under the action of gravity and forces exerted by adjacent units or by fluids in cracks. In fact, it is tilting without collapse.

Table 1-1. Classification of Slope Movements

TYPE OF MOVEMENT			TYPE OF MATERIAL		
				Engineering Soils	
			Bedrock	Predominantly Coarse	Predominantly Fine
Falls			Rock fall	Debris fall	Earth fall
Topples			Rock topple	Debris topple	Earth topple
Slides	Rotational	Few units	Rock slump	Debris slump	Earth slump
	Translational		Rock block slide	Debris block slide	Earth block slide
		Many units	Rock slide	Debris slide	Earth slide
Lateral spreads			Rock spread	Debris spread	Earth spread
Flows			Rock flow (deep creep)	Debris flow (soil creep)	Earth flow (soil creep)
Complex			combination of two or more principal types of movement		

(Varnes 1978, © National Academy of Sciences, Washington, DC. Reproduced with permission of the Transportation Research Board, Washington, DC)

- In slides, the movement consists of shear strain and displacement along one or several surfaces that are visible or may reasonably be inferred, or within a relatively narrow zone. The movement may be progressive; that is, shear failure may not initially occur simultaneously over what eventually becomes a defined surface of rupture, but, rather, it may propagate from an area of local failure. This displaced mass may slide beyond the original surface of rupture onto what had been the original ground surface, which then becomes a surface of separation. Slides are divided into rotational slides and translational slides. This distinction is important because it affects the methods of analysis and control. Furthermore, the presence of a weak sublayer or boundary between weathered and unweathered strata reveals another type of slide known as the compound slide (Cruden and Varnes 1996). These geological anomalies dictate the location of the surface of rupture (Hutchinson 1988).
- In spreads, the dominant form of movement is lateral extension accompanied by shear or tensile fractures. Movements may involve fracturing and extension of coherent material, either bedrock or soil, owing to liquefaction or plastic flow of subjacent material. The coherent upper units may subside, translate, rotate, or disintegrate, or they may liquefy and flow. The mechanism of failure can involve elements not only of rotation and translation but also of flows; hence, some lateral spreading failures may be regarded as complex. The sudden spreading of clay slopes was discussed by Terzaghi and Peck (1967).

- In flows, slope movements may appear in several different forms. In unconsolidated materials, they generally take the form of fairly obvious flows, fast or slow, wet or dry. In saturated soils, they are triggered when the ground becomes completely saturated by infiltration, as a result of intense rainfall or by the rise of the groundwater table. Thus, water is the primary transport agent and the saturated soils resembling a viscous fluid possess no strength and result in shallow slope failures. In bedrock, the movements are extremely slow and are distributed among many closely spaced, noninterconnected fractures that result in folding, bending, or bulging.

According to age, slope movements are divided into contemporary, dormant, and fossil movements. Contemporary movements are generally active and are relatively easily recognizable by their configuration, because the surface forms produced by the mass movements are expressive and not affected by rainwash and erosion. Dormant movements usually are covered by vegetation or are disturbed by erosion so that the traces of their last movements are not easily discernible. However, the causes of their origin still remain and the movement may be renewed. Fossil movements generally developed in the Pleistocene or earlier periods, under different morphological and climate conditions, and cannot repeat themselves at present.

According to stage, slope movements can be divided into initial, advanced, and exhausted movements. At the initial stage, the first signs of the disturbance of equilibrium appear and cracks in the upper part of the slope develop. In the advanced stage, the loosened mass is propelled into motion and slides downslope. In the exhausted stage, the accumulation of slide mass creates temporary equilibrium conditions.

Chowdhury (1980) classified slides according to their causes: (1) exceptional landslides arising from earthquake, intense precipitation, severe flooding, accelerated erosion from wave action, and liquefaction; (2) ordinary landslides resulting from known or usual causes that can be explained by traditional theories; and (3) unexplainable landslides that occur without any apparent cause.

It should be evident from this discussion that the stability of slopes is a complex problem, which may defy any theoretical analysis. In this book only the slide type of mass movements will be discussed, not only because it is more amenable to theoretical analysis but also because it is the predominant type of failures, particularly in human-created slopes.

1.2 Limit Plastic Equilibrium

The primary purpose of most stability analyses is to determine the factor of safety of the slope based on the concept of limit plastic equilibrium. First, a failure surface is assumed. A state of limit equilibrium is said to exist when the shear stress along the failure surface is expressed as

$$\tau = \frac{s}{F} \tag{1-1}$$

in which τ = shear stress, s = shear strength, and F = factor of safety. According to the Mohr-Coulomb failure theory, the shear strength can be expressed as

$$s = c + \sigma_n \tan\phi \tag{1-2}$$

in which c = cohesion, σ_n = normal stress, and ϕ = angle of internal friction. Both c and ϕ are known properties of the soil. Once the factor of safety is known, the shear stress along the failure surface can be determined from Eq. (1-1).

In the limit equilibrium method, only the concept of static equilibrium is applied. Unfortunately, except in the simplest cases, most problems in slope stability are statically indeterminate, because the number of available equations is not sufficient to solve the number of unknowns. In order to determine a unique factor of safety, some simplifying assumptions must be made to increase the number of equations and make it exactly equal to the number of unknowns. Some examples of statically determinate and statically indeterminate problems are discussed as follows.

1.2.1 Statically Determinate Problems

Two cases, one involving a plane failure and the other a cylindrical failure, will be discussed here.

Fig. 1-1 shows a fill on a sloping ground. The failure surface is assumed to be a plane at the bottom of the fill along the sloping ground. The weight of the fill is W, the force normal to the failure plane is N, and the shear force, T, along the failure plane can be expressed as

$$T = \frac{C + N\tan\phi}{F} \tag{1-3}$$

in which C = total cohesion resistance, which is equal to the unit cohesion, c, multiplied by the area of failure surface. There are a total of three unknowns: the

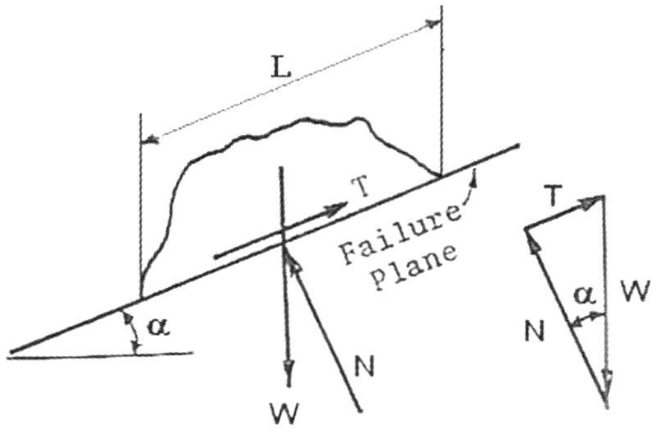

Fig. 1-1. Statically determinate plane failure

factor of safety, F, the magnitude of N, and the point of application of N. According to statics, there are also three equilibrium equations; specifically, the sum of forces in the normal direction is zero, the sum of forces in the tangential direction is zero, and the sum of moments about any given point is zero. The moment requirement implies that W, T, and N must intersect at the same point. Knowing the magnitude and direction of W and the direction of N and T, the magnitude of N and T can be determined from the force diagram shown in Fig. 1-1 and the factor of safety from Eq. (1-3). This case is statically determinate, because all forces applied to the failure mass, including their magnitude, direction, and point of application, can be determined by statics. To find the factor of safety algebraically, from Eq. (1-3),

$$F = \frac{cL + W\cos\alpha\tan\phi}{W\sin\alpha} \tag{1-4}$$

in which L = length of failure surface, which is also the area of failure surface (note that the width is equal to unity, thus the area and length are equivalent), and α = angle of the sloping ground. It can be seen from Eq. (1-4) that, for statically determinate problems, the factor of safety is defined as the ratio between the resisting force and the driving force, both applied on the failure surface.

Example 1.1 Fig. 1-2 shows a triangular fill placed on a sloping surface. It is assumed that the plane failure is along the slope surface, because a thin layer of weak material with a cohesion, c, of 250 psf (12 kN/m^2) and a friction angle, ϕ, of 15° exists at the bottom of the fill. If the fill has a unit weight, γ, of 125 pcf (19.7 kN/m^3), determine its factor of safety.

Solution The triangular fill has a height and a base of 50 ft so, assuming the width of fill as 1 ft, the weight of the fill $W = 0.5 \times 125 \times 50 \times 50 = 156{,}250$ lb, and the area of failure surface $L = [(150)^2 + (50)^2]^{0.5} = 158.1$ ft^2. The angle of slope $\alpha = \tan^{-1}(50/150) = 18.4°$. From Eq. (1-4),

$$F = \frac{250 \times 158.1 + 156{,}250\cos 18.4°\tan 15°}{156{,}250\sin 18.4°} = \frac{39{,}525 + 39{,}727}{49{,}320} = 1.607$$

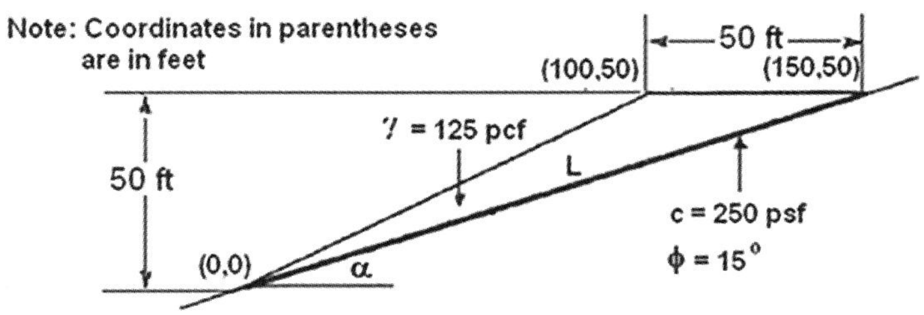

Fig. 1-2. Example 1.1

The second case involves a cylindrical surface cutting through a soil with $\phi = 0$, as shown in Fig. 1-3. When $\phi = 0$, the shear resistance along the failure arc is dependent on the cohesion only, independent of the normal force. By assuming that the cohesion, c, is distributed uniformly along the failure arc, it can be seen from Fig. 1-3(a) that the cohesion resistance along the failure surface can be resolved into two components, one parallel to the chord and the other perpendicular to the chord. The component parallel to the chord is in the same direction and can be added together, while that perpendicular to the chord is in opposite directions and neutralizes each other. Consequently, the resultant cohesion resistance is parallel to the chord with a magnitude of cL_c, where L_c is the chord length, and the shear force, T, is equal to

$$T = \frac{cL_c}{F} \tag{1-5}$$

The distance, d, from the center of circle to the shear force, T, can be obtained by taking moment about point O, i.e., $cL_c d = cL_a R$ or

$$d = \frac{RL_a}{L_c} \tag{1-6}$$

in which L_a = arc length. Given the exact location of weight, W, and shear force, T, their intersection O′ can be found graphically, as shown in Fig. 1-3(b). To satisfy the moment equilibrium, the normal force, N, must also pass through point O′. Because each of the normal forces distributed over the failure arc passes through point O, the resultant of all normal forces, N, must also pass through point O, so the direction of N can be determined graphically by connecting the two points O and O′. Knowing the magnitude and direction of W and the direction of T and N, the magnitude of T can be determined by the force diagram, as

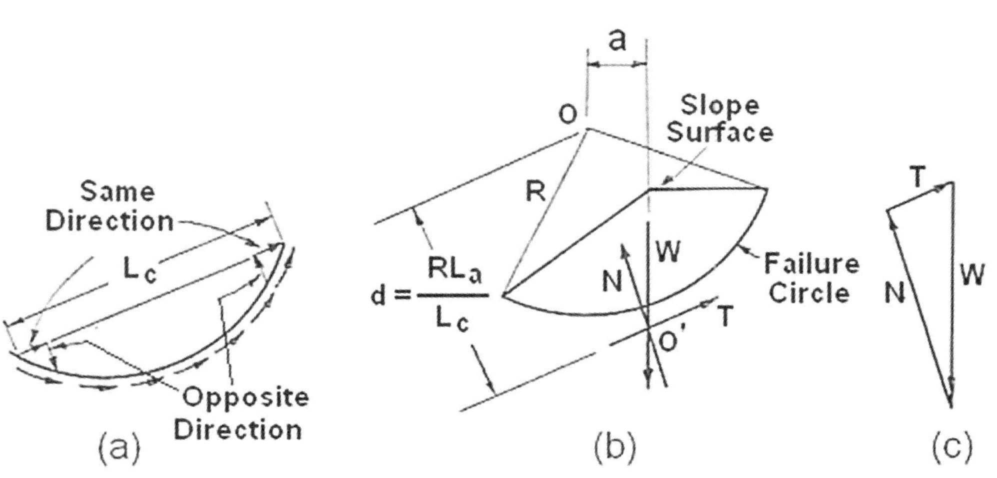

Fig. 1-3. Statically determinate cylindrical failure with $\phi = 0$

shown in Fig. 1-3(c), and the factor of safety obtained from Eq. (1-5). This case is statically determinate, because the magnitude, direction, and point of application of all the forces applied to the failure mass can be determined by statics alone.

To solve the factor of safety algebraically, it is more convenient to take the moment about point O:

$$Td = Wa \tag{1-7}$$

Substituting Eqs. (1-5) and (1-6) into Eq. (1-7), the following equation is obtained:

$$F = \frac{cL_aR}{Wa} \tag{1-8}$$

For statically determinate problems, Eq. (1-8) defines the factor of safety as the ratio between the resisting moment and the overturning moment. The difficulty in applying Eq. (1-8) is how to determine the driving moment, *Wa*. Fig. 1-4 illustrates the method for computing the area of failure mass and the overall driving moment by dividing the total area into several small areas, or

$$\text{Area ABCG} = \text{Area OABC} - \Delta\text{OFC} - \Delta\text{OEF} - \Delta\text{AGD} + \Delta\text{AED} \tag{1-9}$$

Area OABC = $\pi R^2\theta/360$, where R = radius, θ = central angle, or the angle between the two radii in degrees. The center of gravity of Area OABC is on the dashed bisector of the central angle at a distance of b from the center, or

$$b = \frac{4}{3}R\left[\frac{\sin\left(\frac{\theta}{2}\right)}{\theta}\right] \tag{1-10}$$

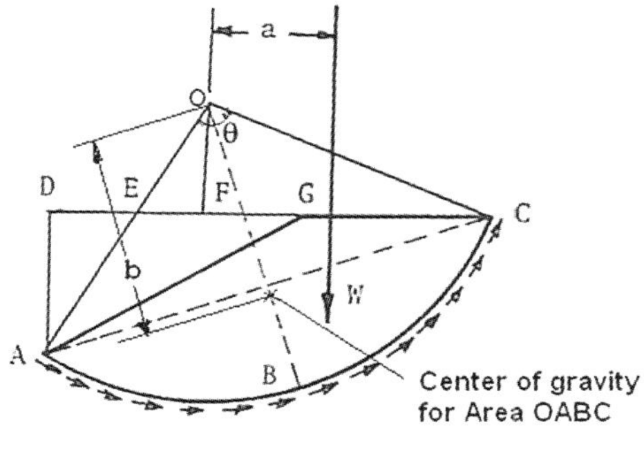

Fig. 1-4. Computation of driving moment

in which θ = central angle in radians. The area of each triangle can be easily computed, and the distance from its center of gravity to the vertical side of each triangle is equal to one-third of the length of the horizontal side.

Example 1.2 Fig. 1-5 shows a circular failure surface. The coordinates at the center, O, and at three points, A, G, and C, on the slope surface are given. If the soil has a unit weight of 125 pcf (19.7 kN/m^3), a cohesion of 1,600 psf (7.7 kN/m^2), and a friction angle of 0°, determine the factor of safety.

Solution All dimensions shown in the figure are based on the four sets of coordinates in parentheses. Radius: $R = [(120 - 30)^2 + (90 - 60)^2]^{0.5} = 94.9$ ft. Chord length: $L_c = [(120)^2 + (60)^2]^{0.5} = 134.2$ ft. Central angle: From geometry, $\sin(\theta/2) = 0.5 \times 134.2/94.9 = 0.707$, or $\theta = 2 \times \sin^{-1}(0.707) = 90°$. Distance EF: $\angle COF = \tan^{-1}(90/30) = 71.6°$, or $\angle EOF = 90 - 71.6 = 18.4°$, so EF $= 30 \times \tan 18.4° = 10$ ft.

The graphical method is more complex and will be discussed first. The weight of failure mass can be determined from Area ABCG shown in Fig. 1-5 multiplied by the unit weight of the soil. The area can be computed by Eq. (1-9), so $W = 125 \times [\pi(94.9)^2 \times 90/360 - 30 \times 90/2 - 30 \times 10/2 - 60 \times 90/2 + 60 \times 20/2] = 125 \times (7{,}073 - 1{,}350 - 150 - 2{,}700 + 600) = 434{,}125$ lb. From Eq. (1.10), $b = (4/3) \times 94.9 \times \sin 45°/(\pi/2) = 56.96$ ft. The overturning moment is $Wa = 125 \times [7{,}073 \times 56.96 \times \sin 26.6° - 1{,}350 \times 30 - 150 \times (-3.333) - 2{,}700 \times 0 + 600 \times (-23.3)] = 15{,}464{,}034$ ft-lb,

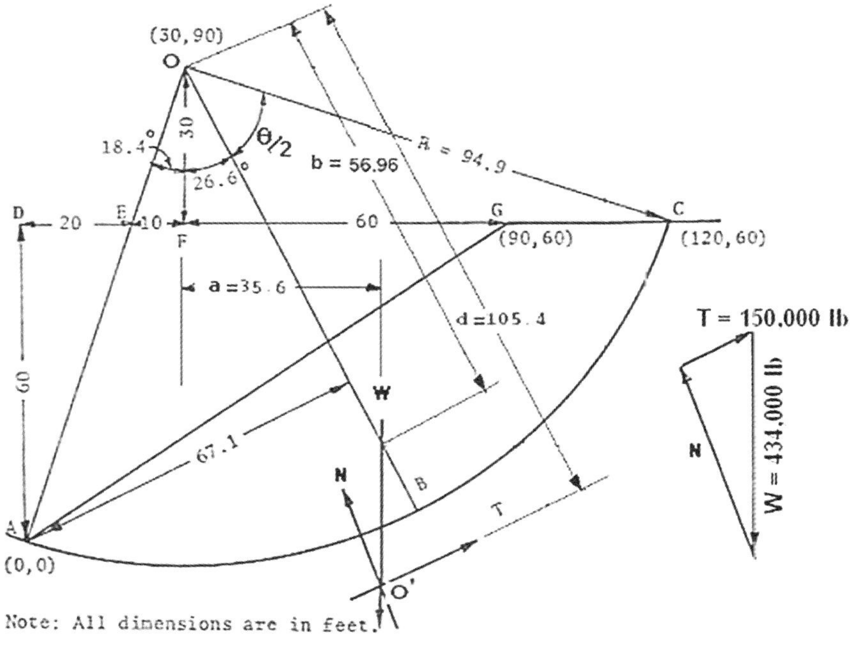

Fig. 1-5. Example 1.2

or $a = 15{,}464{,}034/434{,}125 = 35.6$ ft. The arc length is $L_a = \pi R\theta/180 = 3.1416 \times 94.9 \times 90/180 = 149.1$ ft. From Eq. (1-6), $d = 94.9 \times 149.1/134.2 = 105.4$ ft. With the distances a and d known, the exact location of W and T can be drawn on the figure, and the direction of N, which must pass through points O and O′, can be determined. From the force diagram on the right side of the figure, $T = 150{,}000$ lb. From Eq. (1-5), $F = cL_c/T = 1{,}600 \times 134.2/150{,}000 = 1.43$. The purpose of this graphical method is to show that the magnitude, direction, and line of application of all three forces can be determined by statics. If only the factor of safety is required, it is much quicker and more accurate by the algebraic method using Eq. (1-8), or $F = 1{,}600 \times 149.1 \times 94.9/15{,}464{,}034 = 1.46$, which checks with the 1.43 by the graphical method.

1.2.2 Statically Indeterminate Problems

Except for the simple cases shown in Figs. 1-1 and 1-3, most problems encountered in engineering practice are statically indeterminate. Fig. 1-6(a) shows the free-body diagram of a fill with weight, W, and the normal and shear forces on the two failure planes at the bottom. If both moment and force equilibrium are considered, there are five unknowns but only three equations. The five unknowns are the factor of safety, F, the magnitude and point of application of N_1, and the magnitude and point of application of N_2. The shear forces, T_1 and T_2, on the failure surface are known once the factor of safety is determined. This problem is statically indeterminate, because the number of equations is less than the number of unknowns.

If only force equilibrium is considered, there are three unknowns (factor of safety and magnitudes of N_1 and N_2) but only two equations. To make the problem statically determinate, it is necessary to divide the fill into two sliding blocks, as shown by Fig. 1-6(b), and arbitrarily assume the inclination of the force, P, acting between the two blocks. Because each block can have two equilibrium equations, the number of equations is four, which is equal to the number of unknowns (factor of safety and magnitudes of P, N_1, and N_2). If P is assumed horizontal, or there is no friction between the two blocks, the factor of safety is the minimum. An increase in the inclination of P also will increase the factor of safety. Therefore, a judicious selection of an inclination for P is needed to ensure

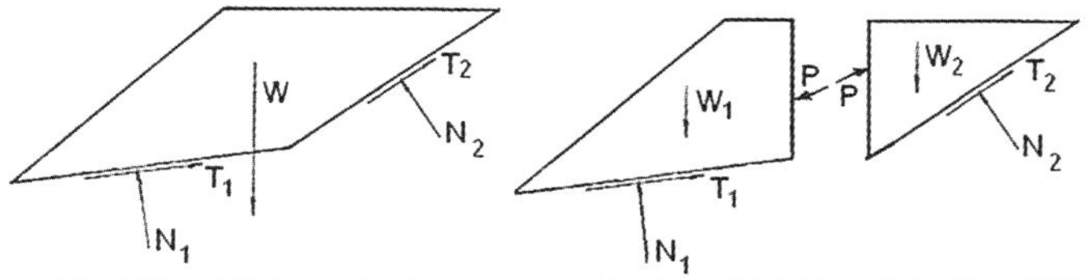

(a) FREE BODY DIAGRAM (b) DIVISION INTO TWO BLOCKS

Fig. 1-6. Statically indeterminate problem

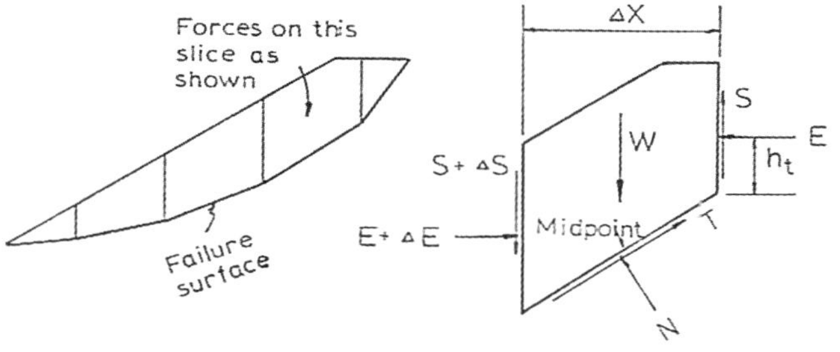

Fig. 1-7. Method of slices

that a reasonable factor of safety can be obtained. Details about sliding block analyses are presented in Sections 6.4 and 6.5.

A very powerful method, which can be applied to either circular or noncircular failure surfaces, is the method of slices. Fig. 1-7 shows an arbitrary failure mass divided into a number of slices. The forces applied on a slice are shown in the free-body diagram. If the failure mass is divided into a sufficient number of slices, the width, Δx, will be small, and it is reasonable to assume that the normal force, N, is applied at the midpoint of the failure surface. In the free-body diagram, the known forces are the weight, W, and the shear force, T, which depends on the factor of safety, as indicated by Eq. (1-3). The unknowns are the factor of safety, F, the shear force on the vertical side, S, the normal force on the vertical side, E, the vertical distance, h_t, and the normal force, N. If there are a total of n slices, the number of unknowns is $4n - 2$, as tabulated:

Unknown	Total Number
F (related to T)	1
N	n
E	$n - 1$
S	$n - 1$
h_t	$n - 1$
Total	$4n - 2$

Because each slice can have three equations by statics—two with respect to force equilibrium and one with respect to moment equilibrium—the total number of equations is $3n$. Therefore, there is an indeterminacy of $n - 2$. The problem can be solved statically only by making assumptions on the forces between two slices.

1.3 Methods of Stability Analysis

The method presented in this book is called the limit equilibrium method, because the factor of safety is based on statics by considering the force and/or moment equilibrium. Another procedure for stability analysis is the finite element method, which is based on solid mechanics by considering not only the equations of equilibrium but also those of compatibility. The advantage of the finite element method is its ability to determine not only the factor of safety but also the displacements. However, the displacements depend strongly on the elastic modulus, which is difficult to ascertain. The finite element method will not be discussed in this book.

Due to the large number of limit equilibrium methods available, it is neither possible nor desirable to review each of them. Therefore, only the most popular and well-known methods will be discussed in this book. Hopkins et al. (1975) and Duncan (1996) presented a review on several methods. The methods are based on the method of slices and can be divided broadly into four categories, depending on the number of equilibrium equations to be satisfied.

1.3.1 Methods that Satisfy Overall Moment Equilibrium

These methods are applicable only to circular failure surfaces and only consider the overall moment equilibrium. The overall force equilibrium is neither considered at all nor satisfied in both directions. Included in this category are the Fellenius method (Fellenius 1936), the normal method (Bailey and Christian 1969), and the simplified Bishop method (Bishop 1955).

The Fellenius method, usually referred to as the ordinary method of slices, has been used extensively for many years, because it is applicable to nonhomogeneous slopes and is very amenable to hand calculations. When pore pressures are present, a modified version of the Fellenius method—based on the concept of submerged weight and hereafter called the normal method—can be used. In both methods, the interslice forces are ignored. The Fellenius and normal methods are used in this book to generate stability charts for practical use. The factor of safety obtained by the normal method is usually smaller than that given by the simplified Bishop method. Details about the normal method are presented in Section 8.2 for two-dimensional (2D) analysis and Section 9.4 for three-dimensional (3D) analysis.

In the simplified Bishop method, the interslice forces are assumed horizontal. Although the overall moment equilibrium and the vertical force equilibrium are satisfied, neither moment nor horizontal force equilibrium is satisfied for each individual slice. Even though equilibrium conditions are not satisfied completely, the method is, nevertheless, a satisfactory procedure and is recommended for most routine work where the failure surface can be approximated by a circle. Bishop compared the factor of safety obtained from his simplified method with that from a more rigorous method in which all equilibrium equations are satisfied. He found that the vertical interslice force, S, could be assumed zero without introducing significant error, typically less than 1%. The simplified Bishop method cannot be used for noncircular failure surfaces, where an arbitrary moment center is assumed, because it only considers the force equilibrium in the vertical direction but not in the horizontal direction, so two moment centers at two different elevations will result in two different factors of safety. Details about the simplified Bishop method are presented in Sections 8.3 and 9.5.

1.3.2 Methods that Satisfy Overall Moment and Overall Force Equilibrium

Included in this category is the original Spencer method (Spencer 1967), which assumes that all interslice forces are parallel and incline at an angle δ with the horizontal, where δ is an unknown to be determined. It considers the overall moment equilibrium, the overall force equilibrium in the δ direction, and the force equilibrium of each slice in the direction perpendicular to δ. Because the overall force equilibrium in two perpendicular directions is satisfied fully, the method also can be used for noncircular failure surfaces, where a moment center must be selected arbitrarily. The original Spencer method was later refined and is hereafter called the Spencer method (Spencer 1973), in which moment equilibrium also is satisfied for each slice. The original Spencer method has the advantage that the factor of safety always converges, whereas the Spencer method sometimes may encounter convergence problems. Details about the original Spencer method are presented in Sections 8.4 and 9.6.

1.3.3 Methods that Satisfy Force Equilibrium of Each Slice

These methods only consider the force equilibrium in each slice. Once the force equilibrium is satisfied in each slice, the overall force equilibrium will be satisfied automatically. Although moment equilibrium is not considered explicitly, these methods may yield accurate results if the inclination of interslice forces is assumed in such a manner that the moment equilibrium is satisfied implicitly. Arbitrary assumptions on the inclination of interslice forces may have a large influence on the factor of safety. Depending on the inclination of interslice forces, a range of safety factors may be obtained in many problems. Force equilibrium methods should be used cautiously, and the user should be well aware of the particular interslice force assumptions employed. Included in this category are the procedures suggested by Janbu (Janbu et al. 1956; Janbu 1973), Lowe and Karafiath (1959), and the U.S. Army Corps of Engineers (1970).

The force equilibrium method proposed by Janbu also is called the simplified Janbu procedure, in contrast to his more rigorous method that also considers moment equilibrium of each slice. In the simplified procedure, the interslice forces are assumed horizontal, so the factor of safety thus obtained is always smaller than that by the more rigorous methods. To increase the factor of safety, Janbu et al. (1956) proposed the use of correction factors based on the depth-to-length ratio of the sliding mass and the type of soils. These correction factors are only approximate and were determined from a number of slope stability calculations by comparing the factors of safety obtained from the simplified procedure with those from the more rigorous procedure. Lowe and Karafiath (1959) suggested that the interslice forces act at the average of the inclination of the slope surface and the failure surface, whereas the U.S. Army Corps of Engineers (1970) recommended that the interslice forces be parallel to the slope surface. Of the three discussed methods, it appears that Lowe's and Karafiath's assumption on the inclination of the interslice forces is most reasonable and generally yields factors of safety closest to those obtained by the more rigorous methods where the moment equilibrium also is satisfied. Due to the negligence of moment equilibrium, force equilibrium methods are not used in the LEAME software presented in the companion volume to this book, *LEAME Software and User's Manual: Analyzing Slope Stability by the Limit Equilibrium Method*.

1.3.4 Methods that Satisfy Moment and Force Equilibrium of Each Slice

Included in this category are the Spencer method (Spencer 1973), the Janbu method (Janbu 1954,1973), and the method by Morgenstern and Price (1965). All these methods consider the moment and force equilibrium in each slice. If the moment and force equilibrium is satisfied in each slice, the overall moment and force equilibrium will be satisfied automatically. The basic concept in these methods is the same; the difference lies in the assumption of the interslice forces. If both moment and force equilibrium are satisfied, the assumption on interslice forces should have only small effect on the factor of safety obtained. All these methods can be applied to both circular and noncircular failure surfaces.

Similar to the original Spencer method, the Spencer method also assumes that the interslice forces are parallel and incline at an angle δ with the horizontal. However, instead of overall moment equilibrium, it considers the moment equilibrium of each slice. Force equilibrium is used to compute the factor of safety, F, and the moment equilibrium is used to compute δ. Because F and δ are interdependent, both need continuous adjustments until the results converge. Details about the Spencer method are presented in Section 8.5.

In the Janbu method, the location of interslice normal forces, or the line of thrust, must be assumed arbitrarily. The method is easy to use and requires less computer time than the Spencer method. Because the number of equations is one less than the number of unknowns and the factor of safety may be difficult to converge to the required tolerance, the Janbu method is not used in LEAME.

In the Morgenstern and Price method, an assumption is made regarding the relationship between interslice shear and normal forces. After obtaining the computer output based on this assumption, all the computed quantities, including the interslice forces, must be examined to determine whether they seem reasonable. If not, a new assumption must be made. Bishop (1955) indicated that the range of equally correct values of safety factor might be quite narrow and that any assumption leading to reasonable stress distributions and magnitudes would give practically the same factor of safety.

1.3.5 Methods Discussed in this Book and Incorporated in LEAME

The following four well-known methods are selected: normal, simplified Bishop, original Spencer, and Spencer. The reasons for their selection are explained as follows.

The normal method is exactly the same as the Fellenius method if there is no seepage. If there is seepage, the normal method applies the concept of submerged weight, instead of the pore water pressure at the bottom of slices, and thus gives a greater factor of safety than the Fellenius method. For circular failure surfaces, generally it is agreed that the simplified Bishop method gives a reasonable factor of safety and the Fellenius method yields a factor of safety smaller than that by the simplified Bishop method. The use of the normal method will draw the factor of safety closer to that by the simplified Bishop method. Although the normal method is not recommended for final design, it can be used to develop stability charts for preliminary design. Because the factor of safety can be determined easily without iterations only by the normal method, it can be used as the first trial factor of safety in all other methods where iterations are required.

The simplified Bishop method is recognized by the engineering profession as a valid method for circular failure surfaces. Most computer programs have included this method, which is also used by LEAME, for both 2D and 3D circular failure surfaces.

The original Spencer method is used for 3D analysis, because it is much simpler than the Spencer method. All the sample problems and examples for 2D analysis presented in this book or on the computer screen indicate that the factors of safety obtained by the original Spencer method check very closely with the more refined Spencer method. Because the 3D analysis is a simple extension of the 2D analysis with the same assumptions, the original Spencer method should be applicable to 3D analysis as well. This can be proved by comparing the 3D factors of safety obtained by LEAME with other available solutions, as reported in Section 3.10 of the companion volume. An advantage of the original Spencer method is that it always converges, whereas the Spencer method may have convergence problems.

The Spencer method satisfies all equations of equilibrium by assuming that the interslice forces in all slices incline at the same angle δ. It is a special case of the Morgenstern and Price method where the inclination of interslice forces can vary from slice to slice. Duncan (1996) evaluated various methods for limit equilibrium analysis of slopes and concluded that factors of safety for solutions with reasonable and unreasonable interslice force distributions were not significantly different. Because slope stability analyses are performed to calculate factors of safety, and not interslice forces, it does not matter in the end whether the interslice force distribution is reasonable, provided the method satisfies all equations of equilibrium. For this reason, the much simpler Spencer method is selected instead of the Morgenstern-Price method.

Summary

1. There exist a number of classification systems for mass movements. The most widely used system was devised by Varnes (1978), who grouped the movements into falls, topples, slides, spreads, and flows. In this book only the slide type of movement will be discussed in detail, not only because it is more amenable to theoretical analysis but also because it is the predominant type of failures, particularly in human-created slopes.

2. Slope movements also can be classified according to age as contemporary, dormant, and fossil; according to stage as initial, advanced, and exhausted; and according to cause, as exceptional, ordinary, and unexplainable.

3. The most commonly used method for stability analysis is the limit equilibrium method in which the shear stress along a failure surface is expressed as a quotient of the shear strength over an unknown factor of safety, and then the factor of safety is solved by using the equilibrium equations from statics.

4. For statically determinate problems, where the number of equilibrium equations is equal to the number of unknowns, the factor of safety can be expressed directly as a ratio between the shear resistance and the shear force, both applied over the failure surface.

5. For statically indeterminate problems, where the number of equations is less than the number of unknowns, some simplifying assumptions must be made to reduce the number of unknowns.

6. A number of limit equilibrium methods that satisfy only overall moment equilibrium, overall moment and overall force equilibrium, force equilibrium of each slice, or both moment and force equilibrium of each slice are described and the assumptions made in each method are briefly discussed.

7. Four well-known limit equilibrium methods (normal, simplified Bishop, original Spencer, and Spencer) are discussed and incorporated in the LEAME computer software. The reasons for their inclusion are explained.

Problems

1.1 Derive Eq. (1-10).
[Hint: moment at center by integrating $(2/3r\cos\theta)(1/2r^2d\theta)$]

1.2 A triangular fill with the dimension shown in Fig. P1-2 is placed on a natural slope of 20°. The fill has a total unit weight of 19.7 kN/m^3. If a thin layer of weaker soil with a cohesion of 7.7 kN/m^2 and a friction angle of 24° exists at the bottom of the fill, determine the factor of safety.
[Answer: 1.319]

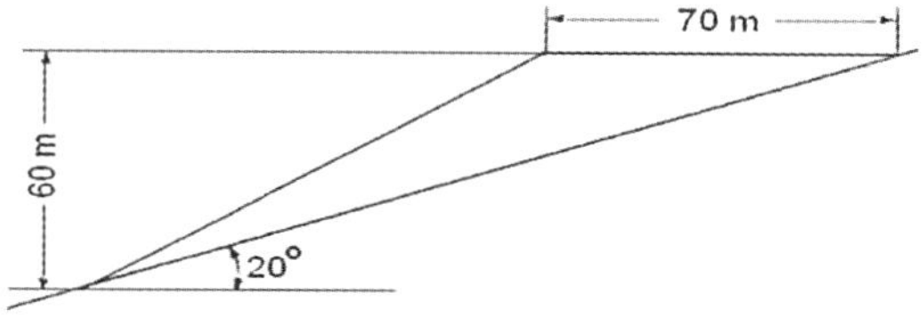

Fig. P1-2.

1.3 An engineer needs to place a new fill onto an existing slope, as shown in Fig. P1-3, to accommodate a roadway widening. The new fill is a granular material of high quality with a unit weight of 145 pcf. The natural soil has a cohesion of 100 psf and a friction angle of 28°. It is assumed that a surcharge load of 300 psf is placed on the new fill and that the most critical failure surface is a plane along the bottom of the new fill. What is the factor of safety against sliding for the proposed widening?
[Answer: 2.073]

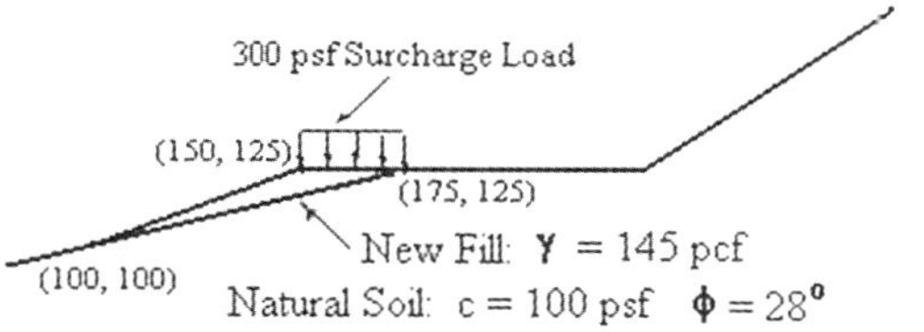

Fig. P1-3.

1.4 Fig. P1-4 shows the coordinates of a circular failure surface with $\phi = 0$. If the soil has a cohesion of $100\,kN/m^2$ and a total unit weight of $18\,kN/m^3$, determine the factor of safety.
[Answer: 1.299]

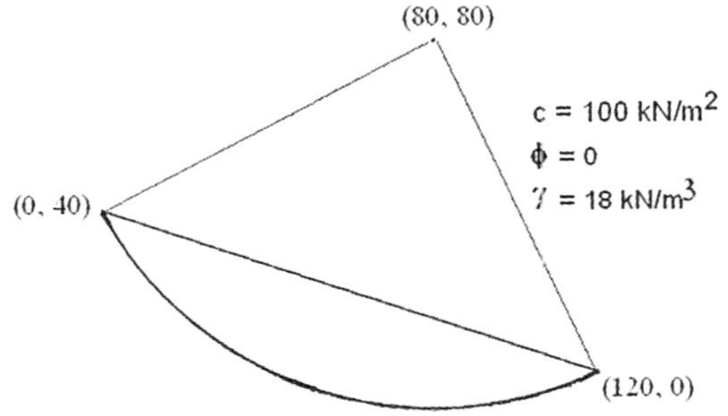

Note: All coordinates are in meters

Fig. P1-4.

1.5 Fig. P1-5 shows the coordinates of a circular failure surface with $\phi = 0$. If the soil has a total unit weight of $19\,kN/m^3$, what should be the minimum cohesion required to achieve a safety factor of 1.5?
[Answer: $150.7\,kN/m^2$]

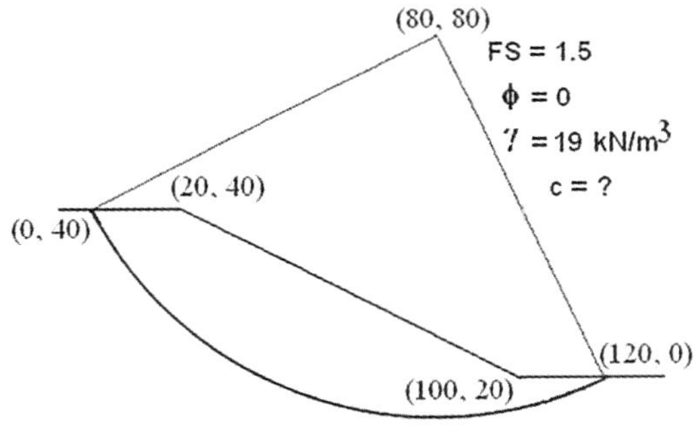

Note: All coordinates are in meters

Fig. P1-5.

1.6 Fig. P1-6 shows a circular failure surface. The soil has a cohesion of 1,300 psf, a friction angle of 0°, and a total unit weight of 120 pcf. Determine the factor of safety.
[Answer: 1.211]

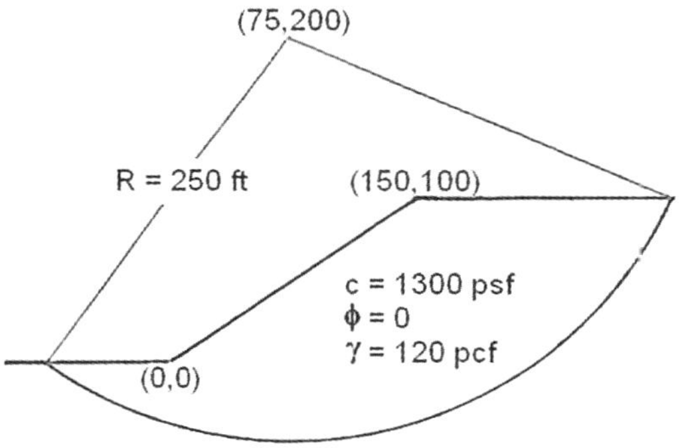

Fig. P1-6.

1.7 Fig. P1-7 shows a 31.5-ft vertical cut in a clay with an unconfined compressive strength of 2,100 psf, or c = 1,050 psf, and a total unit weight of 120 pcf. For the circular failure surface shown in the figure, determine the factor of safety.
[Answer: 1.151]

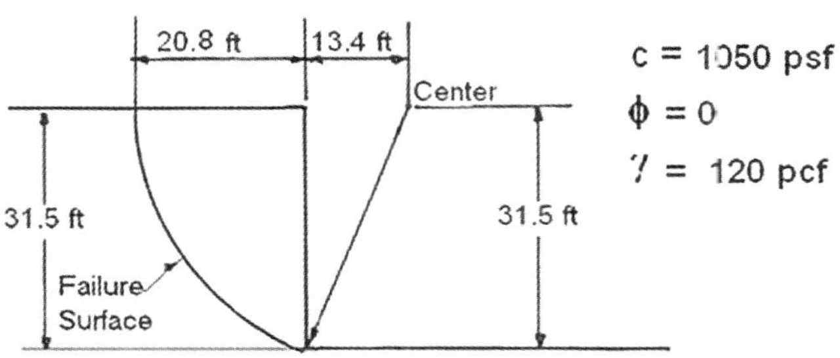

Fig. P1-7.

1.8 The slope is the same as in the previous problem, but the location of the failure surface is shown in Fig. P1-8. Because the center of the circle is 14.9 ft below the ground level and, when the method of slices is used, the circle cannot curve backward, LEAME assumes the failure surface to be a circular arc and a vertical line. Determine the factor of safety.
[Answer: 1.586]

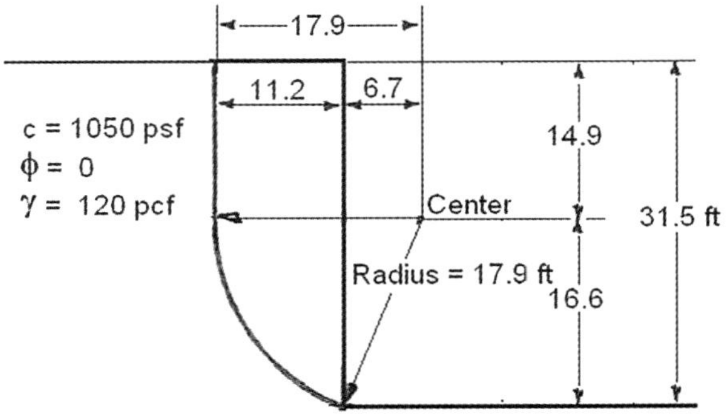

Fig. P1-8.

1.9 What is the factor of safety if the circle is allowed to curve back, as shown in Fig. P1-9. Is this type of failure surface possible?
[Answer: 2.301]

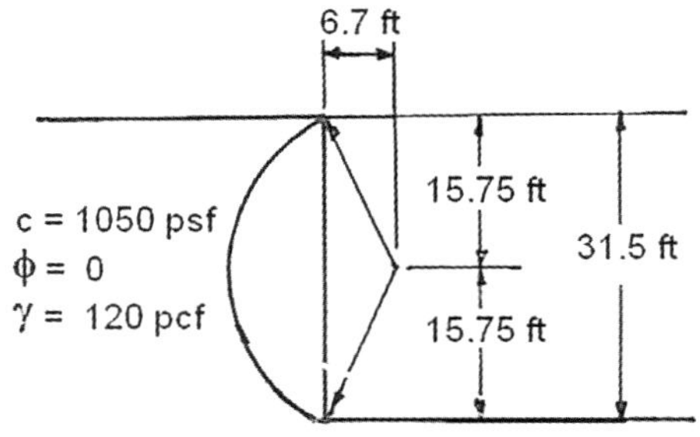

Fig. P1-9.

Chapter 2

Mechanics of Slides

This chapter describes three types of failure surface: circular, noncircular, and composite. The use of the method of slices and the sliding-block analysis for determining the factor of safety is illustrated. The drawback of the well-known Fellenius method is discussed, and the normal method based on the concept of submerged weight is introduced. The differences between the total stress analysis for short-term stability and the effective stress analysis for long-term stability are delineated. The minimum factors of safety required under various conditions are discussed.

2.1 Types of Failure Surface

The purpose of stability analysis is to determine the factor of safety of a potential failure surface. The factor of safety is defined as a ratio between the resisting force and the driving force, both applied along the failure surface, or

$$\text{Factor of Safety} = \frac{\text{Resisting force along failure surface}}{\text{Driving force along failure surface}}$$

When the driving force due to the weight and loading is equal to the resisting force due to the shear strength, the factor of safety is equal to 1 and failure is imminent. The shape of the failure surface may be quite irregular, depending on the homogeneity of the materials in the slope. This is particularly true in natural slopes, where the relic joints and fractures dictate the locus of the failure surface. Fig. 2-1 shows three types of failure surface: circular, noncircular, and composite.

The circular failure surface also is called the cylindrical failure surface, because the actual failure surface is part of a cylinder. If the materials in the slope are relatively homogeneous with no apparent weak layers, the most critical failure surface will be cylindrical, because a cylindrical failure surface has the least surface area per failure mass. Because the surface area is more related to the resisting force and the failure mass is more related to the driving force, a smaller resisting force and a larger driving force will result in a smaller factor of safety. To find the minimum factor of safety, a large number of circles must be tried to determine which is most critical.

The composite failure surface is principally circular, but when the circle cuts a layer of weak material, the failure surface will follow the bottom of the weak layer, so part of the circular failure surface is replaced by the plane failure surface. Fig. 2-2 shows two types of composite failure surface. If the weak layer appears on the surface of the slope, the circular arc is on one end, as shown in Fig. 2-2(a). If the weak layer is buried inside the slope, the circular arc will appear on both ends, as shown in Fig. 2-2(b). The composite failure surface can be analyzed by the method of slices. The method used to generate composite failure surfaces will be discussed in Section 2.16 of the companion volume.

The noncircular failure surface also is called the plane failure surface, because the failure surface consists of a series of planes. The noncircular failure surface may occur if there are weak layers or seams that start and end at or near the

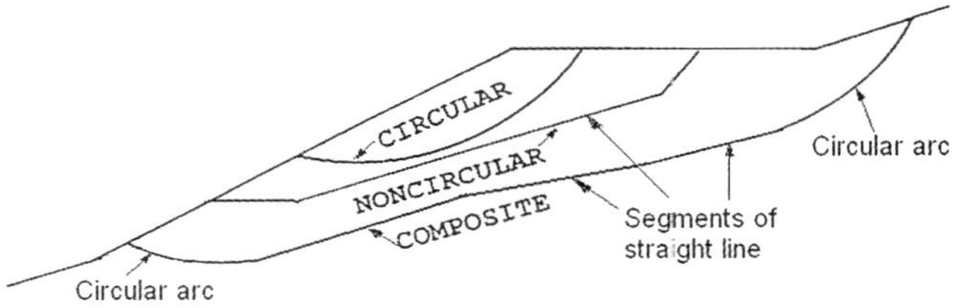

Fig. 2-1. Three types of failure surface

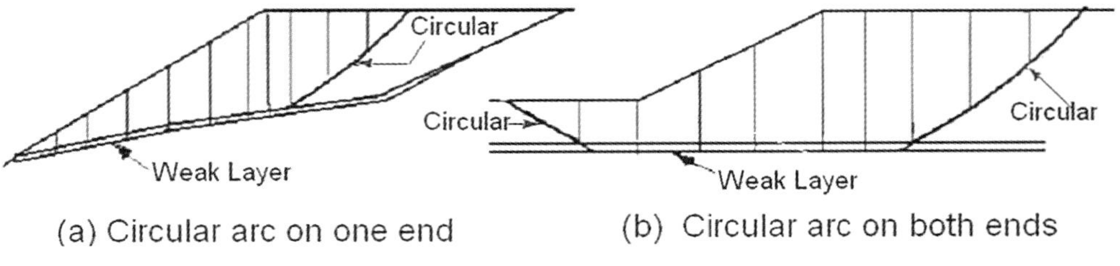

Fig. 2-2. Two types of composite failure surface

slope surface, so the most critical failure surface will be located along the bottom of the weak layers. Sliding-block analysis, which only satisfies force equilibrium, can be used for simple cases of noncircular failures consisting of no more than three failure planes. However, for more accurate results, the method of slices, which satisfies both force and moment equilibrium, should be used.

2.2 Total Stress versus Effective Stress

There are two methods for analyzing the stability of slopes: total stress analysis and effective stress analysis.

Total stress analysis is based on the undrained shear strength and also is called c_u, ϕ_u-analysis. If a saturated soil is undergoing undrained loading, ϕ_u can be assumed 0, so a special case of total stress analysis, commonly called s_u-analysis, with an undrained shear strength, s_u, equal to c_u, can be used. In the s_u- or c_u, ϕ_u-analysis, pore pressure should be taken as 0 along any failure surface where undrained strength is specified. This step does not imply that pore pressures are actually zero, but, rather, is done to be consistent with the fact that the undrained strength determined from tests already has included the effect of pore pressure and needs not be considered again in the stability analysis.

Effective stress analysis is based on the drained shear strength and is called c', ϕ' analysis. In engineering practice, it is not necessary to perform both total and effective stress analyses for every slope. In many cases, the total stress analysis is not needed unless some wet, fine-grained soils with a degree of saturation in the range of 70% or higher are encountered. At a low degree of saturation, the difference between total and effective envelopes is insignificant, so only the effective stress analysis with steady-state seepage needs to be considered for long-term stability.

The total stress analysis based on undrained shear strength usually is used for determining short-term stability during or at the end of construction and the effective stress analysis based on the drained shear strength for long-term stability after the construction. The major difference between a total stress analysis and an effective stress analysis is that the former does not require knowledge of the pore pressure, whereas the latter does. In principle, short-term stability also can be analyzed in terms of effective stress and long-term stability in terms of total stress. However, this would require extra testing efforts and is therefore not recommended. Depending on the type of soils, the following methods of analysis are suggested:

1. For saturated soils, use s_u-analysis with $\phi = 0$ and $c = s_u$ for short-term stability; c', ϕ'-analysis with 0 or steady-state pore pressure for long-term stability; and c', ϕ'-analysis with actual or estimated pore pressure for intermediate times.
2. For partially saturated soils, use c_u, ϕ_u-analysis from undrained tests for short-term stability; c', ϕ'-analysis with 0 or steady-state pore pressure for

long-term stability; and c', ϕ'-analysis with actual or estimated pore pressure for intermediate times.

3. For granular soils or soils with a degree of saturation less than 70%, only the long-term c', ϕ'-analysis with 0 or steady-state pore pressure is needed.

It should be pointed out that, in certain cases, a failure surface may pass partly through a free-draining soil, where strength is expressed appropriately in terms of effective stress, and partly through a clay-like soil, where undrained strength should be used. In such cases, the parameters c' and ϕ', together with appropriate pore pressures, apply along one portion of the surface, and the $\phi = 0$ or c_u, ϕ_u-analysis with zero pore pressure applies along the other part.

For fill slopes, the total stress analysis for short-term stability is more critical than the effective stress analysis for long-term stability because of the increase in effective stress with time, so the total stress analysis should be made before the effective stress analysis. For cut slopes, the effective stress analysis for long-term stability is more critical because, when the weight removes during excavation, the soil expands and generates a negative pore pressure, which gradually dissipates with time, so the effective stress analysis should be conducted first. Another explanation, much easier to understand, is to relate the shear strength to the void ratio; that is, the greater the void ratio, the smaller the shear strength. For fill slopes, the soil consolidates, and void ratio decreases with time, so the short-term stability is more critical. For cut slopes, the soil expands, and the void ratio increases with time, so the long-term stability is more critical.

2.3 Total Stress Analysis

Two examples are given to illustrate the total stress analysis for circular and noncircular failure surfaces. The circular failure surface is analyzed by the well-known Fellenius method and the noncircular failure surface by the sliding-block method.

2.3.1 Fellenius Method

This method was original formulated by Fellenius (1936), a professor of hydraulics at the (Swedish) Royal Institute who was appointed as the chairman of the Geotechnical Commission of Swedish State Railways to investigate the cause of landslide problems. Fig. 2-3 shows one of the many circles for which the factor of safety is to be determined. The sliding mass is divided into n slices. The ith slice has a weight, W_i, a length of failure surface, L_i, an angle of inclination, θ_i, and a normal force, N_i. According to the Mohr-Coulomb theory, the resisting force in slice i is $c_i L_i + N_i \tan\phi_i$. Note that N_i depends on the forces on the two sides of the slice and is statically indeterminate unless some simplifying assumptions are made. By assuming that there are no forces on the two sides of a slice,

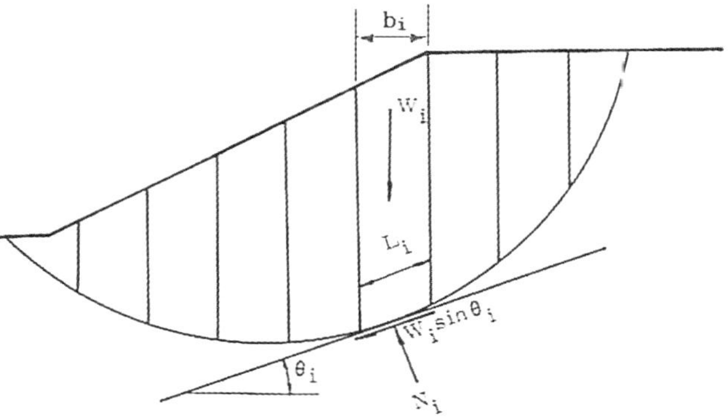

Fig. 2-3. Circular failure surface

the driving force is equal to $W_i \sin\theta_i$, which is the component of weight along the failure surface. The factor of safety is a ratio between the resisting force and the driving force and can be determined by

$$F = \frac{\sum_{i=1}^{n}(c_i L_i + N_i \tan\phi_i)}{\sum_{i=1}^{n}(W_i \sin\theta_i)} \tag{2-1}$$

When the failure surface is circular, the factor of safety also can be defined as the ratio between two moments. Because both the numerator and the denominator in Eq. (2-1) can be multiplied by the same moment arm, which is the radius of the circle, it makes no difference whether the force or moment is used. In the Fellenius method, it is assumed that the forces on the two sides of a slice are zero, so they have no effect on the force normal to the failure surface, or

$$N_i = W_i \cos\theta_i \tag{2-2}$$

Thus Eq. (2-1) becomes

$$F = \frac{\sum_{i=1}^{n}(c_i b_i \sec\theta_i + W_i \cos\theta_i \tan\phi_i)}{\sum_{i=1}^{n} W_i \sin\theta_i} \tag{2-3}$$

The method presented here is called the total stress analysis, because no phreatic surface or pore pressure is considered. Another method, called the effective stress analysis, which involves the application of pore pressure on the failure surface, will be presented in Section 2.4.

Example 2.1 Fig. 2-4 shows a circular failure surface with the coordinates of three points indicated in parentheses. The circle has a radius of 100 ft (30.5 m) and a center at (0, 100). The circle intercepts the slope surface at (0, 0) and (90, 56.4). The failure mass is divided into nine slices, each with a width of 10 ft (3.05 m). The soil in the slope has a cohesion of 800 psf (38.3 kN/m^2), a friction angle of 10°, and a total unit weight of 125 pcf (19.6 kN/m^3). Compute the factor of safety by the Fellenius method.

Solution The equation of the circle is $x^2 + (100 - y)^2 = R^2$, or $y = 100 - (10{,}000 - x^2)^{0.5}$. To determine θ, differentiate the above equation with respect to x as follows: $\tan\theta = dy/dx = -0.5x(10{,}000 - x^2)^{-0.5}(-2) = x/(10{,}000 - x^2)^{0.5}$ or $\theta = \tan^{-1}[x/(10{,}000 - x^2)^{0.5}]$. The height of slice $h = 0.627x + (10{,}000 - x^2)^{0.5} - 100$ and the weight of slice $W = 1{,}250h$. At the centerline of each slice, as indicated by the vertical dashed line in the right figure, Table 2-1 presents the values of x, h, θ, and W tabulated in columns (2) to (5) and other calculated values are in columns (6) to (9). The factor of safety can be obtained by dividing the sum in column (9) with that in column (6), or $F = 109{,}803/66{,}998 = 1.639$.

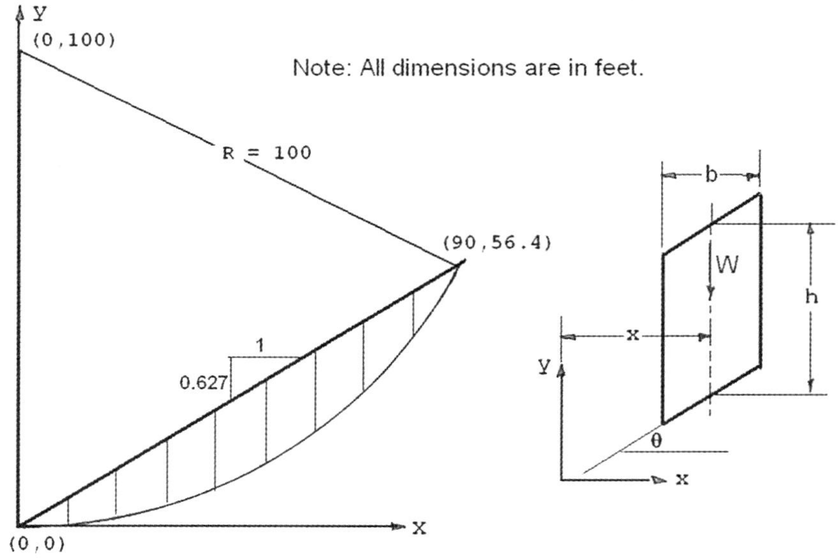

Fig. 2-4. Example 2.1

In Table 2-1, all calculations were done by hand with a pocket calculator. Because the calculations used for each slice are all the same, it is much easier and more accurate to solve this problem by a spreadsheet. A sample spreadsheet is presented in Table 2-2. The formulas used for computing columns C to H of the first slice, cell F14, cell I14, and cell D15 are:

Table 2-1. Computations of Safety Factor by Fellenius Method

No. (1)	x (2)	h (3)	θ (4)	W (5)	$W\sin\theta$ (6)	$cb\sec\theta$ (7)	$W\cos\theta\tan\phi$ (8)	(7) + (8) (9)
1	5	3.1	2.9	3,875	196	8,010	682	8,693
2	15	8.3	8.6	10,375	1,551	8,091	1,809	9,900
3	25	12.5	14.5	15,625	3,912	8,263	2,667	10,930
4	35	15.6	20.5	19,500	6,829	8,541	3,221	11,762
5	45	17.5	26.7	21,875	9,829	8,955	3,446	12,401
6	55	17.9	33.4	22,500	12,386	9,583	3,312	12,895
7	65	16.7	40.5	21,000	13,638	10,521	2,816	13,337
8	75	13.1	48.6	16,375	12,283	12,097	1,909	14,006
9	85	6.0	58.2	7,500	6,374	15,182	697	15,879
Sum					66,998			109,803

Note: θ is in degrees; $b = 10$ ft; $c = 800$ psf; $\phi = 10°$.

Table 2-2. Spreadsheet for Computing Factor of Safety by Fellenius Method

	A	B	C	D	E	F	G	H	I
3	No.	x	h	θ in radian	W	W sin θ	cb sec θ	W cos θ tan ϕ	(7) + (8)
4	(1)	(2)	(3)	(4)	(5)	(6)	(7)	(8)	(9)
5	1	5	3.010	0.050	3,762	188	8,010	663	8,673
6	2	15	8.274	0.151	10,342	1,551	8,092	1,803	9,894
7	3	25	12.500	0.253	15,624	3,906	8,262	2,668	10,930
8	4	35	15.620	0.358	19,525	6,834	8,540	3,225	11,765
9	5	45	17.518	0.467	21,897	9,854	8,958	3,448	12,406
10	6	55	18.001	0.582	22,502	12,376	9,579	3,314	12,893
11	7	65	16.748	0.708	20,936	13,608	10,527	2,805	13,333
12	8	75	13.169	0.848	16,461	12,346	12,095	1,920	14,015
13	9	85	5.973	1.016	7,467	6,347	15,187	694	15,880
14				Sum		67,010			109,788
15	**Factor of safety =**			1.638					

Cell C5: = 0.627*B5 + SQRT(10000 – B5^2)-100
Cell D5: = ATAN(B5/SQRT(10000 – B5^2))
Cell E5: = 1250*C5
Cell F5: = E5*SIN(D5)

Cell G5: = 8000/COS(D5)
Cell H5: = E5*COS(D5)*TAN(RADIANS(10))
Cell F14: = SUM(F5:F13) Cell I14: =SUM(I5:I13)
Cell. D15: = I14/F14

Because of round-off errors, the data computed by hand are somewhat different from those by the spreadsheet. However, the final factor of safety should be nearly the same. It can be seen that the spreadsheet gives a safety factor of 1.638, which is nearly the same as the 1.639 by the manual method.

2.3.2 Sliding-Block Method

Fig. 2-5 shows the free-body diagram of plane failure at the bottom of a fill. The sliding mass is divided into two blocks. Both the normal and the shear forces on each failure plane depend on the interacting force, P, between the two blocks and can be determined only by considering the two blocks jointly. The lower block has a weight, W_1, and a length of failure plane, L_1; the upper block has a weight, W_2, and a length of failure plane, L_2. Note that ϕ_d is the developed friction angle, which must be assumed. When ϕ_d is assumed 0, or the side force is horizontal, the factor of safety is minimum and the design is on the safe side. When the frictional resistance is developed fully, or $\phi_d = \phi$, the factor of safety is maximum. These provide lower and upper bounds within which the factor of safety based on any moment and force equilibrium should lie. As can be seen from the free-body diagram, there are four unknowns: F, P, N_1, and N_2. With four force equilibrium equations, two for each block, these four unknowns can be solved.

In the original Spencer method, the angle δ, which is the same as ϕ_d, is considered as an unknown to be solved. Several values of δ, ranging from 0 to 0.6 rad (35°), are assumed, and the factors of safety based on force equilibrium are

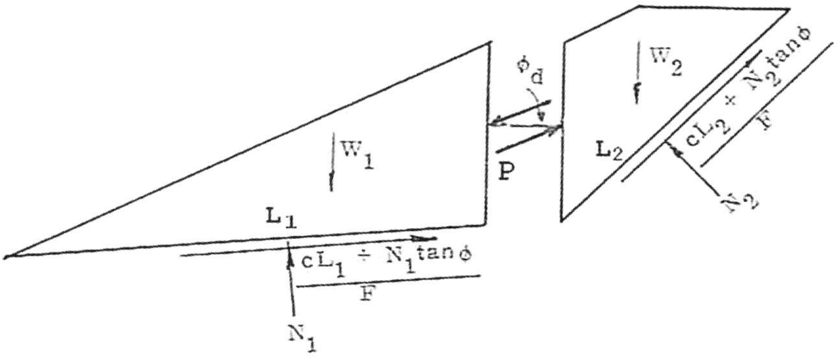

Fig. 2-5. Free-body diagram of sliding blocks

checked against those based on overall moment equilibrium. The value of δ, which gives the same factor of safety for both force and moment equilibrium, is the correct δ to be used. The fact that the correct δ mostly lies between 0 and 35° may indicate that the most reasonable assumption to be used is $\tan\phi_d = (\tan\phi)/F$, or

$$\phi_d = \tan^{-1}\left(\frac{\tan\phi}{F}\right) \tag{2-4}$$

More about the sliding-block analysis is presented in Sections 6.4 and 6.5.

Example 2.2 Fig. 2-6 shows a fill with the coordinates indicated in parentheses. The failure mass is divided into two sliding blocks. To be on the safe side, it is assumed that the force between the two blocks is horizontal. If the soil has a cohesion of 500 psf (23.9 kN/m^2), a friction angle of 15°, and a total unit weight of 120 pcf (18.9 kn/m^3), determine the factor of safety.

Solution From the dimensions shown in the figure, $W_1 = 0.5 \times 45 \times 90 \times 120 =$ 243,000 lb, $L_1 = 90$ ft, $W_2 = 0.5 \times 45 \times 65 \times 120 = 175{,}500$ lb, $L_2 = [(65)^2 + (45)^2]^{0.5} =$ 79.1 ft, and $\alpha = \tan^{-1}(45/65) = 34.7°$.

First consider the left block, or block 1. The equilibrium of forces in the vertical direction requires $N_1 = W_1 = 243{,}000$ lb. In the horizontal direction, $P = T_1 = (cL_1 + N_1\tan\phi)/F = (500 \times 90 + 243{,}000 \times \tan 15°)/F = 110{,}112/F$.

Next, consider the right block, or block 2. The equilibrium of forces in the normal direction requires $N_2 = P\sin\alpha + W_2\cos\alpha = 110{,}112 \times \sin 34.7°/F + 175{,}500 \times \cos 34.7°$, or

$$N_2 = \frac{62{,}685}{F} + 144{,}286 \tag{2-5}$$

In the tangential direction, $P\cos\alpha + T_2 = W_2\sin\alpha$, or

$$\frac{110{,}112 \times \cos 34.7°}{F} + \frac{500 \times 79.1 + N_2 \tan 15°}{F} = 175{,}500 \times \sin 34.7° \tag{2-6}$$

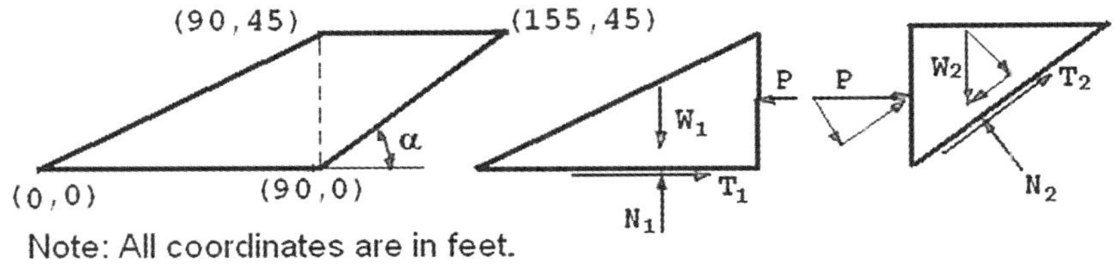

Fig. 2-6. Example 2.2

Substituting N_2 from Eq. (2-5) into Eq. (2-6) and simplifying, the following quadratic equation is obtained:

$$F^2 - 1.689F - 0.168 = 0 \tag{2-7}$$

or

$$F = \frac{1.689 \pm \sqrt{(1.689)^2 + 4 \times 0.168}}{2} = \frac{1.689 \pm 1.877}{2} = 1.783 \quad or -0.094$$

A negative factor of safety of −0.094 is unreasonable and is therefore rejected, so the factor of safety is 1.783.

2.4 Effective Stress Analysis

Eq. (2-1) is based on the total stress analysis, which does not include the effect of pore pressures. In the effective stress analysis, three cases need to be considered: steady-state seepage, rapid drawdown, and earthquake.

2.4.1 Steady-State Seepage

Fig. 2-7 shows the slope with a phreatic surface. The effective normal force, N'_i, is equal to the total normal force, N_i, minus the neutral force, or the force due to

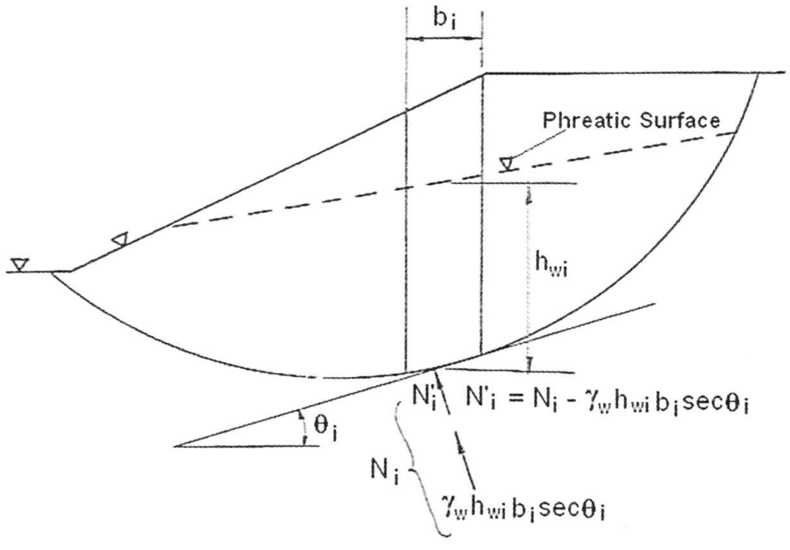

Fig. 2-7. Effect of phreatic surface

water pressure, $\gamma_w h_{iw} b_i \sec\theta_i$. The use of $\gamma_w h_{iw}$ as the pore pressure at the failure surface is an approximation by assuming that the water is stationary with a water level at a distance of h_{iw} above the failure surface. The validity of this assumption will be discussed in Section 4.1.3. In terms of effective stress, the Mohr-Coulomb theory can be represented by

$$s = c' + \sigma_n' \tan\phi' \tag{2-8}$$

in which c' = effective cohesion, σ_n' = effective normal stress, and ϕ' = effective angle of internal friction. Therefore, Eq. (2-1) for a circular failure surface can be written in terms of effective stress as

$$F = \frac{\sum_{i=1}^{n}(c'L_i + N_i' \tan\phi')}{\sum_{i=1}^{n}(W_i \sin\theta_i)} \tag{2-9}$$

and

$$N_i' = W_i \cos\theta_i - \gamma_w h_{wi} b_i \sec\theta_i \tag{2-10}$$

in which h_{wi} = height of phreatic surface above the base of the slice, and b_i = width of the slice. Eq. (2-10) is used in the Fellenius method and sometimes yields unreasonable results, because when θ is large, say greater than 45°, N_i' may become negative.

Example 2.3 Same as Example 2.1 except that the soil has an effective cohesion of 200 psf (9.6 kN/m^2) and an effective friction angle of 35°, and also that a phreatic surface is parallel to the slope surface and located at a distance of 5 ft (1.52 m) below, as shown in Fig. 2-8. Determine the factor of safety by the Fellenius method.

Solution The solution is presented in Table 2-3.

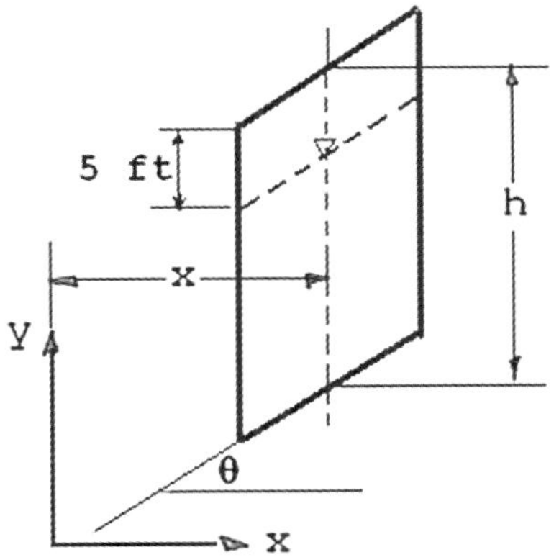

Fig. 2-8. Example 2.3

Table 2-3. Effective Stress Analysis by Fellenius Method

No. (1)	x (2)	h (3)	θ (4)	W (5)	W **sin** θ (6)	cb **sec** θ (7)	N' **tan** ϕ (8)	**(7) + (8)** (9)
1	5	3.1	2.9	3,875	196	2,003	2,710	4,713
2	15	8.3	8.6	10,375	1,551	2,023	5,726	7,749
3	25	12.5	14.5	15,625	3,912	2,066	7,207	9,273
4	35	15.6	20.5	19,500	6,829	2,135	7,845	9,980
5	45	17.5	26.7	21,875	9,829	2,239	7,570	9,809
6	55	18.0	33.4	22,500	12,386	2,396	6,349	8,745
7	65	16.8	40.5	21,000	13,638	2,630	4,401	7,031
8	75	13.1	48.6	16,375	12,283	3,024	2,231	5,255
9	85	6.0	58.2	7,500	6,374	3,795	1,938	5,733
		Sum			66,998			68,288

Note: $b = 10$ ft; $c = 200$ psf; $\phi = 35°$; $W = 1{,}250$h

Except for columns (8) and (9), the table is the same as Table 2-1. The effective normal force, N', is determined by Eq. (2-10). If $h \leq 5$ ft, $N' = W\cos\theta$. If $h > 5$ ft, $N' = W\cos\theta - 624(h - 5)/\cos\theta$. The factor of safety can be obtained by dividing the sum in column (9) with that in column (6), or $F = 68{,}288/66{,}998 = 1.019$. Because the phreatic surface is 5 ft below the slope surface, all values of N' are positive. If the phreatic surface is on the slope surface, for slice 8 with θ of 48.6°, $N' = 16{,}375 \times \cos 48.6° - 624 \times 13.1/\cos 48.6° = -1{,}532$ lb. This clearly indicates the difficulty of using the Fellenius method.

To avoid the negative effective normal stress, N', many engineers have used the more conventional concept of the submerged or effective weight, W_i', by considering

$$N_i' = W_i'\cos\theta_i = (W_i - \gamma_w h_{wi} b_i)\cos\theta_i \qquad (2\text{-}11)$$

The use of Eq. (2-11) for stability analysis is called the normal method, because the normal practice in soil mechanics for computing the effective stress under a soil overburden is to use the total unit weight, γ, if the soil is above the water table, and the submerged unit weight, $\gamma - \gamma_w$, if the soil is below the water table. Based on this concept, $W_i' = b_i(h_i - h_{wi})\gamma + b_i h_{wi}(\gamma - \gamma_w) = \gamma b_i h_i - \gamma_w h_{wi} b_i = W_i - \gamma_w h_{wi} b_i$, which is the same as expressed in Eq. (2-11). It can be seen that the normal method is exactly the same as the Fellenius method when there is no pore pressure. With pore pressures, the normal method generally yields a factor of safety slightly greater than the Fellenius method.

Example 2.4 Same as Example 2.3 but determine the factor of safety by the normal method.

Solution The solution is presented in Table 2-4, in which columns (1) to (7) are the same as in Table 2-1. The effective weight, W', is computed as follows: if $h \leq 5$ ft, $W' = W$; if $h > 5$ ft, $W' = 5b\gamma + b(h-5)(\gamma - \gamma_w) = 6{,}250 + 626(h-5)$. The factor of safety can be obtained by dividing the sum in column (10) with that in column (6), or $F = 78{,}905/66{,}998 = 1.178$, which is slightly greater than the 1.019 by the Fellenius method.

Table 2-4. Effective Stress Analysis by Normal Method

No. (1)	x (2)	h (3)	θ (4)	W (5)	$W\sin\theta$ (6)	$cb\sec\theta$ (7)	W' (8)	$W'\cos\theta\tan\phi'$ (9)	(7) + (9) (10)
1	5	3.1	2.9	3,875	196	2,003	3,875	2,710	4,713
2	15	8.3	8.6	10,375	1,551	2,023	8,315	5,757	7,780
3	25	12.5	14.5	15,625	3,912	2,066	10,945	7,420	9,486
4	35	15.6	20.5	19,500	6,829	2,135	12,886	8,451	10,586
5	45	17.5	26.7	21,875	9,829	2,239	14,075	8,805	11,044
6	55	18.0	33.4	22,500	12,386	2,396	14,388	8,411	10,807
7	65	16.8	40.5	21,000	13,638	2,630	13,637	7,261	9,891
8	75	13.1	48.6	16,375	12,283	3,024	11,321	5,242	8,266
9	85	6.0	58.2	7,500	6,374	3,795	6,876	2,537	6,332
			Sum		66,998				78,905

Note: $b = 10$ ft; $c = 200$ psf; $\phi = 35°$; $W = 1{,}250$ h

2.4.2 Rapid Drawdown

Rapid drawdown is usually the most critical situation in the design of earth dams. The downstream slope is controlled by the case of steady-state seepage, but the upstream slope is controlled by the case of rapid drawdown. As shown in Fig. 2-9, the phreatic surface under rapid drawdown is along the dashed line, as well as along the surface of both slopes. The phreatic surface on the

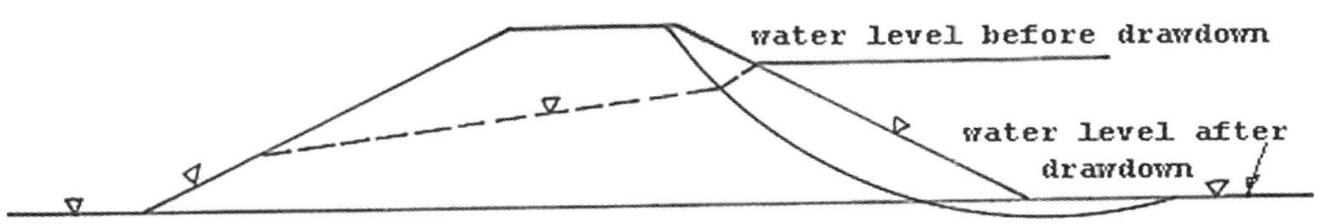

Fig. 2-9. Phreatic surface for rapid drawdown

downstream side is used for steady-state seepage and on the upstream side for rapid drawdown. Because more of the sliding mass is under water, the upstream slope is more critical than the downstream slope.

It should be noted that rapid drawdown occurs only in earth dams with permanent impoundment but not in road embankments subjected to occasional flooding. For clayey soils, it takes many years to develop a steady-state phreatic surface and, without such a phreatic surface inside the embankment, the condition of rapid drawdown never will occur. For granular soils, the phreatic surface inside the embankment will recede as fast as the flood, and no rapid drawdown need be considered.

This method of analyzing rapid drawdown for long-term stability is quite conservative. It is based on the assumption that there is no change in the effective stress during and after the drawdown. In fact, the effective stress will increase because, first, the soils in an earth dam are supposed to be well compacted, so a sudden change in stress because of the lowering of water level in the pond will cause the soils to dilate, thus increasing the effective stress. Second, the soils are located so close to the surface that some consolidation will occur during the period of rapid drawdown.

If rapid drawdown takes place during or at the end of construction, the undrained shear strength used for short-term stability analysis before drawdown also can be used for that after drawdown.

2.4.3 Earthquake Consideration

In the case of earthquake, a horizontal seismic force is applied at the centroid of each slice, as shown in Fig. 2-10. The seismic force is equal to C_sW_i, where C_s is the seismic coefficient and ranges from 0 to 0.4 or more, depending on the geographic location, and W_i is the weight of the slice. With seismic force, Eq. (2-9) can be written as

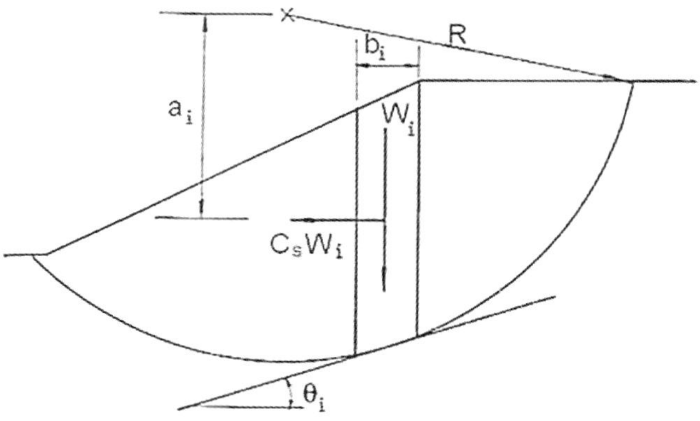

Fig. 2-10. Driving force due to earthquake

$$F = \frac{\sum_{i=1}^{n}(c'b_i \sec\theta_i + N_i' \tan\phi')}{\sum_{i=1}^{n}(W_i \sin\theta_i + C_s W_i a_i / R)} \tag{2-12}$$

in which b_i = width of the slice, a_i = moment arm, and R = radius of the circle. By using the normal method,

$$F = \frac{\sum_{i=1}^{n}[c'b_i \sec\theta_i + W_i' \cos\theta_i \tan\phi']}{\sum_{i=1}^{n}(W_i \sin\theta_i + C_s W_i a_i / R)} \tag{2-13}$$

in which W_i' = submerged weight of the slice.

In the lack of detailed local earthquake information, several maps can be found in the literature to determine the seismic coefficients in the continental United States. The earliest map was presented by Algermissen (1969), as shown in Fig. 2-11.

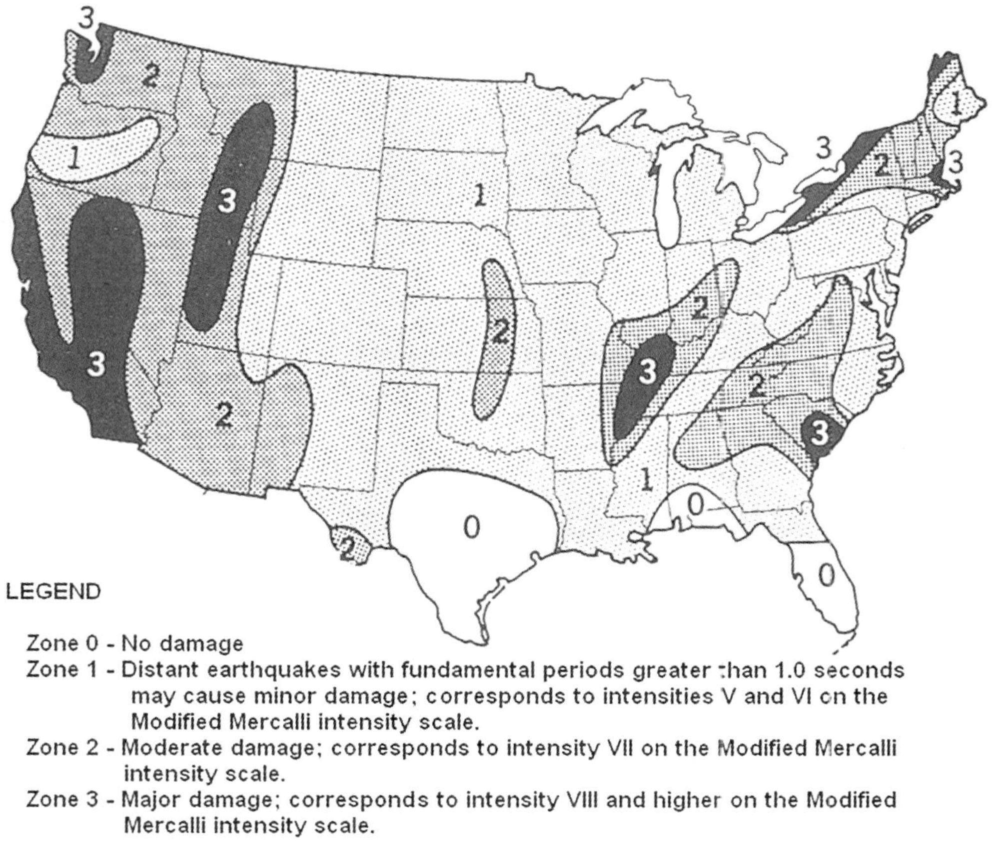

Fig. 2-11. Seismic zone map of the continental United States (Algermissen 1969)

The map is divided into four zones, Zone 0, Zone 1, Zone 2, and Zone 3, with seismic coefficients, C_s, from 0, 0.025, 0.05, to 0.10–0.15 for each zone. Neumann (1954) modified these coefficients, based on an average epicenter of 15 mi (24 km), and used Eq. (2-14) to determine the seismic coefficients, as shown in Table 2-5.

$$C_s = \frac{\log^{-1}[0.267 + (MM - 1) \times 0.308]}{980} \tag{2-14}$$

in which MM is the Modified Mercalli intensity scale.

Fig. 2-12 is a newer map published by the Applied Technology Council (1978). In this publication, a large map, including every county of the 50 states and U.S. territories, is divided into seven areas, each with a different color so the value of seismic coefficients can be found easily.

Table 2-5. *Seismic Coefficients Corresponding to Each Zone*

ZONE	INTENSITY OF MODIFIED MERCALLI SCALE	AVERAGE SEISMIC COEFFICIENT	REMARK
0	—	0	No damage
1	V and VI	0.03 to 0.07	Minor damage
2	VII	0.13	Moderate damage
3	VIII and higher	0.27	Major damage

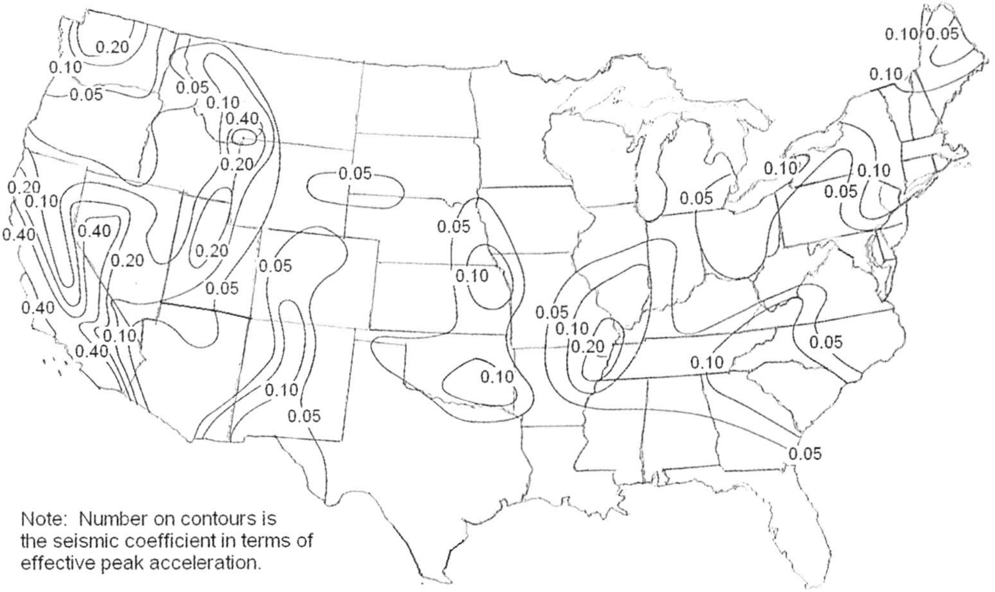

Fig. 2-12. *Seismic coefficients of the continental United States (Applied Technology Council 1978)*

An earthquake probability map was published recently by the U.S. Geological Survey (Petersen et al. 2008). This map uses the seismic coefficient as an indicator for earthquake with a 90% probability not likely to be exceeded in 50 years. The map was printed in color to show the seismic coefficients in different parts of the continental United States, as presented in Section 1.7 of the companion volume.

The earthquake analysis presented here is called the psuedostatic method. For high-risk dams in seismically active regions or with fine-grained soils subjected to liquefaction, more sophisticated dynamic analyses, as suggested by Seed et al. (1975a, 1975b), should be used.

The effective stress analysis is applicable only to granular materials or fine-grained soils with a degree of saturation less than 70%, so there is no change in pore pressure during an earthquake. If the seismic excitation causes a significant change in pore pressure, total stress analysis should be used. Total stress analysis should also be used if the earthquake takes place during or at the end of construction. Following the findings of Makdisi and Seed (1978), most authorities recommend that 80% of the static shear strength should be used for pseudostatic analysis and that the minimum required seismic factor of safety might range from 1 to 1.15.

2.5 Factors of Safety

In the stability analysis of slopes, many design factors cannot be determined with certainty. Therefore, a degree of risk should be assessed in an adopted design. The factor of safety fulfills this requirement. The factor should take into account not only the uncertainties in the design parameters but also the consequence of failure. Where the consequences of failure are slight, a greater risk of failure or a lower factor of safety may be acceptable.

The potential seriousness of failure is related to many factors other than the size of the project. A low dam located above or close to inhabited buildings can pose a greater danger than a high dam in a remote location. Often, the most potentially dangerous types of failure involve soils that undergo a sudden release of energy without much warning. This is true for soils subjected to liquefaction that have a low ratio between the residual and peak strength. For earth slopes composed of intact homogeneous soils, when the strength parameters have been chosen on the basis of good laboratory tests and a careful estimate of pore pressure has been made, a safety factor of at least 1.5 is commonly employed (Lambe and Whitman 1969). With fissured clays and for nonhomogeneous soils, larger uncertainties generally will exist, and more caution is necessary.

The factors of safety suggested by the U.S. Army Corps of Engineers (USACE 1986) are listed in Table 2-6. The factors of safety presented in Table 2-6 may be modified under the following conditions:

1. For slopes where either sliding or large deformations have occurred, and back-analyses have been performed to establish design shear strengths,

Table 2-6. *Factors of Safety Recommended by U.S. Army Corps of Engineers (USACE 1986)*

Types of Slopes	End of Construction	Long-Term Steady Seepage	Rapid Drawdown[a]
Slopes of dams, levees, and dikes, and other embankment and excavation slopes	1.3	1.5	1.0–1.2

[a] $F = 1.0$ applies to drawdown from maximum surcharge pool, for conditions where these water levels are unlikely to persist long enough to establish steady seepage. $F = 1.2$ applies to maximum storage pool level, likely to persist for long periods prior to drawdown. For slopes in pumped storage projects, where rapid drawdown is a normal operating condition, a higher factor of safety (e.g., 1.3 to 1.4) should be used.

a lower factor of safety may be used. In such cases, probabilistic analyses may be useful in supporting the use of lower factors of safety for design. Lower factors of safety also may be justified when the consequences of failure are small.

2. Temporary excavated slopes are sometimes designed only for short-term stability, with knowledge that long-term stability would be inadequate. Special care, and possibly higher factors of safety, should be used in such cases.
3. The factors of safety are based on experience and are applicable only to U.S. Army Corps of Engineers projects, where methods of exploration, testing, and analysis are consistent and the degree of uncertainty does not vary widely. For other situations involving different engineering practices, the factors of safety shown in Table 2-6 may not be appropriate.

Table 2-7 shows the factors of safety suggested by various sources for mining operations (D'Appolonia Consulting Engineers 1975; *Federal Register* 1977; Mines Branch, Canada 1972; National Coal Board 1970). All of these stipulations are based on the assumptions that the most critical failure surface is used in the analysis, that strength parameters are reasonably representative of the actual case, and that sufficient construction control is ensured.

Summary

1. There are three types of failure surface: circular, noncircular, and composite. The composite failure surface is mainly circular, but when the circle intercepts a weak soil layer, the failure surface will follow the bottom of the weak layer, so part of the circular failure surface is replaced by one or more plane failure surfaces. If the weak layer extends to the surface of the slope, the failure surface also will appear on the slope surface.

Table 2-7. *Factors of Safety Suggested for Mining Operations*

	UNITED STATES (*FEDERAL REGISTER*, 1997)	MINIMUM SAFETY FACTOR
I	End of construction	1.3
II	Partial pool with steady seepage saturation	1.5
III	Steady seepage from spillway or decant crest	1.5
IV	Earthquake (cases II and III with seismic loading)	1.0

UNITED STATES (D'APPOLONIA CONSULTING ENGINEERS, INC., 1975)	SUGGESTED MINIMUM FACTORS OF SAFEY WITH HAZARD POTENTIAL		
	High	Moderate	Low
Designs based on shear strength parameters measured in the laboratory	1.5	1.4	1.3
Designs that consider maximum seismic acceleration expected at the site	1.2	1.1	1.0

BRITAIN (NATIONAL COAL BOARD, 1970)	FACTOR OF SAFETY	
	I*	II**
(1) For slip surfaces along which the peak shear stress is used.	1.5	1.25
(2) For slip sufaces passing through a foundation statum that is at its residual shear strength (slip circles wholly within the bank should satisfy (1))	1.35	1.15
(3) For slip surfaces passing along a deep vertical subsidence crack where no shear strength is mobilized and that is filled with water (slip surfaces wholly within intact zones of bank and foundations should satisty (1)).	1.35	1.15
(4) for slip surfaces where both (2) and (3) supply.	1.2	1.1

CANADA (MINES BRANCH, CANADA, 1972)	FACTOR OF SAFETY	
	I*	II**
Design is based on peak shear strength parameters	1.5	1.3
Design is based on residual shear strength parameters	1.3	1.2
Analyses that include the predicted 100-year return period accelerations applied to the potential failure mass	1.2	1.1
For horizontal sliding on base of dike in seismic areas asuming shear strength of the fine refuse in impoundment reduced to zero	1.3	1.3

*where there is a risk of danger to persons or property
**where no risk of danger to persons or property is anticipated

2. Examples are given to illustrate the Fellenius and normal methods for determining the factor of safety, using the method of slices. Both methods can be performed easily by hand calculations with a pocket calculator. Because of the repeated applications of the same formula to each slice, it is more convenient to solve a problem by the use of a spreadsheet.

3. The difference between the Fellenius and the normal methods lies in the treatment of pore pressure for determining the effective normal stress on the failure surface. When the failure surface inclines at a steep angle with the horizontal, the effective normal stress obtained by the Fellenius method may become negative. To avoid this unreasonable occurrence, the normal method uses the submerged unit weight to compute the effective stress, instead of the pore pressure. If there is no seepage, both methods are identical. Otherwise, the factor of safety obtained by the normal method is slightly greater than that by the Fellenius method.

4. Sliding-block analysis, which only satisfies force equilibrium, can be used for simple cases of noncircular failures consisting of no more than three failure planes. The factor of safety depends on the angle of internal friction between the two blocks, and the greater the angle, the greater the factor of safety. The assumption of no friction between the two blocks gives the smallest factor of safety and the most conservative design.

5. Two types of analysis can be performed: total stress analysis for short-term stability and effective stress analysis for long-term stability. The total stress analysis is based on the undrained shear strength and includes s_u-analysis for saturated soils and c_u, ϕ_u-analysis for partially saturated soils. Because the effect of pore pressure has been considered already in determining the shear strength, no pore pressure should be used in the analysis. The effective stress analysis is based on the drained shear strength and also is called c', ϕ'-analysis. Instead of drained tests, the effective shear strength parameters, c' and ϕ', usually are determined by consolidated undrained tests with pore pressure measurements.

6. For fill slopes, the total stress analysis for short-term stability is more critical than the effective stress analysis for long-term stability, so the total stress analysis should be undertaken before the effective stress analysis. For cut slopes, the effective stress analysis for long-term stability is more critical than the total stress analysis for short-term stability, so the effective stress analysis for long-term stability needs to be performed first. In many cases, the total stress analysis is not needed unless some wet, fine-grained soils with a degree of saturation in the range of 70% or higher are encountered.

7. The pore pressure on the failure surface can be obtained by multiplying the depth below the phreatic surface with the unit weight of water. Theoretically, this definition of pore pressure is not correct because the pore pressure should represent the piezometric surface; that is, the water level

in a piezometric tube located at the failure surface However, because of the gentle slope of the phreatic surface, the difference between the phreatic and piezometric surfaces is quite small, and the use of the phreatic surface is on the safe side.

8. The effective stress analysis can be used to analyze rapid drawdown by assuming that there is no change in effective stress after the drawdown. If rapid drawdown takes place during or at the end of construction, the undrained shear strength used for short-term stability also can be used for the analysis of rapid drawdown.

9. The seismic factor of safety is based on the psuedostatic method using a seismic coefficient. For silty soils susceptible to liquefaction, or for high-risk dams in seismically active regions, more sophisticated dynamic analyses should be used. A lower shear strength, say 80% of the static shear strength, and a lower acceptable factor of safety, say 1.1 to 1.15, may be used for pseudostatic analysis.

10. The factor of safety required for a given project should take into account not only the uncertainties in design parameters but also the consequence of failure. Where the consequences of failure are slight, a greater risk of failure or a lower factor of safety may be acceptable. The minimum required factors of safety for several agencies are presented. The minimum factor of safety established by one agency may be different from another because of differences in engineering practices in exploration, testing, and analysis.

Problems

2.1 Fig. P2-1 shows the cross section of a fill. Because a thin layer of weaker material with a cohesion of 160 psf and a friction angle of 24° exists at the bottom of the fill, the potential failure surface is noncircular along the bottom of the fill. If the fill has a total unit weight of 125 pcf and there is no friction between the two sliding blocks, determine the factor of safety. [Answer: 1.472]

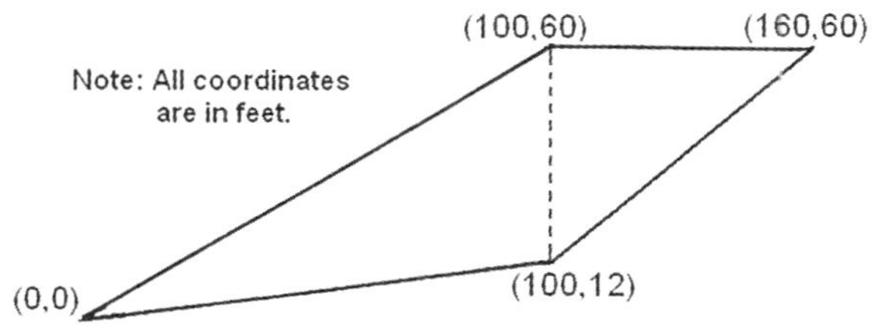

Fig. P2-1.

2.2 Solve Example 2.3 by the use of a spreadsheet.
[Answer: 1.022]

2.3 Solve Example 2.4 by the use of a spreadsheet.
[Answer: 1.180]

2.4 Fig. P2-4 shows the dimensions of the slope and the location of the failure circle and the phreatic surface. The soil has a cohesion of 200 psf, a friction angle of 30°, and a total unit weight of 125 pcf. If the failure mass is divided into eight slices, determine the factor of safety by the Fellenius method.
[Answer: 1.029]

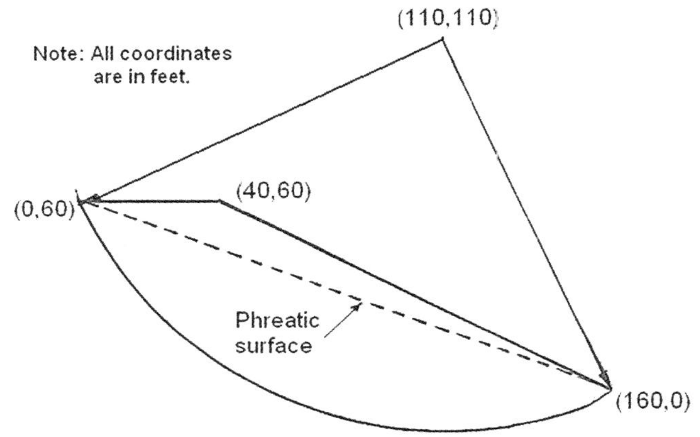

Fig. P2-4.

2.5 Same as Problem 2.4, but determine the factor of safety by the normal method.

[Answer: 1.208]

Chapter 3

Shear Strength

Shear strength is one of the most important factors that affect the factor of safety. A subsurface investigation is needed to determine the nature and extent of materials underground, the location of the sites for field tests, and the collection of representative samples for laboratory tests. The use of field tests, such as standard penetration, Dutch cone, piezocone penetration, and vane shear—together with laboratory direct shear and triaxial compression tests to determine the shear strength of soils—is described. Also presented are the ranges of shear strength for various materials, including municipal wastes, and correlations of effective friction angle or undrained shear strength with other index properties, such as plasticity index, liquid limit, and percent of clay. Other subjects include the shear strength of granular soils with a curved strength envelope, and the back-calculation of shear strength from failed slopes.

3.1 Subsurface Investigations

The shear strength of soils can be determined by field or laboratory tests. No matter what tests are used, it is necessary to conduct an overall geologic appraisal of the site, followed by a planned subsurface investigation. The purpose of the subsurface investigation is to determine the nature and extent of each type of material that may have an effect on the stability of the slope. For simple cases, a detailed knowledge of the slope from toe to crest is essential. For more complex cases, knowledge outside of this zone also should be considered. Fills situated over a deep layer of clays and silts may merit expensive drilling. Auger holes, pits, or trenches will suffice for smaller fills or those with bedrock only a short distance below the surface.

The log of boring forms the permanent record used for design. Either disturbed or undisturbed samples can be taken while boring. To obtain reliable results, the strength parameters should be determined from undisturbed or remolded samples. However, the effective strength parameters of saturated granular soils and silt clays are not affected significantly by the moisture content and density, so disturbed samples may be used for the direct shear test to determine the effective strength. It is difficult to generalize the appropriate number, depth, and spacing of borings required for a project as these will depend on a variety of factors, such as site conditions, size of the project, among others. Often the final location of borings should be made in the field, and additions must be made to the boring program based on the information from the boring already completed. During boring, the depth to the groundwater table also should be determined.

3.2 Field Tests

There are a variety of field tests for determining the shear strength of soils in the field. However, only the standard penetration test, the Dutch cone test, the piezocone test, and the vane shear test will be discussed here. These tests are applicable to soils free from substantial gravel or cobble-sized particles.

3.2.1 Standard Penetration Test

This test also is used to collect soil samples. When a borehole is extended to a given depth, the drill tools are removed and a standard split-spoon sampler is lowered to the bottom of the borehole. The sampler is driven into the soil by a 140-lb (623-N) hammer with a 30-in. (0.76-m) drop. The number of blows required for the spoon to penetrate three 6-in. (152-mm) intervals is recorded. The standard penetration number, generally referred to as the N-value, is obtained by adding the blow counts for the last two intervals. The sampler then is removed, and the soil sample is recovered, placed in a glass jar, and shipped to the laboratory. The method is specified by ASTM D1586 "Standard Test Method for Penetration Test and Split-Barrel Sampling of Soils" (ASTM 2010).

For sands, Fig. 3-1 gives a relationship between effective friction angle, ϕ', and N-value based on the effective overburden pressure, σ'_{vo}, during the field test. If there is no water table, the effective overburden pressure is equal to the depth below the ground surface multiplied by the total unit weight of the soil. If part of the overburden is below the water table, the submerged unit weight should be used for the overburden below the water table.

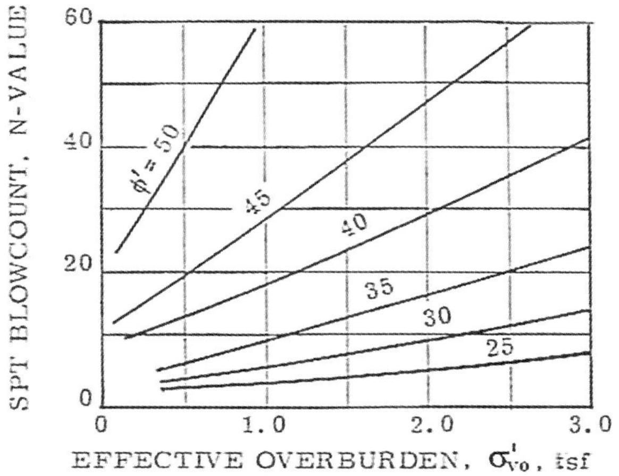

Fig. 3-1. Blow count versus effective friction angle for sand (Schmertmann 1975. Reproduced with permission)
Note: 1 tsf = 95.8 kN/m²

Example 3.1 A standard penetration test was performed on sand at a depth of 15 ft (4.6 m) below the ground surface. The water table is located 10 ft (3.05 m) below the surface. The soil has a total unit weight of 125 pcf (19.6 kN/m^3). If the *N*-value obtained from the test is 14, determine the effective friction angle of the soil.

Solution Effective overburden pressure $\sigma'_{vo} = [10 \times 125 + 5 \times (125 - 62.4)]/2{,}000 = 0.78$ tsf. From Fig. 3.1, $\phi' = 39°$.

For clays, Fig. 3-2 shows the correlation between the standard penetration test and the unconfined compressive strength. The undrained shear strength, s_u, is equal to one-half of the unconfined compressive strength, q_u.

If the clay normally is consolidated with $N <$ depth in feet/5 (depth in meters/1.5), Schmertmann (1975) suggests

$$s_u(tsf) \geq N/15 \tag{3-1}$$

$$s_u(kN/m^2) \geq 6.4N \tag{3-2}$$

in which s_u = undrained shear strength and N = blow count per foot of penetration.

It is worthy of note that blow count is not a reliable method for determining the shear strength of clays. The use of this method should proceed with caution, and check some samples with the laboratory unconfined compression tests.

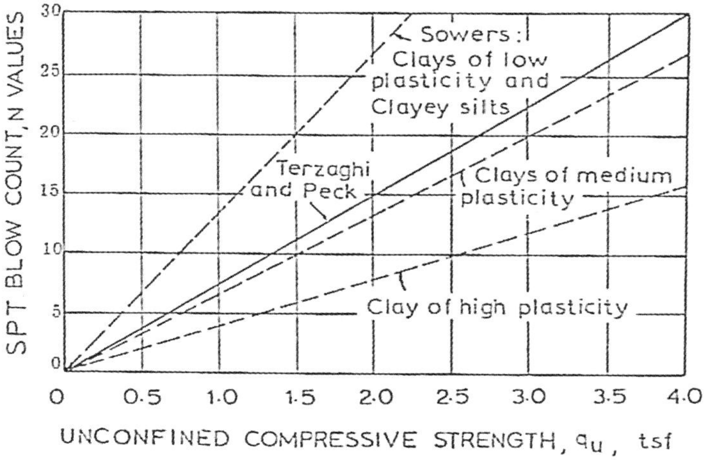

Fig. 3-2. *Blow count versus unconfined compressive strength for clays (U.S. Navy 1971)*
Note: 1 tsf = 95.8 kN/m²

Example 3.2 The standard penetration test on a silty clay gives a blow count of 8 at a depth of 50 ft (15.3 m) from the ground surface. If the silty clay is of medium plasticity, determine the undrained shear strength.

Solution Given $N = 8$ and clay of medium plasticity, from Fig. 3-2, unconfined compressive strength $q_u = 1.3$ tsf. The undrained shear strength is equal to one-half of the unconfined compressive strength, so $s_u = 0.5 \times 1.3 = 0.65$ tsf.

If Eq. (3-1) is used, the depth of the borehole must be greater than 5*N*, or 40 ft. The actual depth is 50 ft, so Eq. (3-1) is valid. From Eq. (3-1), $s_u = 8/15 = 0.53$ tsf, which compares with 0.65 tsf from Fig. 3-2. The discrepancy is expected, because these empirical correlations are approximate at best. As can be seen from Fig. 3-2, the unconfined compressive strength for clays of low plasticity is 3.3 times greater than those of high plasticity, so a small change in plasticity will have a large effect on the undrained shear strength.

3.2.2 Dutch Cone Test

This test, originally developed by the Dutch engineer P. Barentsen (Broms and Flodin 1988), has been used for soil exploration since the early 1930s. A 60° cone with a base area of 1.55 in.^{2} (10 cm^2) is pushed into the ground at a steady rate of about 0.8 in./s (20 mm/s). The resistance to the penetration of the cone and the frictional resistance of the sleeve are measured. The test also is called the cone penetration test (CPT) and does not require the drilling of a borehole. Electrical versions were developed in late 1940s by the Delft Soil Mechanics Laboratory, which offered continuous measurements of tip resistance and sleeve friction with depth and direct strip chart plotting of the sounding record.

Based on the CPT calibration tests from five sands, Robertson and Campanella (1983) proposed the following expression for the peak friction angle of clean quartz sand:

$$\phi' = \tan^{-1}\left[0.1 + 0.38\log\left(\frac{q_c}{\sigma'_{vo}}\right)\right] \tag{3-3}$$

in which q_c = cone resistance and σ'_{vo} = effective overburden pressure. Eq. (3-3) can be used to plot the chart shown in Fig. 3-3.

Another expression derived from calibration tests of 24 sands was proposed by Kulhawy and Mayne (1990):

$$\phi' = 17.6 + 11\left[\log\left(\frac{q_c}{p_a}\right) - 0.5\log\left(\frac{\sigma'_{vo}}{p_a}\right)\right] \tag{3-4}$$

in which p_a = atmospheric pressure, or 1.06 tsf.

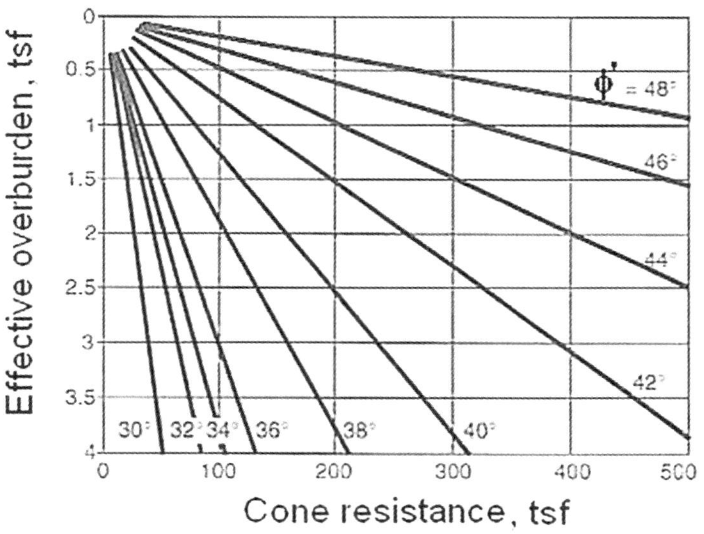

Fig. 3-3. Cone resistance versus effective friction angle for sands (Robertson and Campanella 1983. Reproduced with permission from IOS Press)
Note: 1 tsf = 95.8 kN/m²

Example 3.3 When the Dutch cone penetrates to a sand deposit at a depth of 35 ft (10.7 m) below the ground surface, the recorded cone resistance is 100 tsf (9.6 MN/m^2). If the total unit weight of sand is 130 pcf (20.5 kN/m^3) and the water table is 10 ft (3.1 m) below the ground surface, estimate the effective friction angle of the sand.

Solution The effective overburden pressure $\sigma'_{vo} = 10 \times 130 + 25 \times (130 - 62.5) = 2{,}987.5$ psf or 1.49 tsf. Given q_c = 100 tsf, from Fig. 3-3, ϕ' = 39°, which checks with the 38.5° from Eq. (3-3), or $\phi' = \tan^{-1}[0.1 + 0.38 \log(100/1.49)] = 38.5°$.

If Eq. (3-4) is used, $\phi' = 17.6 + 11 \times [\log(100/1.06) - 0.5 \times \log(1.49/1.06)] = 38.5°$, which is the same as from Eq. (3-3).

For clays, the correlation between the cone resistance and the undrained shear strength depends on the overconsolidation ratio, which is the ratio between the maximum precompression and the existing overburden pressure. An equation used by many engineers for clays that are not highly sensitive with an overconsolidation ratio less than 2 and a plasticity index greater than 10 is (Schmertmann 1975),

$$s_u = \frac{q_c - \sigma'_{vo}}{16} \tag{3-5}$$

Drnevich et al. (1974) showed that

$$s_u = 0.8 \times \text{friction sleeve resistance} \tag{3-6}$$

3.2.3 Piezocone Penetration Test

The curves presented in Fig. 3-3 are designed for cohesionless sand but are also applicable to fine-grained soils with a small cohesion. However, it may not be used for clayey soils because the penetration of the cone will generate considerable excess pore pressures, which affect significantly not only the load applied to the cone but also the effective stress in the soil. The introduction of the piezocone in the mid 1970s permits the simultaneous measurements of cone resistance, q_c, sleeve friction, f_s, and pore pressure, u_t, and provides new possibilities for soil identification and classification and the interpretation of soil parameters. This type of CPT with pore pressure measurements is called CPTU and is specified by ASTM D5778 "Standard Test Method for Performing Electric Friction Cone and Piezocone Penetration Testing of Soils" (ASTM 2010).

Piezocones are normally available in two standard sizes: (1) a 35.7-mm (1.4-in.) diameter version with cone area A_c = 10 cm^2 (1.55 in.^{2}) and sleeve area A_s = 150 cm^2 (23.3 in.^{2}), and (2) a 44-mm (1.75-in.) diameter version with A_c = 15 cm^2 (2.33 in.^{2}) and sleeve area A_s = 200 to 300 cm^2 (31 to 46.5 in.^{2}). Although the 10-cm^2 (1.55-in.^{2}) version is the original standard size, many commercial firms have found that the 15-cm^2 (2.33-in.^{2}) version is stronger for routine profiling and is more easily outfitted with additional sensors for specific needs. Because the rod size is normally 35.7 mm (1.4 in.) in diameter, the 15-cm^2 (2.33-in.^{2}) cone also tends to open a larger hole and thus reduce side rod friction during pushing. Fig. 3-4 shows a schematic of a piezocone and its calibration. This figure actually is divided into four parts and described as follows:

In Fig. 3-4(a), when the rod is pushed downward, the cone and the friction sleeve also will move down, so the direction of the reactive forces P_c beneath the cone and P_f around the sleeve is upward. The two strain gauges at the lower location are used to measure P_c and the other two gauges at the higher location to measure $P_c + P_f$. Knowing P_c and $P_c + P_f$, P_f can be obtained by subtraction. Division of P_c by the area of the cone, A_c, gives the cone resistance, q_c, and the division of P_f by the surface area of sleeve gives the sleeve friction, f_s. Note that pore pressure has not been considered in these definitions of q_c and f_s.

Fig. 3-4(b) illustrates the effect of pore pressure on q_c and f_s. Because the cone and the friction sleeve are two different pieces separated by a small space, pore pressures will exert on the top of the cone and the bottom of the sleeve, as indicated by the two arrows identified by u_t in the figure. The total pore pressure, u_t, consists of the hydrostatic pore pressure, u_o, due to the phreatic surface and the excess pore pressure, Δu, due to penetration of the cone, or

$$u_t = u_o + \Delta u \tag{3-7}$$

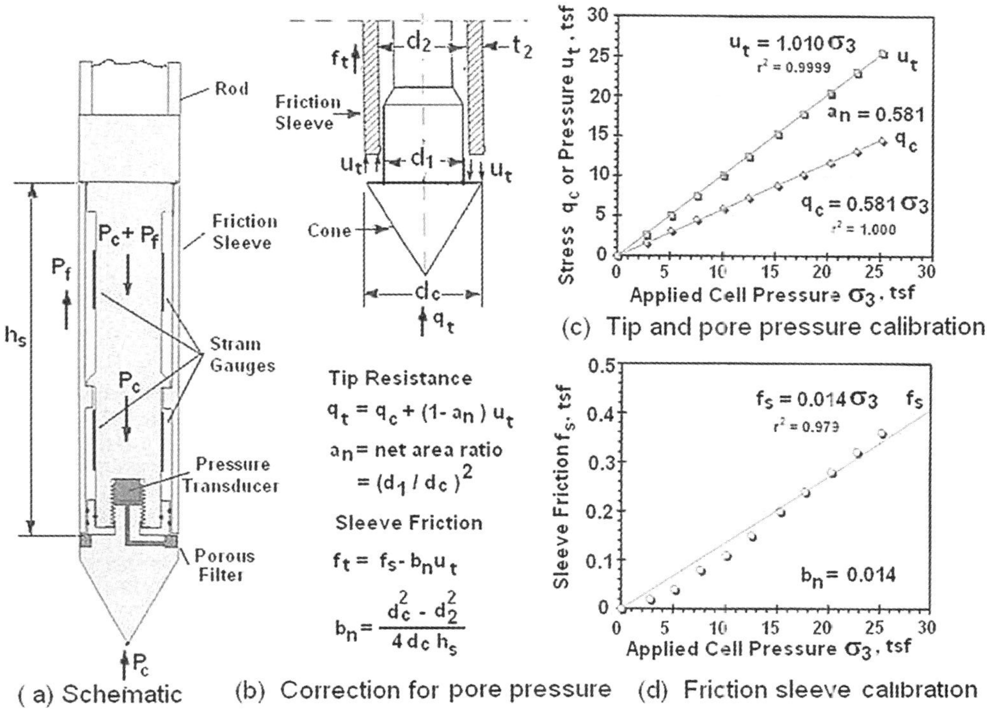

Fig. 3-4. Schematic and calibration of a piezocone
Note: 1 tsf = 95.8 kN/m²

The downward pore pressure on the cone increases the cone resistance from q_c to q_t, or

$$q_t = q_c + \left(\frac{A_c - A_1}{A_c}\right)u_t = q_c + \left(1 - \frac{A_1}{A_c}\right)u_t = q_c + (1 - a_n)u_t \tag{3-8}$$

in which A_c = area of cone with a diameter of d_c, A_1 = area with a diameter of d_1, and a_n = net area ratio.

The upward pore pressure on the friction sleeve decreases the frictional resistance from f_s to f_t, or

$$f_t = f_s - \left(\frac{d_c^2 - d_2^2}{4d_c h_s}\right)u_t = f_s - b_n u_t \tag{3-9}$$

in which h_s = height of friction sleeve and b_n = cross section and surface area ratio. Although values of a_n and b_n can be computed from the dimensions of the piezocone, as shown in Eqs. (3-8) and (3-9), it is more accurate and reliable to calibrate them directly in a triaxial chamber.

Fig. 3-4(c) demonstrates the calibration of u_t and q_c under various chamber pressures, σ_3. The u_t registered by the pressure transducer should be equal to σ_3, as expected. The plot of q_c versus σ_3 results in a straight line with a slope a_n of 0.581, because the q_c registered by the lower strain gauges represents the difference in pore pressure at the top of the cone relative to that at the bottom, expressed as an area ratio A_1/A_c.

Fig. 3-4(d) illustrates the calibration of f_s. The slope of the straight line gives a b_n of 0.014. Given q_c and f_s together with the calibrated a_n and b_n, the corrected q_t and f_t can be computed by Eqs. (3-8) and (3-9), respectively.

The following procedures for determining the effective friction angle of both granular and fine-grained soils were developed by the Norwegian Institute of Technology, as reported by Senneset et al. (1989).

The effective shear strength of fine-grained soils can be approximated by a straight line, as shown in Fig. 3-5. The negative intercept at the normal stress axis is called the effective attraction, a', which is related to the effective cohesion by

$$c' = a' \tan \phi' \tag{3-10}$$

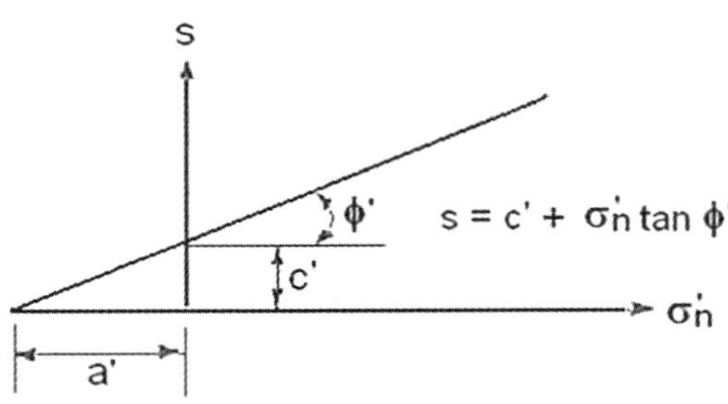

Fig. 3-5. Effective shear strength of fine-grained soils

For a given type of soil, the effective friction angle, ϕ', depends on two dimensionless parameters: cone resistance number, N_m, and pore pressure ratio, B_q, defined as

$$N_m = \frac{q_t - \sigma_{vo}}{\sigma'_{vo} + a'} \tag{3-11}$$

and

$$B_q = \frac{u_t - u_o}{q_t - \sigma_{vo}} \tag{3-12}$$

in which q_t = corrected cone resistance, σ_{vo} = total overburden pressure, σ'_{vo} = effective overburden pressure, u_t = measured pore pressure, u_o i= initial pore pressure due to phreatic surface, and a' = effective attraction. The effective attraction can be obtained from the CPTU records by plotting σ'_{vo} versus $q_t - \sigma_{vo}$, as indicated by Eq. (3-11) and Fig. 3-6. This method is applicable when relatively homogeneous soil deposits or layers are encountered.

In cases where such estimates cannot be obtained, it is suggested that typical values from triaxial tests on similar soils be used, as shown in Table 3-1. Table 3-1 also gives the typical ranges of shear strength parameters for some common soil types and may be useful for evaluating the parameter values interpreted from CPTU data. The soil type can be identified from q_t and B_q by Fig. 3-7.

Fig. 3-8 can be used to determine $\tan\phi'$ for three different types of soil, based on the values of N_m and B_q.

Mayne and Campanella (2005) indicated that the charts for most soils with a small cohesion, as shown in Fig. 3-8(a), can be approximated by

$$\phi'(\text{in degrees}) = 29.5 B_q^{0.121}[0.256 + 0.336 B_q + \log N_m] \tag{3-13}$$

Eq. (3-13) is applicable for $0.1 < B_q < 1.0$ and $20° < \phi' < 45°$.

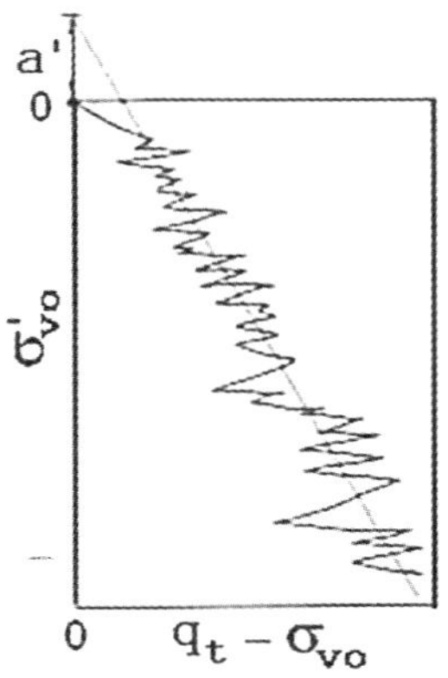

Fig. 3-6. Determination of effective attraction from CPTU data

Table 3-1. Typical Values of Attraction, Friction Angle, and Other Parameters

Soil Type	Consistency	a' (psf)	$\tan\phi'$	ϕ'	N_m	B_q
Clay	Soft	100–200	0.35–0.45	19–24	1–3	0.8–1.0
	Medium	200–400	0.40–0.55	19–29	3–5	0.6–0.8
	Stiff	400–1,000	0.50–0.60	27–31	5–8	0.3–0.6
Silt	Soft	0–100	0.50–0.60	27–31	—	—
	Medium	100–300	0.55–0.65	29–33	5–30	0–0.4
	Stiff	300–600	0.60–0.70	31–35	—	—
Sand	Loose	0	0.55–0.65	29–33	—	—
	Medium	200–400	0.60–0.75	31–37	30–100	<0.1
	Dense	400–1,000	0.70–0.90	35–42	—	—
Hard, stiff soil, over-consolidated, cement		>1,000	0.8–1.0	38–45	100	<0

(Senneset et al. 1989, © National Academy of Sciences, Washington, DC. Reproduced with permission of the Transportation Research Board, Washington, DC)
Note: 1 psf = 0.048 kN/m^2

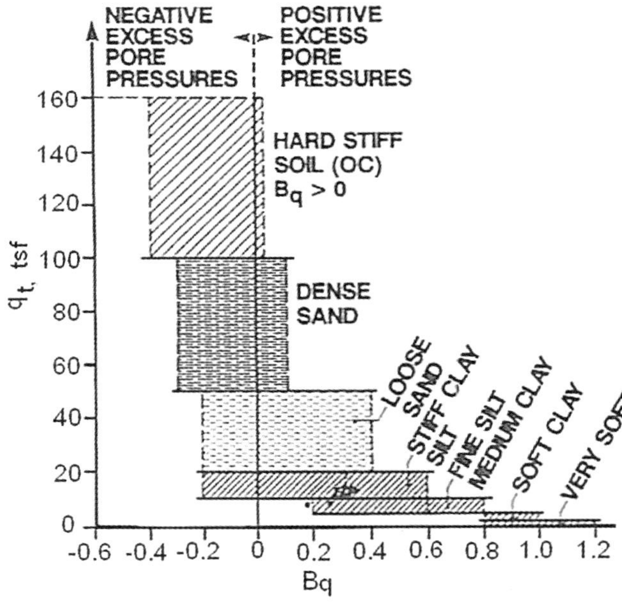

Fig. 3-7. Chart for classification of soil type by CPTU data (Senneset et al. 1989, © National Academy of Sciences, Washington, DC. Reproduced with permission of the Transportation Research Board, Washington, DC)
Note: 1 tsf = 0.1 MN/m^2

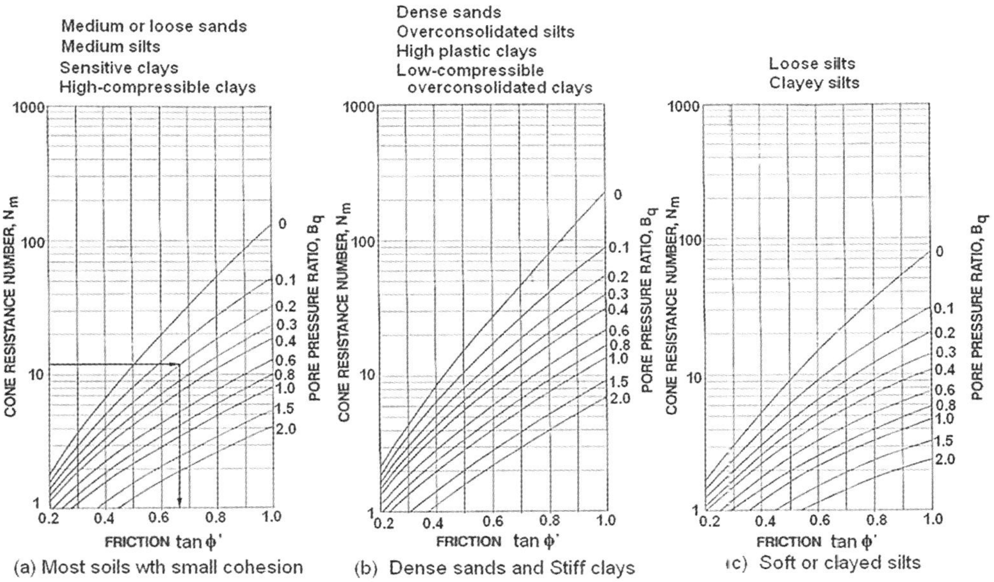

Fig. 3-8. *Charts for determining friction tan φ′ (Senneset et al. 1989, © National Academy of Sciences, Washington, DC. Reproduced with permission of the Transportation Research Board, Washington, DC)*

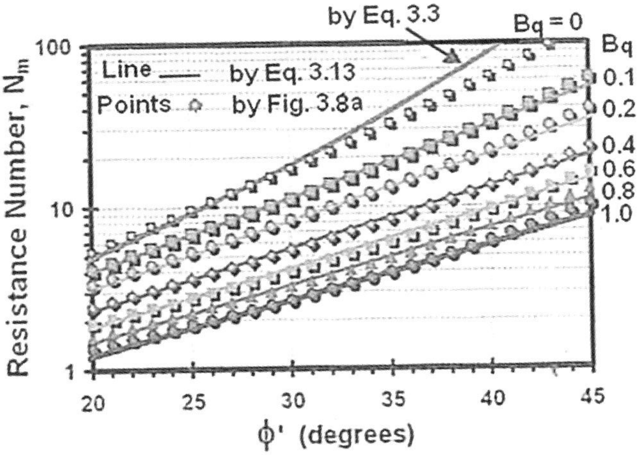

Fig. 3-9. *Comparison of effective friction angle based on Eq. (3-13) and Fig. 3-8(a) (Mayne and Campanella 2005. Reproduced with permission from IOS Press)*

Fig. 3-9 is a comparison of effective friction angles, φ′, computed by Eq. (3-13), as shown by the straight lines, versus those obtained from Fig. 3-8(a), as shown by the individual points. It can be seen that, in the applicable ranges, both check very closely. For granular soils with $B_q = 0$, Eq. (3-13) does not work, so

Eq. (3-3) is plotted instead. The comparison between using Eq. (3-3) and Fig. 3-8(a) indicates that the results check very well when ϕ' is smaller than 30°; however, when ϕ' is greater than 40°, Eq. (3-3) gives a ϕ' value two to three degrees lower than that obtained by Fig. 3-8(a). Because Eq. (3-3) is more conservative, it is suggested that Eq. (3-3) be used for granular material with $B_q = 0$.

Example 3.4 Piezocone penetration tests were conducted on a site with the water table 10 ft (3.1 m) below the surface. At a depth of 20 ft (6.1 m) below the ground surface is a layer of medium clay with a cone resistance, q_c, of 7,500 psf (359 kN/m^2) and a total pore pressure, u_t, of 5,500 psf (263 kN/m^2). If the net area ratio, a_n, of the piezocone is 0.581 and the average unit weight of soil in the top 20 ft (6.1 m) is 125 pcf (19.7 kN/m^3), estimate the effective cohesion and angle of internal friction of the clay.

Solution From Eq. (3-8), $q_t = 7{,}500 + (1 - 0.581) \times 5{,}500 = 9{,}805$ psf. $\sigma_{vo} = 125 \times 20 = 2{,}500$ psf. $\sigma'_{vo} = 2{,}500 - 62.4 \times 10 = 1{,}876$ psf. Assuming $a' = 300$ psf (average for medium clay in Table 3-1), from Eq. (3-11), $N_m = (9{,}805 - 2{,}500)/(1{,}876 + 300) = 3.357$. From Eq. (3-12), $B_q = (5{,}500 - 624)/(9{,}805 - 2{,}500) = 0.667$. From Fig. 3-8(a), $\tan\phi' = 0.55$, or $\phi' = 28.8°$. From Eq. (3-10), $c' = 300 \times \tan 28.8° = 164.9$ psf. If Eq. (3-13) is used, $\phi' = 29.5 \times (0.667)^{0.121}[0.256 + 0.336 \times 0.667 + \log 3.357] = 28.3°$, which compares with the 28.8° obtained from the chart. In this example, the soil type is given as medium clay. If this information is not given, it can be found from Fig. 3-7 that, with $q_t = 4.9$ tsf and $B_q = 0.667$, the soil is classified as medium clay.

It should be noted that, when the effective cohesion is small, the arbitrary assumption of a' has very little effect on ϕ'. For example, if $a' = 200$ psf, $N_m = (9{,}805 - 2{,}500)/(1{,}876 + 200) = 3.52$ and, from Eq. (3-13), $\phi' = 29.3°$; if $a' = 400$ psf, $N_m = (9{,}805 - 2{,}500)/(1{,}876 + 400) = 3.21$ and, from Eq. (3-13), $\phi' = 27.7°$. Even if the assumed a' is not correct, the use of a larger a' will be compensated by a decrease in ϕ', so the net effect on shear strength may not be significant.

The relationship between undrained shear strength, s_u, and cone resistance, q_t, can be expressed as

$$s_u = \frac{q_t - \sigma_{vo}}{N_c} \tag{3-14}$$

in which N_c = cone factor. The cone factor usually is determined from a reference value for s_u, either from a field vane shear test or a laboratory triaxial compression test. Values of N_c seem to range from 10 to 15 for normally consolidated clays, and from about 15 to 19 for overconsolidated clays. Because of the large scatter in values of N_c, local correlations at the site are highly recommended.

3.2.4 Vane Shear Test

This test commonly is used for determining the undrained strength of clays in situ. The test basically consists of placing a four-blade vane in the undisturbed

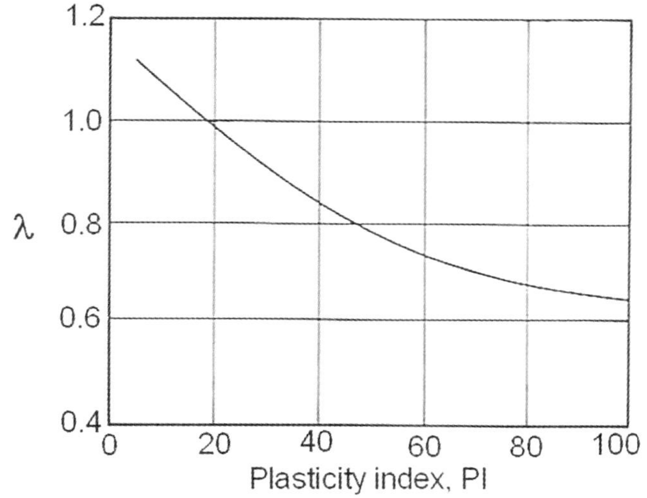

Fig. 3-10. Correction factor for vane shear test (Bjerrum 1972. Reproduced with permission)

soil and rotating it from the surface to determine the torsional force required to cause a cylindrical surface to be sheared by the vane. This force then is converted to the undrained shear strength. Both the peak and residual undrained strength can be determined by measuring the maximum and post-maximum torsional forces. The method is specified by ASTM D2573 "Standard Test Method for Field Vane Shear Test in Cohesive Soils" (ASTM 2010). Because of the difference in failure mode, the results of field vane tests do not always agree with other shear tests. Bjerrum (1972) recommended that, depending on the plasticity index, PI, of the soil, the undrained shear strength obtained from the field vane shear test be multiplied by a correction factor, λ, as shown in Fig. 3-10.

Example 3.5 In the previous example, field vane shear tests on the clay layer give an undrained shear strength of 650 psf (31.1 kN/m^2). If the clay has a plasticity index of 22, determine the cone factor, N_c, in Eq. (3-14).

Solution From Fig. 3-10, $\lambda = 0.9$, so $s_u = 0.9 \times 650 = 585$ psf. From Example 3.4, $q_t = 9{,}805$ psf and $\sigma_{vo} = 2{,}500$ psf. From Eq. (3-14), $N_c = (9{,}805 - 2{,}500)/585 = 12.5$.

3.3 Laboratory Tests

Laboratory tests complement field tests to give a more complete picture of the materials within the slope and their engineering properties. Furthermore, it is possible in the laboratory to establish the changes in soil behavior due to the changes in environment. For example, the construction of an embankment certainly will affect the shear strength in the foundation soils. Field tests before construction cannot establish these changes, whereas laboratory tests can simulate these changes as they occur in the field.

The major laboratory tests for determining the shear strength of soils include the direct shear test, the triaxial compression test, and the unconfined compression test. The direct shear test is very easy to conduct because of its simplicity. Because of the thin specimen used, drained conditions exist for most materials except for the highly plastic clays. Therefore, the direct shear test usually is used to determine the effective shear srrength. The triaxial compression test can be used for determining either the total strength or the effective strength. The unconfined compression test is similar to the triaxial compression test but without the confining pressure and can only be used to determine the undrained shear strength.

3.3.1 Direct Shear Test

Fig. 3-11 shows a schematic of the direct shear box. The soil sample is placed between two porous stones to facilitate drainage. The normal load is applied to the sample by placing weights in a hanger system. The shear force is applied by the piston driven by an electric motor. The horizontal displacement is measured by a horizontal dial gauge and the shear force by a proving ring and load dial, which are not shown in the figure. Instead of the dial gauge, proving ring, and hanging weights, LVDT and load cell with automatic recording devices are available, but the basic arrangement is the same. The method is specified by ASTM D3080 "Direct Shear Test of Soils under Consolidated Drained Condition" (ASTM 2010).

For silty clays and coal refuse, Huang (1978b) found that their effective shear strength can be determined easily by the direct shear tests, which check closely with the results of triaxial compression tests with pore pressure measurements. His suggested procedure is as follows.

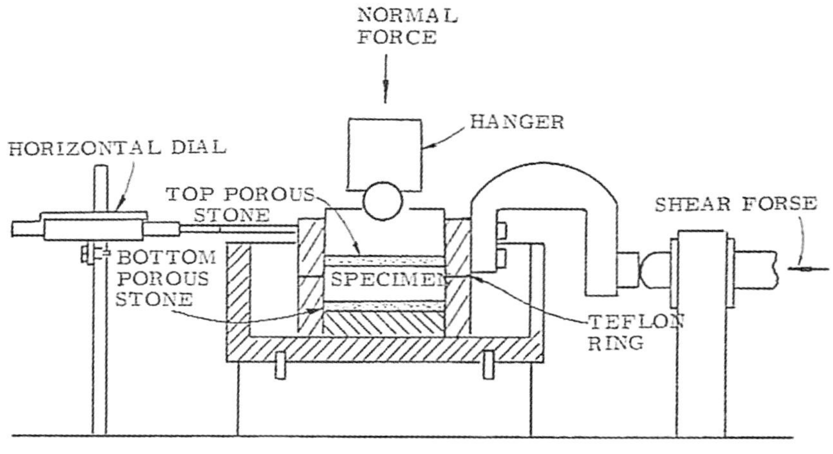

Fig. 3-11. Schematic diagram of direct shear box assembly

The soil is air-dried and sieved through a No. 4 sieve (4.75 mm). The material retained on the sieve is discarded, because the specimen is only 2.5 in. (63.5 mm) in diameter and is not adequate for large particles. The material passing the sieve is mixed with a large amount of water to make it very plastic, and then it is placed in the direct shear box. To prevent the sample from squeezing out, a Teflon ring is used to separate the two halves of the shear box.

After a given normal stress is applied for about 10 min, the shear stress is applied at a rate of 0.02 in. (0.5 mm) per min until the specimen fails, as indicated by a decrease in the reading of the proving ring dial. If the specimen does not fail, the test is stopped at 20 min or a horizontal deformation of 0.5 in. (12.7 mm). At least three tests involving three different normal stresses must be performed.

Fig. 3-12 shows the stress-displacement curves of a fine coal refuse under three different normal stresses: 0.52, 1.55, and 2.58 tsf (49.8, 148.5, and 247.2 kN/m^2). In all three curves, the shear stress increases with the horizontal displacement up to a peak and then decreases until a nearly constant value is obtained. The shear strength at the peak is called the peak shear strength, and that at the constant value is called the residual shear strength. Because of progressive failure, the average shear strength actually developed along a failure surface is somewhere between the peak and the residual strength. If the two strengths are not significantly different, as is the case shown in Fig. 3-12, the peak strength can be used for stability analysis.

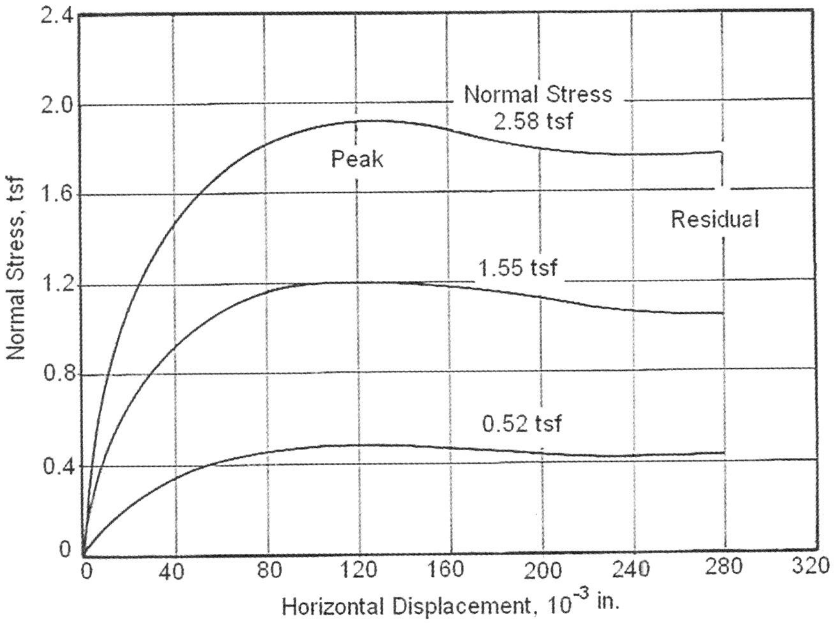

Fig. 3-12. Stress-displacement curves of fine refuse
Note: 1 in. = 25.4 mm; 1 tsf = 95.8 kN/m^2

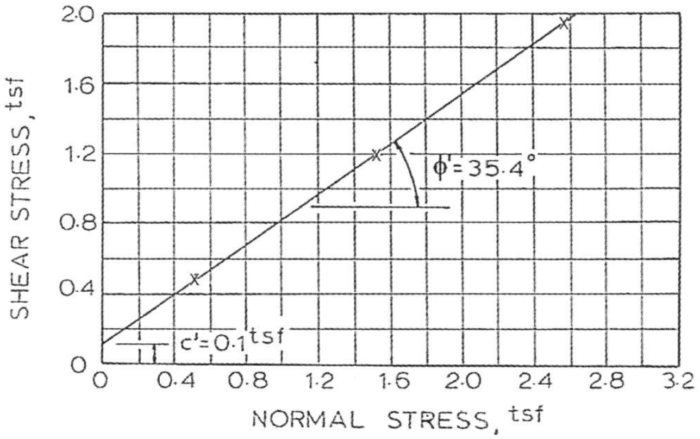

Fig. 3-13. Strength of fine refuse by direct shear test
Note: 1 tsf = 95.8 kN/m²

Fig. 3-13 shows a plot of peak shear stress versus normal stress for fine coal refuse. A straight line passing through the three points is drawn. The vertical intercept at zero normal stress is the effective cohesion, c', and the angle of the straight line with the horizontal is the effective friction angle, ϕ'. The figure shows that the fine coal refuse has an effective cohesion of 0.1 tsf (9.6 kN/m^2) and an effective friction angle of 35.4°.

It is believed that the strength parameters determined from this procedure are quite conservative because: (1) only the fraction passing through the No. 4 sieve is used in the test. If sufficient coarse materials are present, the shear strength may be slightly greater; (2) no compaction is applied to the specimen other than the normal load used in the test. If the material is compacted, a slightly higher strength may be obtained; and (3) the specimen is very wet and completely saturated, which may not occur in the field.

It should be pointed out that this procedure for determining effective shear strength is applicable only to silty clays and coal refuse. It may not be used for highly plastic clays unless the rate of loading is kept exceedingly low.

3.3.2 Triaxial Compression Test

The triaxial compression test can be used to determine either the total strength parameters or the effective strength parameters. Fig. 3-14 shows the schematic of a triaxial chamber. The specimen is covered with a rubber membrane and placed in the triaxial chamber. Water is introduced into the chamber and a given confining pressure is applied. A vertical axial load then is applied gradually until the specimen fails, as indicated by a decrease in reading of the applied load. The deformation of the specimen and magnitude of the applied load can be measured either by dial gauges or other electronic devices. If the specimen does not fail, the test continues until a strain of 15% is obtained.

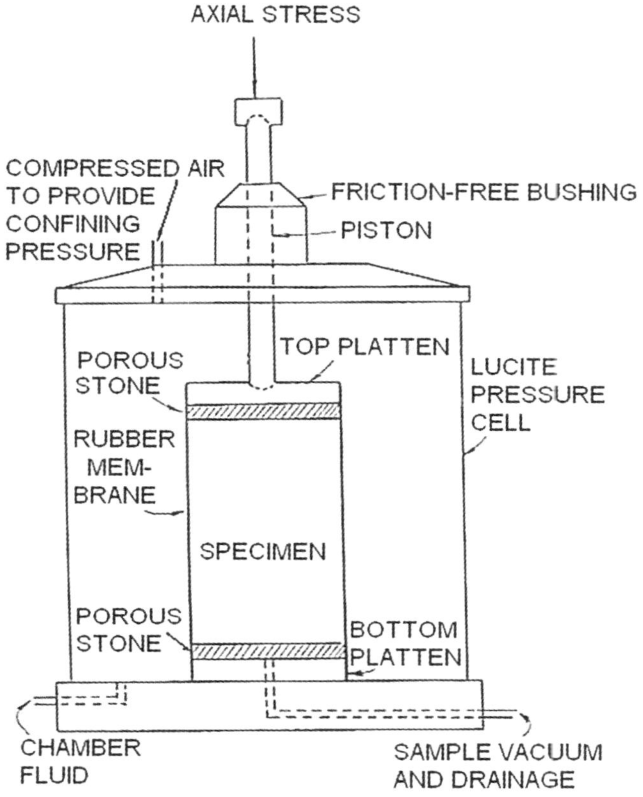

Fig. 3-14. *Schematic of a triaxial chamber*

One simple way to determine the total strength parameters of unsaturated soils is to prepare two identical specimens, and then subject one to the unconfined compression test and the other to the triaxial compression test. The confining pressure used for the triaxial test should nearly equal the maximum overburden pressure expected in the field. The procedure for the unconfined compression test is similar to the triaxial test, except that the specimen is not enclosed in the rubber membrane and no confining pressure is applied.

To prevent any drainage in the triaxial test, the drainage valves must be closed. After both tests are completed, two Mohr's circles are drawn, and a straight line tangent to these two circles is the Mohr's envelope. The vertical intercept of the envelope at zero normal stress is the cohesion, and the angle of the envelope with the horizontal is the friction angle, as shown in Fig. 3-15. The total strength parameters generally exhibit a large cohesion and a small friction angle. If the specimen is saturated completely, the envelope will be horizontal with an angle of internal friction equal to zero. Procedures for undrained triaxial test are specified by ASTM D2850 "Unconsolidated-Undrained Triaxial Compression Test on Cohesive Soils" (ASTM 2010).

The effective strength parameters can be determined by a consolidated drained test or a consolidated undrained test with pore pressure measurements. Instead of using the total normal stress as shown in Fig. 3-15, the shear stress is plotted versus the effective normal stress, and a Mohr's envelope in terms of effective stress is obtained. The vertical intercept of the envelope at zero effective normal stress is the effective cohesion, and the angle of the envelope with the horizontal is the effective friction angle. The effective strength parameters always exhibit a small effective cohesion and a large effective friction angle. The method is specified by ASTM D4767 "Consolidated Undrained Triaxial Compression Test for Cohesive Soils" (ASTM 2010).

Another procedure to obtain the effective strength parameters is by the use of the stress path method (Lambe and Whitman 1969). Fig. 3-16 shows the p' versus q diagram of the fine coal refuse used in the direct shear test shown in Fig. 3-13.

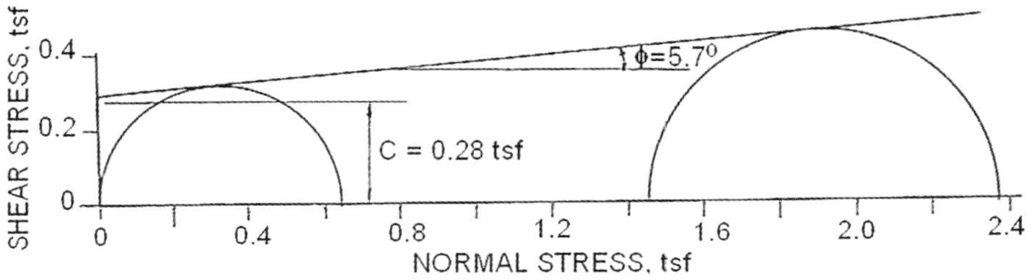

Fig. 3-15. *Total strength parameters of a compacted specimen*
Note: 1 tsf = 95.8 kN/m²; 1 pcf = 157.1 N/m³

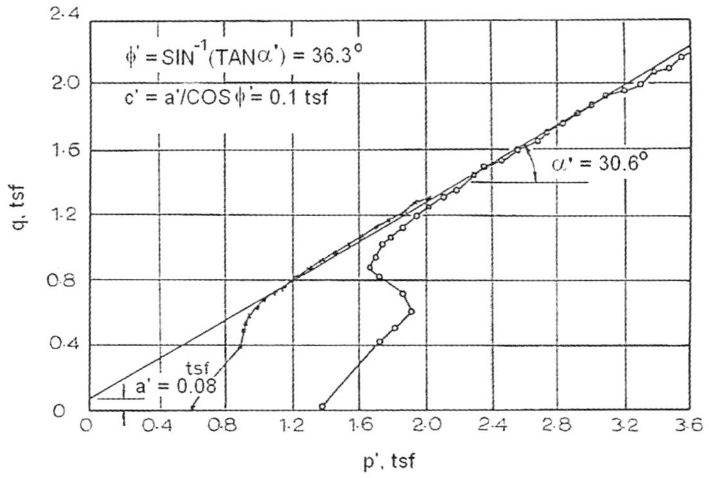

Fig. 3-16. *Triaxial compression test on fine refuse*
Note: 1 tsf = 95.8 kN/m²

Note that

$$p' = \frac{\sigma_1' + \sigma_3'}{2} \tag{3-15}$$

$$q = \frac{\sigma_1 - \sigma_3}{2} = \frac{\sigma_1' - \sigma_3'}{2} \tag{3-16}$$

in which σ_1 = major principal stress and σ_3 = minor principal stress. The corresponding effective stresses, σ_1' and σ_3', can be determined by subtracting the measured pore pressure from the total stresses, σ_1 and σ_3.

Fig. 3-16 shows the effective stress path of triaxial compression tests with pore pressure measurements using the same fine refuse for the direct shear test shown in Fig. 13-13. Each point on the stress path is called the stress point. The tests were made on two specimens, one subjected to an effective confining pressure of 0.6 tsf (58 kN/m^2) and the other to 1.4 tsf (134 kN/m^2). To saturate the specimens, filter strips were placed around the sample and a chamber pressure was applied. The load was increased until the stress path approached a straight line. A line tangent to the failure part of the stress path is called the K_f-line. The angle of the K_f-line with the horizontal is called α', which is related to ϕ' by

$$\sin\phi' = \tan\alpha' \tag{3-17}$$

The intercept of the K_f-line with the q-axis is called a', which is related to c' by

$$c' = \frac{a'}{\cos\phi'} \tag{3-18}$$

Fig. 3-16 shows that the fine coal refuse has an effective cohesion of 0.10 tsf (7.7 kN/m^2) and an effective friction angle of 36.3°, which check closely with the result of direct shear test.

An advantage of using the stress path method is that the effective cohesion and the effective friction angle may be estimated from a single test by approximating a straight line through the stress points at the latter part of the test, whereas two tests are required if the Mohr's circle is used. Furthermore, if several triaxial tests are run and only the σ_1' and σ_3' at the time of failure are measured, it is much easier to approximate a straight line through all the stress points than to plot a straight line tangent to all the circles.

Example 3.6 The principal stresses, σ_1' and σ_3', of three triaxial shear tests at the time of failure are tabulated as follows (where 1 psi = 144 psf = 6.9 kN/m^2):

Test No.	σ_1' (psi)	σ_3' (psi)	p' (psi)	q (psi)
1	46	14	30	16
2	89	29	59	30
3	137	45	91	46

By the use of a p' versus q diagram and the principle of least squares, determine the effective cohesion, c', and the effective angle of internal friction, ϕ', of the soil.

Solution From Eqs. (3-15) and (3-16), values of p' and q are computed, as listed in the preceeding table. The K_f-line, which passes through the three points with coordinates (p', q), is supposed to be a straight line represented by the following linear equation:

$$y_q = a + bx \tag{3-19}$$

in which y_q = computed value of q obtained by substituting p' as the value of x in Eq. (3-19), and a and b are constants to be determined by the principle of least squares. The purpose herein is to obtain a line so that

$$\sum(y - y_q)^2 = \sum(y - a - bx)^2 = \text{minimum} \tag{3-20}$$

This can be achieved by taking the partial derivative of Eq. (3-20) with respect to a and b and setting it to 0, or

$$\begin{aligned} &\frac{\partial}{\partial a}\sum(y-a-bx)^2 = \sum[-2(y-a-bx)] = 0 \\ &\text{or} \sum y = na + b\sum x \end{aligned} \tag{3-21}$$

in which n = number of points, which is 3 in this example.

$$\begin{aligned} &\frac{\partial}{\partial b}\sum(y-a-bx)^2 = \sum[-2(y-a-bx)x] = 0 \\ &\text{or} \sum xy = a\sum x + b\sum x^2 \end{aligned} \tag{3-22}$$

Eqs. (3-21) and (3-22) can be used to solve a and b. In this example, substituting values of q as y and p' as x, Eq. (3-21) becomes

$$\begin{aligned} &16 + 30 + 46 = 3a + (30 + 59 + 91)b \\ &\text{or } 92 = 3a + 180b \end{aligned} \tag{3-23}$$

From Eq. (3-22),

$$\begin{aligned} &30 \times 16 + 59 \times 30 + 91 \times 46 = 180a + [(30)^2 + (59)^2 + (91)^2]b \\ &\text{or } 6{,}436 = 180a + 12{,}662b \end{aligned} \tag{3-24}$$

From Eqs. (3-23) and (3-24), $a = 1.147$ psi and $b = 0.492$. From Eq. (3-17), $\sin\phi' = 0.492$, or $\phi' = \sin^{-1}(0.492) = 29.5°$. From Eq. (3-18), $c' = 1.147/\cos 29.5° = 1.318$ psi. Fig. 3-17 is the p' versus q diagram and the best fit least square line.

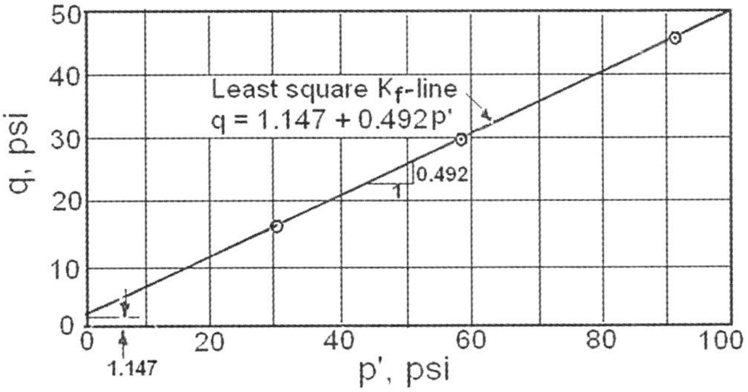

Fig. 3-17. *Example 3.6*
Note: 1 psi = 144 psf = 6.9 kN/m²

The discussed procedures for determining the cohesion and internal friction may not be applicable to granular soils. Theoretically, granular materials are cohesionless, and the cohesion should be zero. However, when two triaxial tests under two different confining pressures are run, the failure envelope—which is a line tangent to both circles—does not pass through the origin, because the Mohr's failure envelope is actually a curve, and the angle of internal friction decreases with the increase in confining pressure due to the increasing breakdown of particles under higher stresses. Consequently, a line tangent to both circles may result in a cohesion intercept, which does not exist in reality. To approximate the failure envelope by a straight line, the line must be forced to pass through the origin. When a cohesionless material is placed in an embankment, the most critical failure surface obtained by LEAME is a very shallow circle, which barely scratches the surface of the slope. This type of failure surface is impossible, because the higher friction angle near the surface will prevent the formation of a very shallow failure surface. As the depth increases, the friction angle decreases, but the normal stress increases, and there is a critical depth at which the factor of safety is minimum. A theoretical method for analyzing curved envelope is presented in the next section.

3.4 Shear Strength of Granular Materials

In all the methods of stability analysis discussed so far, it is assumed that the Mohr's failure envelope is a straight line, as indicated by Eq. (1-2) or (2-8). Because granular materials are cohesionless and fully drained, the shear strength can be expressed in effective stress by

$$s = \sigma' \tan \phi' \tag{3-25}$$

in which σ' = effective normal stress on the failure plane at the time of failure.

Fig. 3-18 shows the Mohr's circles for four triaxial tests on the rockfill materials used for the shell of Oroville Dam in California, as reported by Marachi

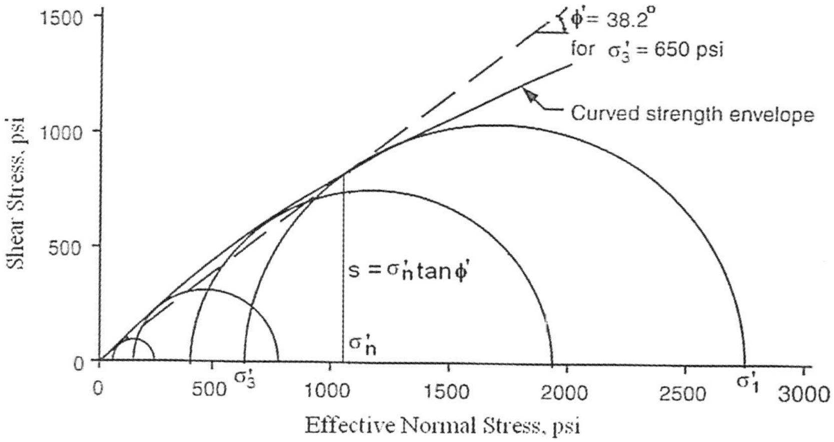

Fig. 3-18. Mohr's circles and failure envelope for a rockfill material
Note: 1 psi = 6.9 kN/m²

Table 3-2. Stresses at Time of Failure and Values of ϕ'

Test	σ_3' (psi)	σ_1' (psi)	ϕ' (deg)	σ_3'/p_a
1	30	193	46.9	2.04
2	140	754	43.4	9.52
3	420	1,914	39.8	28.6
4	650	2,770	38.2	44.2
Note: 1 psi = 6.9 kN/m²				

et al. (1969). The principal stresses used to plot the Mohr's circles are shown in Table 3-2. The strength envelope tangent to the circles is a curve, and each point on the curve represents the normal and shear stresses on a failure plane at the time of failure. With no cohesion, a line passing through the origin and tangent to each circle gives the value of ϕ' for each circle. The dashed line in the figure is the tangent to the largest circle. It can be seen that the values of ϕ' decrease with the increase in σ_3'. Given the major principal stress, σ_1', and the minor principal stress, σ_3', the angle of internal friction can be computed by

$$\phi' = \sin^{-1}\left(\frac{\sigma_1' - \sigma_3'}{\sigma_1' + \sigma_3'}\right) \tag{3-26}$$

Experimental evidences show that the value of ϕ' for curved envelope can be expressed by

$$\phi' = \phi_o - \Delta\phi \log\left(\frac{\sigma_3'}{p_a}\right) \tag{3-27}$$

in which p_a = atmospheric pressure, or 14.7 psi (101 kN/m^2), ϕ_o is the value of ϕ' when $\sigma'_3 = p_a$, and $\Delta\phi$ is the reduction in ϕ' for a 10-fold increase in σ'_3. Eq. (3-27) implies that a plot of ϕ' versus $\log(\sigma'_3)$ should result in a straight line.

Table 3-2 is the numerical values of the principal stresses at the time of failure and the computed ϕ' and σ'_3/p_a for the Oroville Dam shell. To determine ϕ_o and $\Delta\phi$, values of ϕ' are plotted versus $\log(\sigma'_3/p_a)$, as shown in Fig. 3-19.

In Fig. 3-19, a straight line is drawn through these four points and the value of ϕ' at $\sigma'_3/p_a = 1$ is ϕ_o, and the difference in ϕ' between $\sigma'_3/p_a = 1$ and 10 is $\Delta\phi$. The plot gives a ϕ_o of 49.5° and a $\Delta\phi$ of 6.5°.

In the method of slices, the shear strength is based on the normal stress, which is the effective normal stress on the failure plane at the time of failure, σ'_{ff}. It can be proved by geometry that

$$\sigma'_3 = \frac{\sigma'_{ff}}{1+\sin\phi'} \tag{3-28}$$

Values of ϕ' may range from 30° to 50°, so the ratio σ'_3/σ'_{ff} may range from 0.667 to 0.566. With $\Delta\phi$ = 6.5°, the use of 0.667 as the σ'_3/σ'_{ff} ratio will increase ϕ' by –6.5 × log(0.667), or 1.14°, whereas the use of 0.566 will increase by –6.5 × log(0.566), or 1.61°, so the difference is only 0.47°. It is therefore concluded that the σ'_3/σ'_{ff} ratio has very little effect on the computed ϕ' and that a conservative ratio of 0.667 can be used regardless of ϕ'.

In stability analysis, $\sigma'_3 = 0.667\sigma'_n$, where σ'_n is the normal stress on the failure plane at the bottom of the slice, so Eq. (3-27) can be written as

$$\phi' = \phi_o - \Delta\phi\log\left(\frac{0.667\sigma'_n}{p_a}\right) \tag{3-29}$$

Using English units with stress in psf, Eq. (3-29) becomes

$$\phi' = \phi_o - \Delta\phi\log\left(\frac{\sigma'_n}{3,174}\right) \tag{3-30a}$$

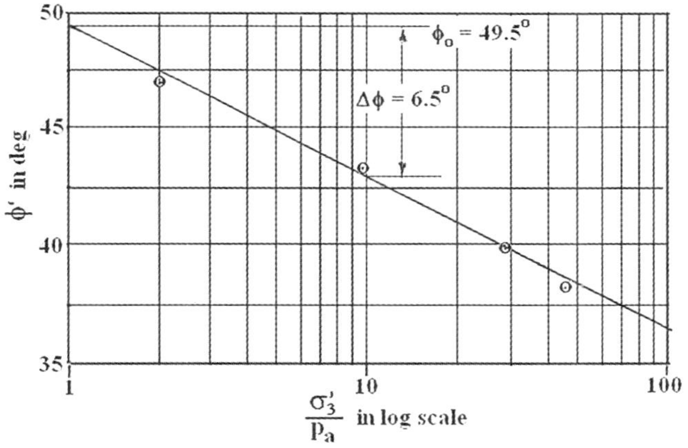

Fig. 3-19. Effect of confining pressure on ϕ'

Using SI units with stress in kN/m^2, Eq. (3-29) becomes

$$\phi' = \phi_o - \Delta\phi \log\left(\frac{\sigma'_n}{152}\right) \quad (3\text{-}30b)$$

Eq. (3-29) was incorporated in LEAME to determine the effective friction angle at the bottom of each slice based on the effective force normal to the failure surface.

It should be noted that the most critical failure surface for granular materials is not a deep circle. Even though the Oroville Dam is more than 770 ft (235 m) high, the tallest slice in the most critical circle is less than 200 ft (61 m). It really is not necessary to use very high confining pressures to define the failure envelope unless some extraordinary conditions dictate the formation of deep failure surfaces. Assuming a unit weight of 150 pcf (23.7 kN/m^3), the second circle with a confining pressure of 140 psi (965 kN/m^2) is equivalent to a normal stress of 210 psi or an overburden of not more than 200 ft (61 m).

Table 3-3 lists the typical values of ϕ_o and $\Delta\phi$, as reported by Wong and Duncan (1974). In Table 3-3, *RC* is the relative compaction defined by

$$RC = \frac{\gamma_d}{\gamma_{d\max}} \quad (3\text{-}31)$$

in which γ_d = in situ dry density, and $\gamma_{d\max}$ = maximum dry density determined in the laboratory according to ASTM D698 "Standard Test method for Laboratory

Table 3-3. *Values of ϕ and $\Delta\phi$ for Gravels and Sands*

Unified Classification	**Standard Proctor, *RC* (%)**	**Relative Density, D_r (%)**	**ϕ_o (deg)**	**$\Delta\phi$ (deg)**
GW, SW	105	100	46	10
	100	75	43	8
	95	50	40	6
	90	25	37	4
GP, SP	105	100	42	9
	100	75	39	7
	95	50	36	5
	90	25	33	3
SM	100	—	36	8
	95	—	34	6
	90	—	32	4
	85	—	30	2

(Wong and Duncan 1974)

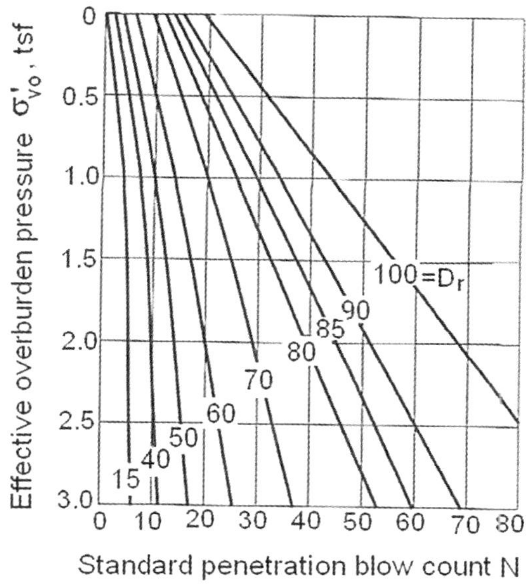

Fig. 3-20. Use of standard penetration test to determine relative density of sands (USBR 1974)
Note: 1 tsf = 95.8 kN/m²

Compaction Characteristics of Soil Using Standard Effort" (ASTM 2010). The relative density, D_r, is defined by

$$D_r = \frac{e_{\max} - e}{e_{\max} - e_{\min}} \tag{3-32}$$

in which $e_{\max}$ = void ratio for minimum dry density in the loosest state, $e_{\min}$ = void ratio for maximum dry density in the densest state, and e = in situ void ratio. The values of minimum dry density can be determined in the laboratory according to ASTM D4257 "Standard Test Methods for Minimum Index Density and Unit Weight of Soils and Calculation of Relative Density" (ASTM 2010).

For well-graded (SW) or poorly-graded (SP) sands, the most important factor that affects the shear strength is the relative density. When the standard or cone penetration tests are used for subsurface exploration, the blow count, N, of the standard penetration test or the cone resistance, q_c, of the cone penetration test can be used to determine the relative density, as shown in Figs. 3-20 and 3-21, so the shear strength, ϕ_o and $\Delta\phi$ can be estimated from Table 3-3.

3.5 Shear Strength of Municipal Solid Waste

In the design of landfills, it is necessary to know the shear strength of solid waste. The strength of solid waste varies considerably depending on the amount of soil

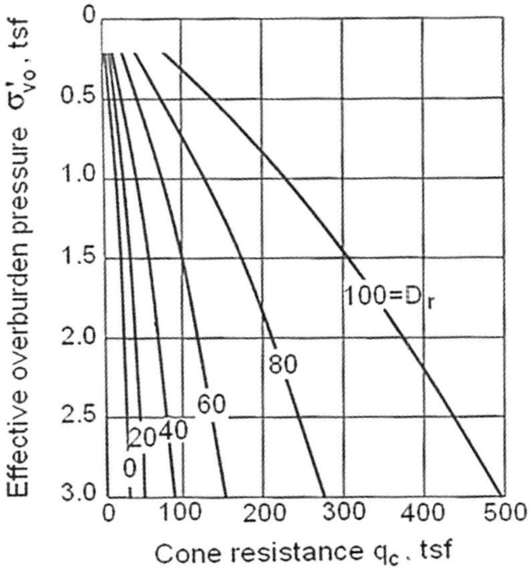

Fig. 3-21. *Use of cone penetration test to determine relative density of sands (Schmertmann 1975. Reproduced with permission)*
Note: 1 tsf = 95.8 kN/m²

and sludge and the proportion of plastic and other materials that cause interlock action and increase in shear strength. Although solid waste tends to decompose and degrade with time, Kavazanjian (2001) indicated that the degradation did not have significant effect on shear strength.

Kavazanjian et al. (1995) developed the lower-bound strength envelope of municipal solid waste (MSW), using direct shear test data and back-analysis of seven stable landfills, as shown in Fig. 3-22. When the normal stress is less than 0.39 tsf (37 kN/m^2), the envelope is horizontal with c = 0.25 tsf (24 kN/m^2) and $\phi = 0$. When the normal stress is greater than 0.39 tsf (37 kN/m^2), the envelope is inclined at = 33° with $c = 0$.

Based on the results of large-scale direct shear tests and back-analysis of a few failed slopes, Eid et al. (2000) developed the range of strength envelopes for MSW, as shown in Fig. 3-23. All three envelopes are inclined at a friction angle of 35°. The cohesion ranges from 0 to 0.52 tsf (50 kN/m^2) with an average of 0.26 tsf (25 kN/m^2).

Bray et al. (2009) conducted a comprehensive testing program using direct shear, triaxial, and simple shear tests to determine the shear strength of MSW. The results of their studies indicate that the direct shear test is appropriate to evaluate the shear strength along the weakest orientation of the large fibrous particles and, thus, gives the most conservative shear strength. More than 100 test results from their and other studies indicate that the static shear strength is best characterized by a stress-dependent Mohr-Coulomb strength criterion with

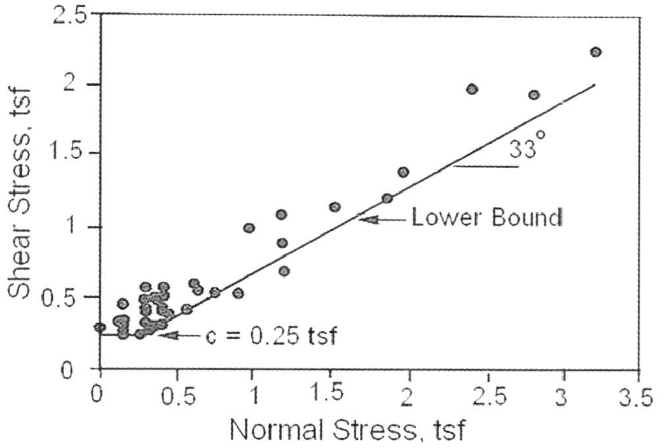

Fig. 3-22. Shear strength envelope for municipal solid waste (Kavazanjian et al. 1995. Reproduced with permission)
Note: 1 tsf = 95.8 kN/m²

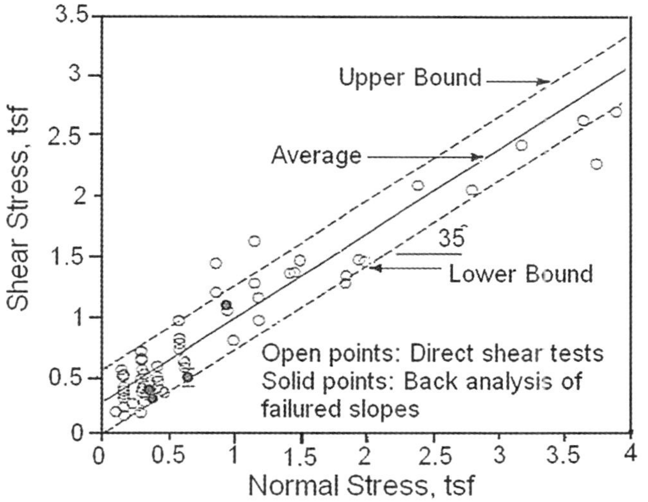

Fig. 3-23. Range of shear envelopes for municipal solid waste (Eid et al. 2000. Reproduced with permission)
Note: 1 tsf = 95.8 kN/m²

$c = 1.5\,\text{kN/m}^2$ (310 psf), $\phi_o = 36°$, and $\Delta\phi = 5°$. Without considering the undrained behavior of saturated waste and strength loss due to pore pressure generation resulting from cyclic loading, the dynamic shear strength of MSW can be estimated to be a minimum of 20% greater than its static strength.

The following two equations, similar to Eqs. (3-25) and (3-29) based on triaxial tests, can be used to determine the shear strength of MSW by direct shear tests:

$$s = c' + \sigma'_n \tan\phi' \tag{3-33}$$

$$\phi' = \phi_o - \Delta\phi \log\left(\frac{\sigma'_n}{p_a}\right) \tag{3-34}$$

For granular materials, c' is equal to 0, and the input parameters, ϕ_o and $\Delta\phi$, must be obtained from triaxial tests. For MSW, c' is not 0, because of their fibrous fragments, and the input parameters, c', ϕ_o, and $\Delta\phi$, must be obtained from direct shear tests.

3.6 Typical Ranges and Correlations

When actual test data are not available, empirical correlations between shear strength and soil classification or index properties are available. However, these correlations should be used cautiously, because there is substantial scatter in the data to establish these correlations.

3.6.1 Effective Shear Strength

Kenney (1959) presented the relationship between $\sin\phi'$ and the plasticity index, PI, for normally consolidated soils, as shown in Fig. 3-24. Although there is considerable scatter in the data, a definite trend toward decreasing ϕ' with increasing

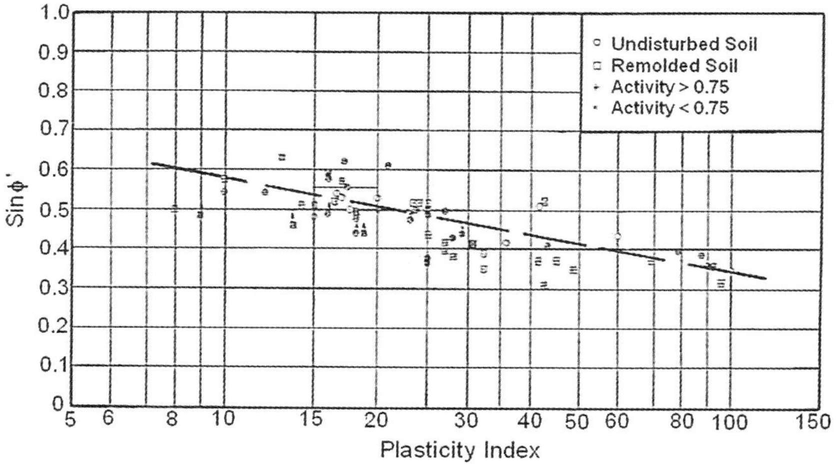

Fig. 3-24. Plasticity index versus $\sin\phi'$ *for normally consolidated soils (Kenny 1959. Reproduced with permission)*

plasticity is apparent. Note that normally consolidated soil is a soil that has not been subjected to precompression or to a previous effective overburden pressure greater than the present effective overburden pressure. If the soil has been precompressed, the effective angle of internal friction should be slightly higher.

Bjerrum and Simons (1960) presented a similar relationship for both undisturbed and remolded soil, as shown in Fig. 3-25. The relationship by Kenney is plotted in the dashed curve for comparison.

Skempton (1964) presented a correlation between the residual effective angle of internal friction and percent of clay, as shown in Fig. 3-26. The friction angle based on the residual stress is smaller than that based on the peak stress.

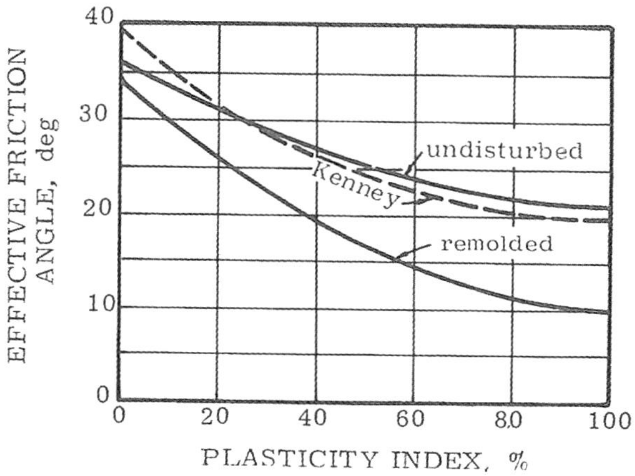

Fig. 3-25. Plasticity index versus effective friction angle (Bjerrum and Simons 1960. Reproduced with permission)

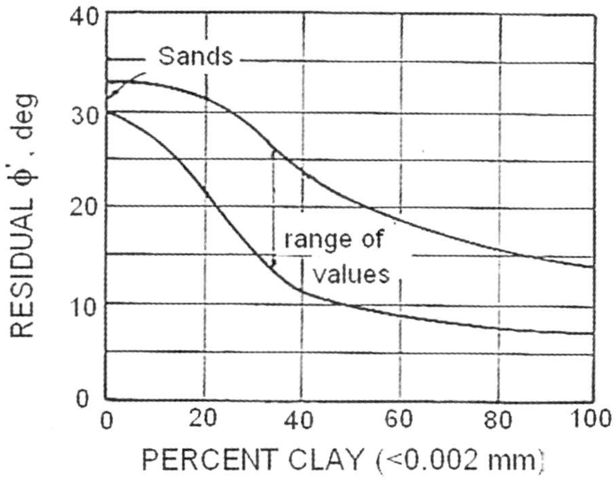

Fig. 3-26. Percent clay versus residual effective friction angle (Skempton 1964. Reproduced with permission from Geotechnique)

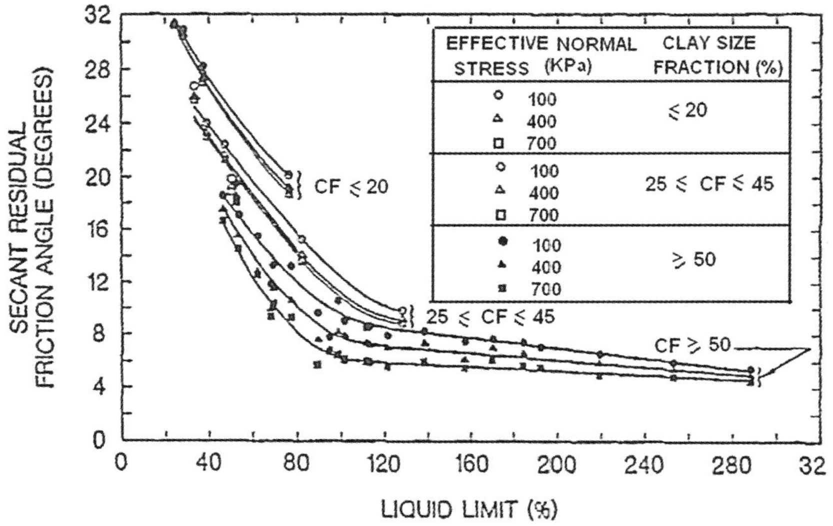

Fig. 3-27. Relationship between effective friction angle and liquid limit in terms of clay size friction and effective normal stress (Stark and Eid 1994. Reproduced with permission)

There are many correlations available today for residual friction angle of clayed soils. Stark and Eid (1994) found that the residual friction angle was related to the liquid limit as an indicator of clay mineralogy, the percent of clay size fraction (CF) smaller than 0.002 mm, and the effective normal stress, as shown in Fig. 3-27.

Table 3-4 shows the average effective shear strength of soils compacted to the Proctor maximum dry density at optimum moisture content, as suggested by the U.S. Bureau of Reclamation (1973).

If the soil is subject to saturation, then $c' = c_{sat}$. If the soil is at the optimum moisture content and the maximum density, $c' = c_o$. The shear strength listed in Table 3-4 is for compacted soils. For natural soils, the effective cohesion may be larger or smaller than the listed values, depending on whether the soil is overly or normally consolidated, but the effective angle of internal friction should be affected to a much lesser degree.

Table 3-5 shows the typical ranges of effective friction angle for granular materials and silts, as suggested by Bowles (1984).

Example 3.7 According to the plasticity chart of the Unified Soil Classification system, the silty and clayey soil with the dual classification ML-CL should have a plasticity index, PI, between 4 and 8 and a liquid limit, LL, between 12 and 30. If such a soil is normally consolidated with an average plastic index of 6, an average liquid limit of 21, and a clay content of 5%, estimate its effective friction angle by the various tables and figures presented in this section.

Table 3-4. Average Effective Shear Strength of Compacted Soils

UNIFIED CLASSIFICATION	SOIL TYPE	PROCTOR COMPACTION Maximum Dry Density pcf	PROCTOR COMPACTION Optimum Moisture Content %	As Compacted Cohesion c_o tsf	Saturated Cohesion c_{sat} tsf	Friction Angle ϕ' deg
GW	well graded clean graves, gravel-sand mixture	>119	<13.3	*	*	>38
GP	poorly graded clean gravels, gravel sand mixture	>110	<12.4	*	*	>37
GM	silty gravels, poorly graded gravel-sand-silt	>114	<14.5	*	*	>34
GC	clayed gravels, poorly graded gravel-sand-clay	>115	<14.7	*	*	>31
SW	well graded clean sands, gravelly sands	119 ± 5	13.3 ± 2.5	0.41 ± 0.04	*	38 ± 1
SP	poorly graded clean sands, sand-gravel mixture	110 ± 2	12.4 ± 1.0	0.24 ± 0.06	*	37 ± 1
SM	silty sands, poorly graded sand-silt mixture	114 ± 1	14.5 ± 0.4	0.53 ± 0.06	0.21 ± 0.07	34 ± 1
SM-SC	sand-silt-clay with slightly plastic fines	119 ± 1	12.8 ± 0.5	0.21 ± 0.07	0.15 ± 0.06	33 ± 3

Table 3-4. *(Continued)*

UNIFIED CLASSIFICATION	SOIL TYPE	PROCTOR COMPACTION Maximum Dry Density pcf	PROCTOR COMPACTION Optimum Moisture Content %	As Compacted Cohesion c_o tsf	Saturated Cohesion c_{sat} tsf	Friction Angle ϕ' deg
SC	clayey sands, poorly graded sand-clay mixture	115 ± 1	14.7 ± 0.4	0.78 ± 0.16	0.12 ± 0.06	31 ± 3
ML	inorganic silts and clayey silts	103 ± 1	19.2 ± 0.7	0.70 ± 0.10	0.09 ± *	32 ± 3
ML-CL	mixtures of inorganic silts and clays	109 ± 2	16.8 ± 0.7	0.66 ± 0.18	0.23 ± *	32 ± 2
CL	inorganic clays of low to medium plasticity	108 ± 1	17.3 ± 3	0.91 ± 0.11	0.14 ± 0.02	28 ± 2
OL	organic silts and silty clays of low plasticity	*	*	*	*	*
MH	inorganic clayed silts, elastic silts	82 ± 4	36.3 ± 3.2	0.76 ± 0.31	0.21 ± 0.09	25 ± 3
CH	inorganic clays of high plasticity	94 ± 2	25.5 ± 1.2	1.07 ± 0.35	0.12 ± 0.06	19 ± 5
OH	organic clays and silty clays	*	*	*	*	*

*denotes insufficient data, > is greater than, < is less than
(U.S. Bureau of Reclamation 1973; 1 pef = 157.1 N/m^3, 1 tsf = 95.8 kPa)

Table 3-5. Typical Range of Effective Friction Angle for Soils Other than Clays

SOIL	EFFECTIVE FRICTION ANGLE, deg	
	LOOSE	DENSE
Sand, crushed (angular)	32–36	35–45
Sand, bank run (subangular)	30–34	34–40
Sand, beach (well rounded)	28–32	32–38
Gravel, crushed	36–40	40–50
Gravel, bank run	34–38	38–42
Silty sand	25–35	30–36
Silt, inorganic	25–35	30–35
(Bowles 1984. Reproduced with permission from McGraw-Hill, Inc.)		

Solution By Table 3-4: For the ML-CL classification, $\phi' = 32°$, or ranging from 30° to 34°. Because the specimen is not subject to any precompression before the strength test, other than the Proctor compactive effort, it is considered as a normally consolidated remolded soil.

By Table 3-5: There is no silty and clayey soil type in the table. The one closest to it is the inorganic silt. If compacted, the soil should be dense with ϕ' ranging from 30° to 35°, or an average of 32.5°. If some clay is added to the silt to fill the voids, ϕ' should increase slightly.

By Fig. 3-24: With PI = 6, $\sin\phi'$ is outside the range of the chart. By extrapolation, $\sin\phi' = 0.63$, or $\phi' = 39°$, which certainly is too high. Note the two test points in the figure, one with $\sin\phi' = 0.5$ when PI = 8 and the other with $\sin\phi'$ slightly smaller when PI = 9. When PI = 6, by a straight-line extrapolation, $\sin\phi' = 0.53$, or $\phi' = 32°$.

By Fig. 3-25: With PI = 6, $\phi' = 35°$ if undisturbed and 32° if remolded. The remolded case is the same as in Table 3-4 and yields the same average of 32°.

By Fig. 3-26: With 5% of clay, or particles smaller than 0.002 mm, $\phi' = 29°$ to 33°, with an average of about 31°.

By Fig. 3-27: With LL = 21 and CF = 5% < 20%, $\phi' = 32°$ regardless of normal stress.

It can be seen that most of the tables and charts give an effective friction angle of 32°.

Table 3-6 shows the relationship among relative density, STP blow count, CPT cone resistance, and effective friction angle for sands, as suggested by Meyerhof (1956).

3.6.2 Undrained Shear Strength

For normally consolidated soils, Bjerrum and Simons (1960) show that the ratio between the undrained shear strength and the effective overburden pressure,

Table 3-6. *Correlation of Effective Friction Angle to Relative Density, STP Blow Count, and CPT Cone Resistance for Sand*

State of Packing	Relative Density, D_r (%)	SPT blow count, N	CPT cone resistance, q_c (tsf)	Effective friction angle, ϕ' (deg)
Very loose	<20	<4	<20	<30
Loose	20–40	4–10	20–40	30–35
Medium	40–60	10–30	40–120	35–40
Dense	60–80	30–50	120–200	40–45
Very dense	>80	>50	>200	>45

(Meyerhof 1956. Reproduced with permission)
Note: 1 tsf = 95.8 kN/m^2

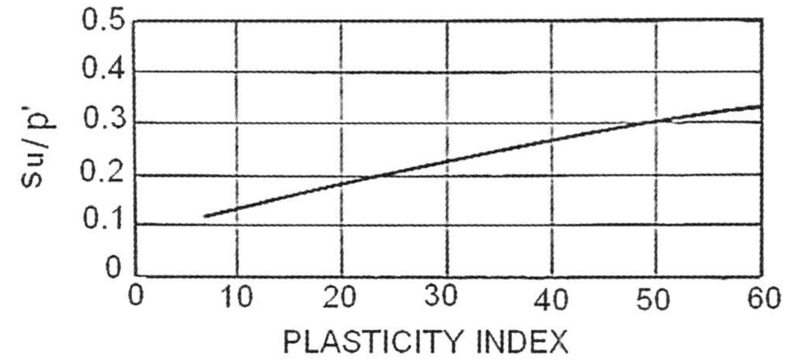

Fig. 3-28. *Plasticity index versus s_u/p' (Bjerrum and Simons 1960. Reproduced with permission)*

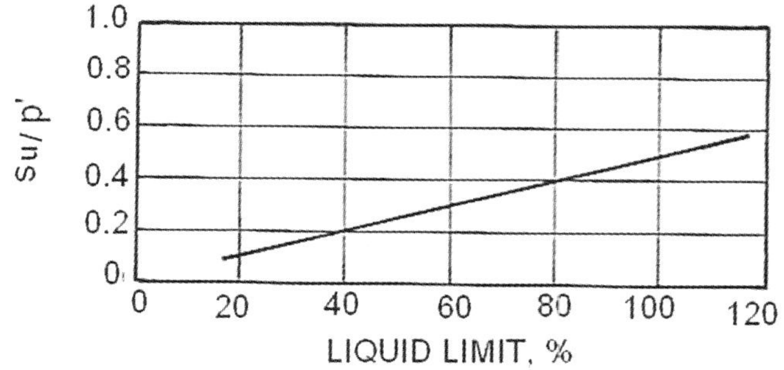

Fig. 3-29. *Liquid limit versus s_u/p' (Karlsson and Viberg 1967. Reproduced with permission from Norwegian Geotechnical Institute)*

s_u/p', is related to the plasticity index, as shown in Fig. 3-28. Karlsson and Viberg (1967) presented the relationship between s_u/p' and liquid limit, as shown in Fig. 3-29.

For overconsolidated soils, Ladd and Foott (1974) presented the relationship between the overconsolidated ratio and the ratio between the overconsolidated

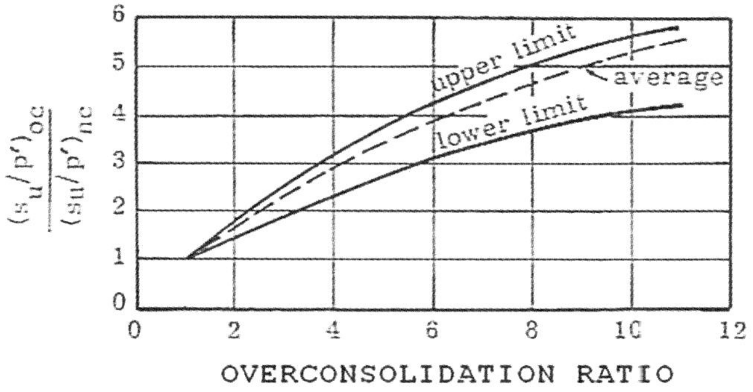

Fig. 3-30. *Relationship between overconsolidated and normally consolidated s_u/p' (Ladd and Foott 1974. Reproduced with permission)*

Table 3-7. *Undrained Shear Strength of Soils*

CONSISTENCY	UNDRAINED SHEAR STRENGTH, tsf	FIELD TEST
Very soft	0–1	Squeezed between fingers when fist is closed
Soft	1–2	Easily molded by fingers
Firm	2–4	Molded by strong pressure of fingers
Stiff	4–6	Dented by strong pressures of fingers
Very stiff	6–8	Dented only slightly by finger pressure
Hand	8^+	Dented only slightly by pencil point
(Adapted from Sowers 1979; 1 tsf = 95.8 kPa)		

and normally consolidated s_u/p', as shown in Fig. 3-30. The overconsolidated ratio is the ratio between the maximum effective overburden pressure, in which the soil has ever been subjected to during the past, and the present effective overburden pressure. The maximum effective overburden pressure can be determined from a consolidation test.

The undrained shear strength of soils varies a great deal depending on the moisture content and density. Table 3-7 shows the range of undrained shear strength of soils and a simple method of identification.

Example 3.8 The ML-CL soil in Example 3.7 with a plasticity index of 8 is normally consolidated and located at a distance of 40 ft (12.2 m) below the ground surface. The groundwater table is 15 ft (4.6 m) below the surface, or 25 ft (7.6 m) above the soil. If the total wet unit weight of the soil is 130 pcf (20.4 kN/m^3), determine the undrained shear strength of the soil. If the soil has an overconsolidation ratio of 4, what should be its undrained shear strength?

Solution To determine the effective overburden pressure, the total unit weight should be used for the soil above the water table and the submerged unit weight for the soil below the water table, so $p' = 15 \times 130 + 25 \times (130 - 62.4) = 3{,}640\,\text{psf}$. With PI = 8, from Fig. 3-26, $s_u/p' = 0.12$, or $s_u = 0.12 \times 3{,}640 = 436.8\,\text{psf} = 0.218\,\text{tsf}$. If the soil is overconsolidated with an overconsolidation ratio of 4, from Fig. 3-27,

$$\frac{(s_u/p')_{oc}}{(s_u/p')_{nc}} = 2.9 \quad \text{or} \quad (s_u/p')_{oc} = 2.9 \times 0.218 = 0.632\ \text{tsf}$$

It was mentioned in the previous example that ML-CL soil has a liquid limit between 12 and 30. From Fig. 3-26, a value of 0.12 for s_u/p' is equivalent to a liquid limit of 25, which is within the given range of 12 to 30.

3.7 Back-Analysis of Shear Strength

When failure occurs in a simple slope involving only one type of soil, one way to determine the shear strength of the soil is by back-calculation, using the failure site as a large-scale model test. When a slope fails, the factor of safety should be equal to 1. Based on the dimensions of the fill and the groundwater conditions at the time of failure, the shear strength that results in a safety factor of 1 can be back-calculated using stability charts or a computer program such as LEAME. In view of the fact that the factor of safety depends on not only the shear strength but also on so many other factors, which are difficult to evaluate, the use of the back-calculation method should proceed with caution, preferably in collaboration with other field or laboratory tests.

3.7.1 Total Stress Analysis with $\phi = 0$

If $\phi = 0$, the only shear strength parameter to be determined is the cohesion, c. First, a value of c is assumed and the factor of safety, F, is determined. Because the factor of safety is proportional to c, the developed cohesion, c_d, that produces a safety factor of 1 can be computed by

$$c_d = \frac{c}{F} \tag{3-35}$$

3.7.2 Total Stress Analysis with both c and ϕ

For total stress analysis with both cohesion and internal friction, a safety factor of 1 can be produced by various combinations of c and ϕ. It is well known that the depth of the failure surface for soils with a large cohesion is greater than that with a small cohesion, so the depth of a failure surface can be used to determine the amount of cohesion relative to the internal friction. The depth of the failure surface is related to the dimensionless parameter, $\lambda_{c\phi}$, defined as

$$\lambda_{c\phi} = \frac{\gamma H \tan\phi}{c} \tag{3-36}$$

in which H = height of a simple slope and γ = total unit weight of the soil. Eq. (3-36) indicates that, for a given slope, the relationship between $\tan\phi$ and c is represented by the parameter $\lambda_{c\phi}$. A smaller $\lambda_{c\phi}$ implies that the soil has a larger cohesion and a smaller friction angle, so the depth of the failure surface is greater. In the back-calculation, several values of $\lambda_{c\phi}$ and c are assumed. For each pair of c and ϕ, the factor of safety, F, can be determined by LEAME. The developed cohesion, c_d, to produce a safety factor of 1 can be computed by Eq. (3-35), and the developed angle of internal friction, ϕ_d, by

$$\phi_d = \tan^{-1}\left(\frac{\tan\phi}{F}\right) \tag{3-37}$$

The LEAME computer program also gives the height of tallest slice, which is the same as the depth of the failure surface. The pair of c_d and ϕ_d that gives a height of the tallest slice nearly equal to the actual depth of the failure surface is the back-calculated shear strength to produce a safety factor of 1.

Example 3.9 Fig. 3-31 shows a 2:1 fill slope with a height of 50 ft (15.2 m). The fill and foundation are of the same soil, which is a partially saturated silty clay with a total unit weight of 125 pcf (19.7 kN/m^3). The slope failed immediately

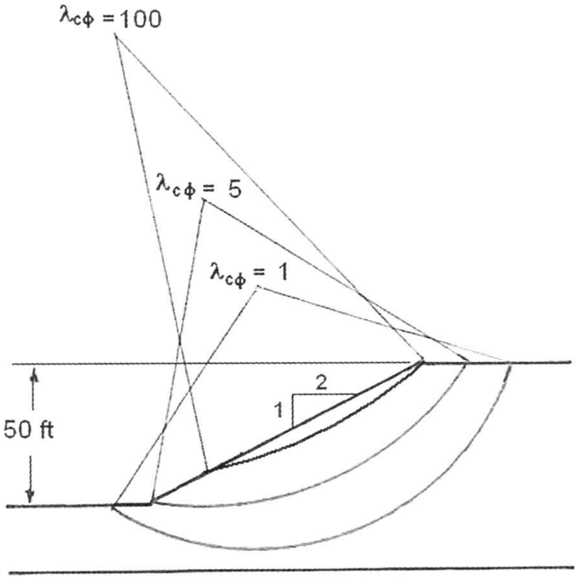

Fig. 3-31. Example 3.9
Note: 1 ft = 0.305 m

after construction, and the depth of failure surface is 30 ft (9.2 m). Back-calculate the soil parameters c and ϕ at the time of failure.

Solution Procedures for determining the pair of c_d and ϕ_d to produce a safety factor of 1 can be best illustrated by the following table:

$\lambda_{c\phi}$ (1)	c (psf) (2)	ϕ (deg) (3)	F (4)	c_d (psf) (5)	ϕ_d (deg) (6)	Height of tallest slice (ft) (7)
1	1,000	9.1	1.613	620	5.7	45.8
5	700	29.2	2.411	290	13.1	25.6
500	50	38.7	1.860	27	23.3	8.4

Each column is explained as follows:

(1) At least three values of $\lambda_{c\phi}$ must be assumed, starting from a small $\lambda_{c\phi}$ for a deep circle to a large $\lambda_{c\phi}$ for a shallow circle.
(2) The assumed value of c is immaterial. It is reasonable to assume a large c for the smaller $\lambda_{c\phi}$ and a small c for the large $\lambda_{c\phi}$. However, any value of c can be used. Even if the same c of 1,000 psf is assumed for all three cases, the same results will be obtained.
(3) Compute ϕ from Eq. (3-36), or $\phi = \tan^{-1}(c\lambda_{c\phi}/6{,}250)$.
(4) Use LEAME to determine the minimum factor of safety, F.
(5) Compute c_d by Eq. (3-38).
(6) Compute ϕ_d by Eq. (3-39).
(7) For each pair of c and ϕ, LEAME also gives the depth of the failure surface, which is the same no matter whether c and ϕ or c_d and ϕ_d are used.

Finally, plot c_d and ϕ_d versus the depth of the failure surface, as shown in Fig. 3-32. When the depth of the failure surface is 30 ft, the cohesion and

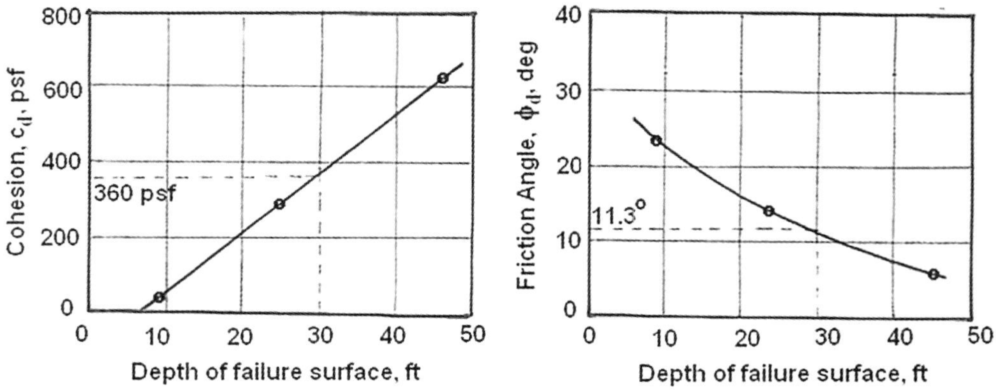

Fig. 3-32. Total shear strength versus depth of failure surface
Note: 1 ft = 0.305 m; 1 psf = 47.9 N/m²

friction angle to produce a safety factor of 1 are 360 psf and 11.3°, respectively, as indicated by the dashed lines. A rerun of LEAME using the stated shear strength resulted in a safety factor of 1.006 and a height of tallest slice of 29.151 ft, which check with the expected values of 1 and 30 ft.

3.7.3 Effective Stress Analysis with c' and ϕ'

The same procedures for total stress analysis can be applied to the effective stress analysis, except that Eqs. (3-35) and (3-37) be changed to

$$c_d = \frac{c'}{F} \tag{3-38}$$

and

$$\phi_d = \tan^{-1}\left(\frac{\tan\phi'}{F}\right) \tag{3-39}$$

Because the effective shear strength has a small cohesion and a large friction angle, $\lambda_{c\phi}$ cannot be too small. Too small a $\lambda_{c\phi}$, such as $\lambda_{c\phi} = 1$, may result in a large cohesion and a small friction angle, which fall outside the range of the effective shear strength.

Example 3.10 Fig. 3-33 shows a 1.5:1 slope with a height of 50 ft (15.2 m) and a total unit weight of 125 pcf (19.7 kN/m^3). The location of the phreatic surface at the time of failure is indicated by the dashed line. If the slope fails and the depth of the failure surface is 22 ft (6.7 m), what should be the effective cohesion and friction angle at the time of failure?

Solutions The results are tabulated in the following table and plotted in Fig. 3-34:

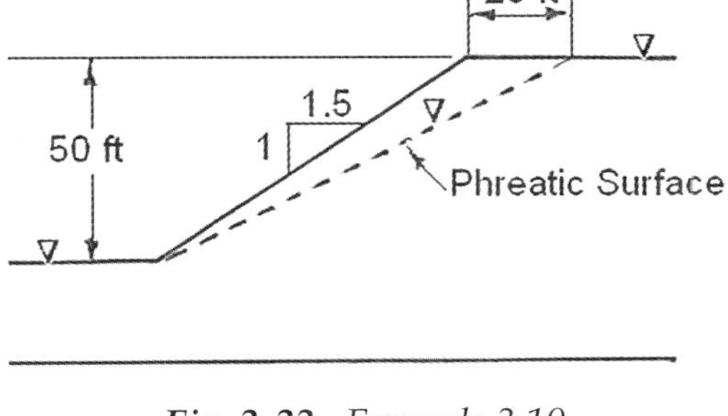

Fig. 3-33. *Example 3.10*
Note: 1 ft = 0.305 m

$\lambda_{c\phi}$ (1)	c'(psf) (2)	ϕ'(deg) (3)	F (4)	c'_d (psf) (5)	ϕ'_d (deg) (6)	Height of tallest slice (ft) (7)
10	100	9.1	0.329	304	25.9	28.5
50	100	38.7	1.132	88	35.3	19.8
300	100	25.6	5.939	17	38.9	15.8
20	100	17.9	0.537	186	30.7	24.0

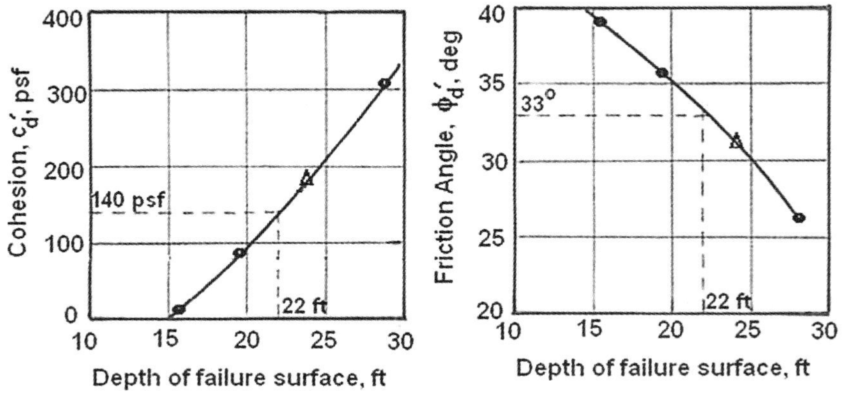

Fig. 3-34. Effective shear strength versus depth of failure surface
Note: 1 ft = 0.305 m; 1 psf = 47.9 N/m²

Originally, three values of 10, 50, and 300 were assumed for $\lambda_{c\phi}$. It was later found that the spacing between $\lambda_{c\phi} = 10$ and $\lambda_{c\phi} = 50$ was too far apart, so an additional $\lambda_{c\phi}$ of 20 was added, as shown by the small triangles in the figure. With a depth of the failure surface equal to 22 ft, the effective cohesion at the time of failure is 140 psf, and the effective friction angle is 33°. A rerun of LEAME using the stated shear strength resulted in a safety factor of 1.007 and a height of the tallest slice of 23.331 ft, which compare with the expected values of 1 and 22 ft, respectively.

The two examples shown indicate that the back-calculated shear strength obtained from LEAME gives a factor of safety very close to 1, but the height of the tallest slice may be somewhat different from the expected depth of the failure surface, indicating that the height of the tallest slice obtained by LEAME might not be very accurate. Two circles may produce the same minimum factor of safety, but the height of the tallest slice may be quite different. As a result, the depth of the failure surface may have two values depending on which circle is selected as the most critical. It is therefore suggested that more than three points be used to plot the curve relating c_d or ϕ_d to the depth of the failure surface. If a point is slightly out of line, the error can be minimized by smoothing out the curve through all points.

Summary

1. A subsurface investigation is needed to determine the nature and extent of materials from toe to crest of the slope. The log of boring forms the permanent record and should be kept as complete as possible. The final location of borings should be made in the field, and additions must be made to the boring program based on the information from the borings already completed. During boring, the depth to the groundwater table also should be recorded.

2. The charts relating the effective friction angle, ϕ', to the blow count, N, of standard penetration tests or to the cone resistance, q_c, of Dutch cone tests are designed for cohesionless sands but are also applicable to fine-grained soils with a small cohesion. However, if the soils are overconsolidated with a large effective cohesion or the penetration causes a significant change in pore pressure or effective stress, piezocone penetration tests, which take into account the effect of effective cohesion and excess pore pressure, should be used.

3. For clays, the correlation between the cone resistance and the undrained shear strength can be evaluated by Eq. (3-14), where the empirical cone factor, N_c, can be determined by correlating the cone resistance with the undrained shear strength obtained from field vane or laboratory triaxial compression tests.

4. In both field and laboratory tests, it is generally assumed that the Mohr's failure envelope is a straight line with a slope equal to the angle of internal friction and an intercept at the origin equal to the cohesion. For granular materials with no cohesion, the Mohr's envelope should pass through the origin. However, when two triaxial compression tests under two different confining pressures are run, the failure envelope, which is a line tangent to both circles, does not pass through the origin, because the Mohr's failure envelope is actually a curve with the angle of internal friction decreasing with the increase in confining pressure owing to the increasing breakdown of particles under higher stresses. Consequently, a line tangent to both circles may result in a cohesion intercept, which does not exist in reality. The current practice is to tilt the envelope slightly and force it to pass through the origin. When a cohesionless material is placed in an embankment, the most critical failure surface obtained by LEAME is a very shallow circle, which barely scratches the surface of the slope. This type of failure surface is impossible, because the higher friction angle near to the surface will prevent the formation of very shallow failure surfaces. As the depth increases, the friction angle decreases, but the normal stress increases, and there is a critical depth at which the factor of safety is at minimum. A theoretical method for analyzing a curved envelope is presented, in which the

conventional soil parameters, c and ϕ, are replaced by a new pair of parameters, ϕ_o and $\Delta\phi$.

5. When actual test data are not available, empirical correlations between shear strength and soil classification or index properties are available. However, these correlations should be used cautiously, because there is substantial scatter in the data to establish these correlations. Charts are presented for normally consolidated clays showing the relationship between the effective friction angle and the plasticity index or the percent of clay. There are also many correlations available today for relating the residual friction angle of clayey soils to the liquid limit, the percent of clay size fraction smaller than 0.002 mm, and the effective normal stress for overconsolidated clays; the undrained shear strength depends on the overconsolidation ratio.

6. In the design of landfills, it is necessary to know the shear strength of solid waste. The strength of municipal solid wastes varies considerably depending on the amount of soil and sludge and the proportion of plastic and other materials that cause interlock action and increase in shear strength. Based on the results of large-scale direct shear tests and the back-analysis of some failed slopes, it was found that solid waste has an effective cohesion from 0 to 0.52 tsf (49.8 kN/m^2) and an effective friction from 33° to 35°. The most recent studies indicate that the shear strength is best characterized by a curved strength envelope with $c = 15\,kN/m^2$ (310 psf), $\phi_o = 36°$, and $\Delta\phi = 5°$. These strength parameters can be determined more readily by the large-scale direct shear test than by the triaxial text. Although solid waste tends to decompose and degrade with time, it also was found that the degradation did not have significant effect on shear strength.

7. When failure occurs in a simple slope involving only one type of soil, one way to determine the shear strength of the soil is by back-calculation, using the failure site as a large-scale model test. When a slope fails, the factor of safety should be equal to 1. Based on the dimensions of the fill and the groundwater conditions at the time of failure, the shear strength that results in a safety factor of 1 can be back-calculated. It is well known that the most critical failure surface for a soil with a large cohesion and a small friction angle is a deep circle, whereas that with a small cohesion and a large friction angle is a shallow circle. Based on the actual depth of the failure surface, a method is presented to back-calculate the cohesion and friction angle of the soil and results in a safety factor of 1. In view of the fact that the factor of safety depends on not only the shear strength but also on so many other factors, which may be difficult to evaluate, the use of back-analysis should be done cautiously, preferably in collaboration with other field or laboratory tests.

Problems

3.1 A standard penetration test was performed on sand at a depth of 30 ft below the ground surface. The water table is located 15 ft below the surface. The soil has a total unit weight of 130 pcf. If the *N*-value obtained from the test is 30, determine the effective friction angle of the soil. If the sand is well graded, estimate its ϕ_o and $\Delta\phi$.
[Answer: 43°, 43°, 8°]

3.2 Based on Fig. 3-2, derive equations showing the relationship between undrained shear strength and blow count for clay of low plasticity and also for clay of high plasticity. For the same blow count, what is the strength ratio between the two soils?
[Answer: 3.3]

3.3 When a Dutch cone penetrates to a sand deposit at a depth of 20 ft below the ground surface, the recorded cone resistance is 68 tsf. If the total unit weight of the sand is 125 pcf and the water table is 10 ft below the ground surface, estimate the effective friction angle of the sand using Fig. 3-3, Eq. (3-3), and Eq. (3-4). If the sand is poorly graded, estimate its ϕ_o and $\Delta\phi$.
[Answer: 39°, 38.9°, 37.8°, 37°, 6°]

3.4 In Fig. 3-4, the standard size piezocone has a cone diameter, d_c, of 35.7 mm and a sleeve height, h_s, of 134 mm. Theoretically, what should be the dimension of d_1 to obtain a calibrated a_n of 0.581? What should be the dimension of d_2 for a calibrated b_n of 0.014?
[Answer: 27.2 mm, 31.7 mm]

3.5 The cone resistance, q_c, of a dense sand at a depth of 40 ft below the ground is 300 tsf. The total unit weight of the sand is 130 pcf, and the groundwater table is at a depth of 20 ft below the ground. If there is no excess pore pressure, determine the effective angle of internal friction of the sand by Fig. 3-3, and compare with that by Fig. 3-8.
[Answer: 42.5°, 42°]

3.6 The piezocone data indicates that a loose silt with a total unit weight of 120 pcf has a cone resistance, q_c, of 6.6 tsf, and a total pore pressure, u_t, of 1.6 tsf at a depth of 25 ft below the ground surface. If the calibrated net area ratio, a_n, is 0.581 and the water table is 5 ft below the ground surface, estimate the effective cohesion and the effective friction angle of the silt by Fig. 3-8.
[Answer: 28 psf, 29.2°]

3.7 Derive Eqs. (3-17) and (3-18).

3.8 For the granular material with the four sets of principal stresses shown in Table 3-2, plot the p' versus q diagram and determine the K_f-line by the principle of least squares. What are the effective cohesion and the effective friction angle of the material?
[Answer: 30.5 psi, 37.7°]

3.9 If the K_f-line in Problem 3.8 is forced to pass through the origin, plot the p' versus q diagram and the K_f-line. Determine the slope of the K_f-line by least squares and the effective friction angle of the granular material.
[Answer: 0.629, 39°]

3.10 Derive Eq. (3-28).

Chapter 4

Phreatic Surfaces

Seepage in embankments and dams is one of the important factors affecting stability, and many dam failures are caused by seepage. Seepage can be best represented by a phreatic surface. Theoretically, the pore water pressure along the failure surface should be determined from the piezometric surface rather than the phreatic surface. However, unless the phreatic surface is steeply inclined, the difference between the two surfaces is insignificant. Furthermore, the use of the phreatic surface always gives a lower and more conservative factor of safety. This chapter discusses various methods for determining the phreatic surface. Once a phreatic surface is known, its coordinates can be read by a computer program, and its effect on the factor of safety can be evaluated. A simplified but approximate method to characterize seepage is by the use of a pore pressure ratio. Pore pressure ratios also can be used for short-term stability analysis to predict the excess water pressure because of consolidation. These applications of pore pressure ratios also will be briefly described.

4.1 Flownets

In the stability analysis of slopes, particularly those related to earth dams, it is necessary to estimate the location of the phreatic surface, or the line of seepage.

In the case of an existing slope, the phreatic surface can be determined from subsurface investigation by observing the water level in bore holes with adjustments for seasonal changes. If the slope has not been constructed and is quite complex in configuration, the phreatic surface can be obtained by drawing a flownet or using a finite-element computer program. For simple cases, such as a homogeneous dam, the phreatic surface can be obtained by simple charts. The differences between phreatic and piezometric surfaces will be explained in Section 4.1.3. In this section, only cases involving homogeneous cross sections will be discussed. Details about flownets in nonhomogeneous cross sections consisting of soils with different coefficients of permeability can be found in Cedergren (1977).

4.1.1 Isotropic Cross Section

Fig. 4-1 shows several flownets for homogeneous and isotropic cross sections. An isotropic soil has the same permeability in both the horizontal and vertical directions. A flownet consists of a number of flow lines, as shown by the solid curves in the figure, and a series of equipotential lines, as indicated by the dashed curves. To construct the flownet, first a trial phreatic surface must be drawn. At

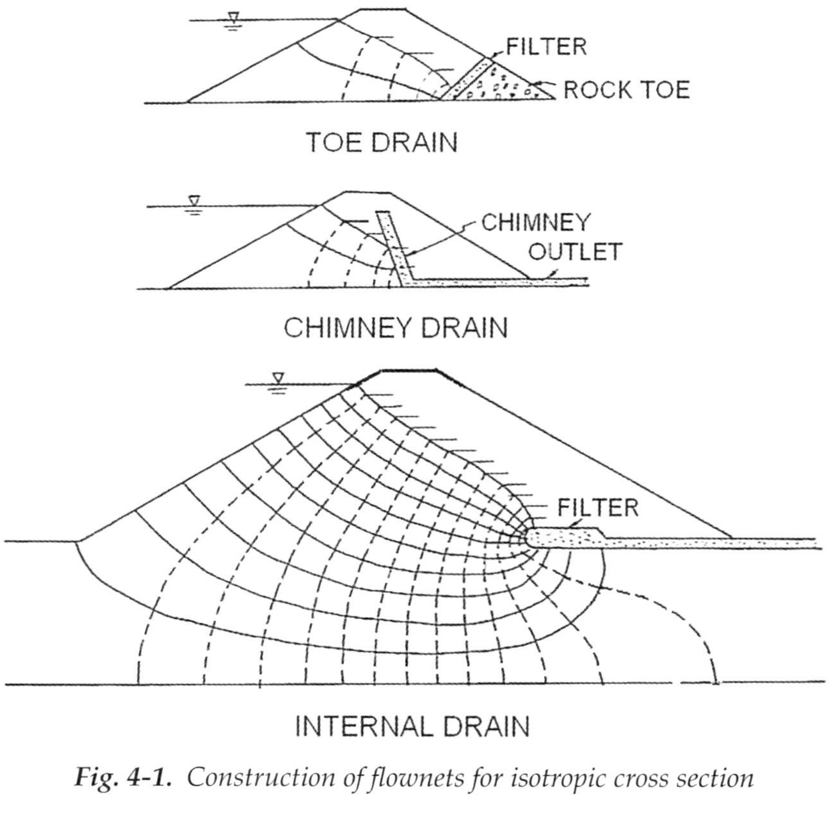

Fig. 4-1. Construction of flownets for isotropic cross section

equal intervals on the phreatic surface, as indicated by the short horizontal lines, the equipotential lines can be drawn, followed by the flow lines. If the flow lines and equipotential lines are perpendicular and form curvilinear squares, at which the average distance between the two flow lines is equal to that between the two equipotential lines, the assumed phreatic surface is correct; otherwise, the phreatic surface must be changed until a satisfactory flownet is obtained. The use of squares is for convenience only. At the impervious boundary, rectangles also may be used as long as they are all of the same shape. For example, the toe drain at the top of Fig. 4-1 has only three flow lines, including the impervious boundary, and two flow channels. The bottom channel should not be considered a full channel but only a 0.6 channel, because it is formed by rectangles with the same width-to-length ratio of 0.6.

If the rate of seepage is desired, the following equation can be used:

$$q = kh\frac{N_f}{N_d} \tag{4-1}$$

in which q = rate of flow per unit width, k = coefficient of permeability, h = total head loss, N_f = number of flow channels, and N_d = number of equipotential drops.

4.1.2 Anisotropic Cross Section

An anisotropic soil has a coefficient of permeability in the horizontal direction different from that in the vertical direction. When the fill is compacted in layers, the coarse particles may be broken down under the high-contact pressure of the roller, so a thin layer of fine materials may be formed at the top of each compacted layer. These thin layers will reduce greatly the permeability in the vertical direction, but the permeability in the horizontal direction will be affected to a much lesser degree. To draw the flownet, an anisotropic cross section must be transformed to an isotropic section by changing the x-coordinates while the y-coordinates remain unchanged, or

$$X = \left(\sqrt{\frac{k_y}{k_x}}\right)x \tag{4-2}$$

in which X is the coordinate after transformation, x is the original coordinate before transformation, k_x is the coefficient of permeability in the x direction, and k_y is the coefficient of permeability in the y direction. Because k_y is usually smaller than k_x, the horizontal distance is reduced. Once the section is transformed, a flownet can be drawn and the location, or X-coordinates, of the phreatic surface can be determined. The transformed X-coordinates of the phreatic surface then can be transformed back to the original x-coordinates, which are the coordinates to be used for stability analysis.

Eq. (4-1) also can be used to determine the rate of seepage for the anisotropic cross section. However, in the equation, k is an equivalent permeability defined by

$$k = \sqrt{k_x k_y} \tag{4-3}$$

Example 4.1 Fig. 4-2(a) shows the original cross section of a refuse dam constructed of two different materials, silty clay and coarse coal refuse. The starter dam was built of silty clay, which has a permeability several orders of magnitude smaller than that of coarse coal refuse, so the surface of the starter dam can be considered as an impervious boundary, or the bottom-most flow line. The coarse coal refuse is anisotropic with a horizontal permeability of 3.28×10^{-6} ft/s (1×10^{-4} cm/s) and a vertical permeability of 8.2×10^{-7} ft/s (2.5×10^{-5} cm/s). A drainage blanket is provided under the coarse coal refuse to lower the phreatic surface. Determine the coordinates of the phreatic surface and compute the rate of seepage through the dam.

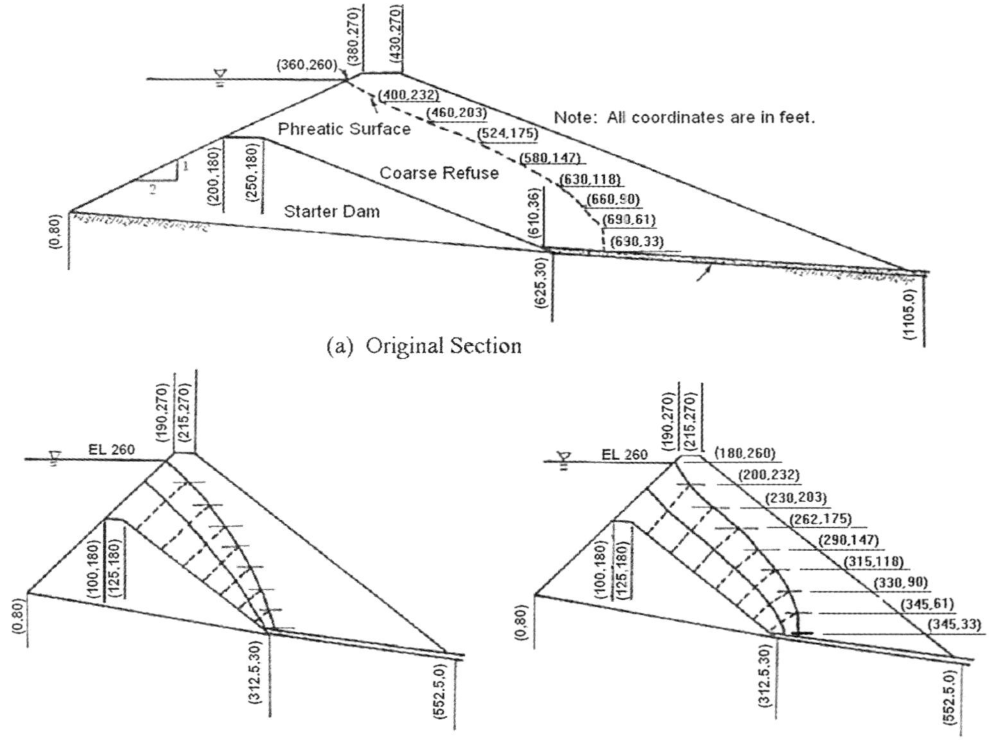

(a) Original Section

(b) Transformed Section: unacceptable flownet (c) Transformed Section: acceptable flownet

Fig. 4-2. Determination of phreatic surface for anisotropic cross section
Note: 1 ft = 0.305 m

Solution From Eq. (4-2), $X = x\sqrt{8.2\times10^{-7}/3.28\times10^{-6}} = 0.5x$, so the transformed section can be drawn by reducing the x-coordinates by one-half, while the y-coordinates remain the same. Fig. 4-2(b) is the transformed section with an unacceptable flownet, because it is not formed by squares. Fig. 4-2(c) is an acceptable flownet with the coordinates of the phreatic surface shown in parentheses. The section is transformed back to the original by doubling the x-coordinates, and the coordinates of the phreatic surface are shown in Fig. 4-2(a). To determine the rate of seepage, first from Eq. (4-3), $k = \sqrt{3.28\times8.2\times10^{-13}} = 1.64\times10^{-6}$ ft/s; then from Eq. (4-1), $q = 1.64\times10^{-6}\times(260-33)\times2/8 = 9.31\times10^{-5}$ cfs per foot of dam.

If the flownet in the transformed section is transformed back to the original section, the flownet will become rectangles instead of squares, and the equipotential lines no longer will be perpendicular to the flow lines.

4.1.3 Phreatic Surface versus Piezometric Surface

Theoretically, the pore pressure along the failure surface under steady-state seepage should be determined by drawing a flownet, or, more accurately, by using a finite-element program for seepage analysis. Fig. 4-3 shows the phreatic surface and the equipotential line passing through point A at the bottom of a slice. If point B on the phreatic surface, which lies directly above point A, is at a distance of h_w above point A, the pressure head at point A is $h_w\cos^2\beta$, where β is the slope of the phreatic surface at point B. When a piezometer is placed at point A, the water table in the piezometric tube will rise to point C at a distance of $h_w\cos^2\beta$ above point A, so point C is on the piezometric surface. The pore pressure at the bottom of slice can be written as

$$u = \gamma_w h_w \cos^2\beta \tag{4-4}$$

in which γ_w = unit weight of water. A simplification is to consider the phreatic surface as a piezometric surface, so if a piezometer is placed a point A, the water level in the piezometer will rise to the elevation at point B. Thus, the pore pressure at point A can be expressed simply as

$$u = \gamma_w h_w \tag{4-5}$$

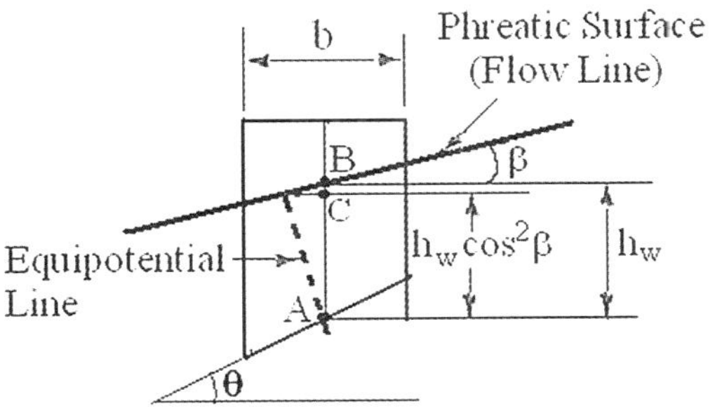

Fig. 4-3. Pore pressure at bottom of slice

Compared to Eq. (4.4), Eq. (4-5) is a more conservative estimate of the pore water pressure. However, the differences between the two representations of pore water pressure are typically small because the slope angle, β, for most phreatic surfaces is quite small. Even if β is large in some local regions, the difference between the two surfaces occurs only over a limited area, and its effect on the factor of safety is quite small. In view of the fact that the phreatic surface is difficult to estimate and is at times unconservative, the use of Eq. (4-5) to represent the pore pressure at the bottom of the slice is just as suitable as Eq. (4-4). The same conclusion was drawn by Duncan and Wright (2005). For this reason, Eq. (4-4) is used to compute the pore water pressure throughout this book.

Based on $\beta = 0$, the Fellenius method assumes that the neutral force normal to the failure surface is $\gamma_w h_w b \sec\theta$. It is more reasonable to assume $\beta = \theta$, so the neutral force by the Fellenius method should be multiplied by $\cos^2\theta$, which is the same as $\gamma_w h_w b \cos\theta$ by the normal method. Therefore, the use of the normal method to replace the Fellenius method is theoretically sound.

4.2 Earth Dams without Filter Drains

Fig. 4-4 shows an earth dam on an impervious base. The downstream face of the dam has a slope of S:1 (horizontal:vertical). If no drainage system is provided, the downstream slope should be relatively flat, generally not steeper than 1.5:1, or 30°. In such a case, Dupuit's assumption that the hydraulic gradient in every point on a vertical line is constant and equal to the slope, dy/dx, is valid. The seepage through the dam can be expressed by Darcy's law as

$$q = k(y - x\tan\alpha)\frac{dy}{dx} \tag{4-6}$$

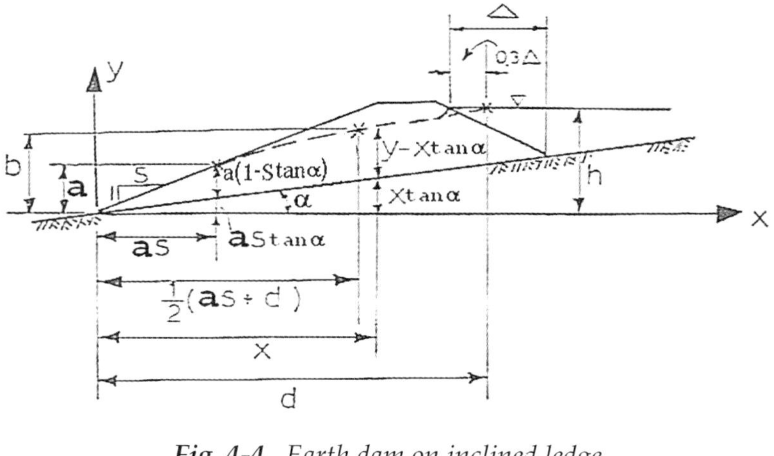

Fig. 4-4. Earth dam on inclined ledge

in which q = discharge per unit time, k = permeability, α = angle of inclination of the base, and x and y are the coordinates. At the point of exit,

$$q = \frac{ka(1 - S\tan\alpha)}{S} \tag{4-7}$$

in which a is the y-coordinate of the exit point.

Equating Eqs. (4-6) and (4-7) and integrating, an equation of the following form is obtained:

$$\text{Function}(x, y, c_1) = 0 \tag{4-8}$$

in which c_1 is a constant of integration.

Following the procedure originally suggested by Casagrande (1937), it is assumed that the theoretical line of seepage starts from the pool level at a distance of 0.3Δ from the dam, where Δ is the horizontal distance shown in Fig. 4-4. Therefore, when the toe of the downstream slope is used as the origin of coordinates, one point on the line of seepage with $x = d$ and $y = h$ is known. Substituting this x, y pair into Eq. (4-8) allows the evaluation of the constant of integration, c_1. Assuming the x- and y-coordinates of the exit point as aS and a, and substituting this x, y pair into Eq. (4-8), an equation of the following form is obtained:

$$\text{Function}(a) = 0 \tag{4-9}$$

Eq. (4-9) was solved by Huang (1981) using a numerical method, and the results are presented in Fig. 4-5.

Next, assume that the x- and y-coordinates of the midpoint are $(aS + d)/2$ and b, respectively. Substituting this x, y pair into Eq. (4-8) and bearing in mind that the value of a has been determined by Eq. (4-9), an equation of the following form is obtained:

$$\text{Function}(b) = 0 \tag{4-10}$$

The solution of Eq. (4-10) is presented in Fig. 4-6. Knowing the three points on the phreatic surface, i.e., the starting point, the midpoint, and the exit point, a curve can be drawn, which is the theoretical line of seepage. Because the actual line of seepage must be perpendicular to the upstream slope and tangent to the downstream slope, the theoretical line can be adjusted slightly to fulfill these boundary requirements. This adjustment is not really necessary when the phreatic surface is inputted into a computer program. The phreatic surface within a dam can be represented simply by two straight lines: one from the water entrance point (point of intersection between the upstream water table and the upstream slope) to the midpoint, and the other from the midpoint to the exit point. There is no need to draw a flownet or use a finite-element seepage program.

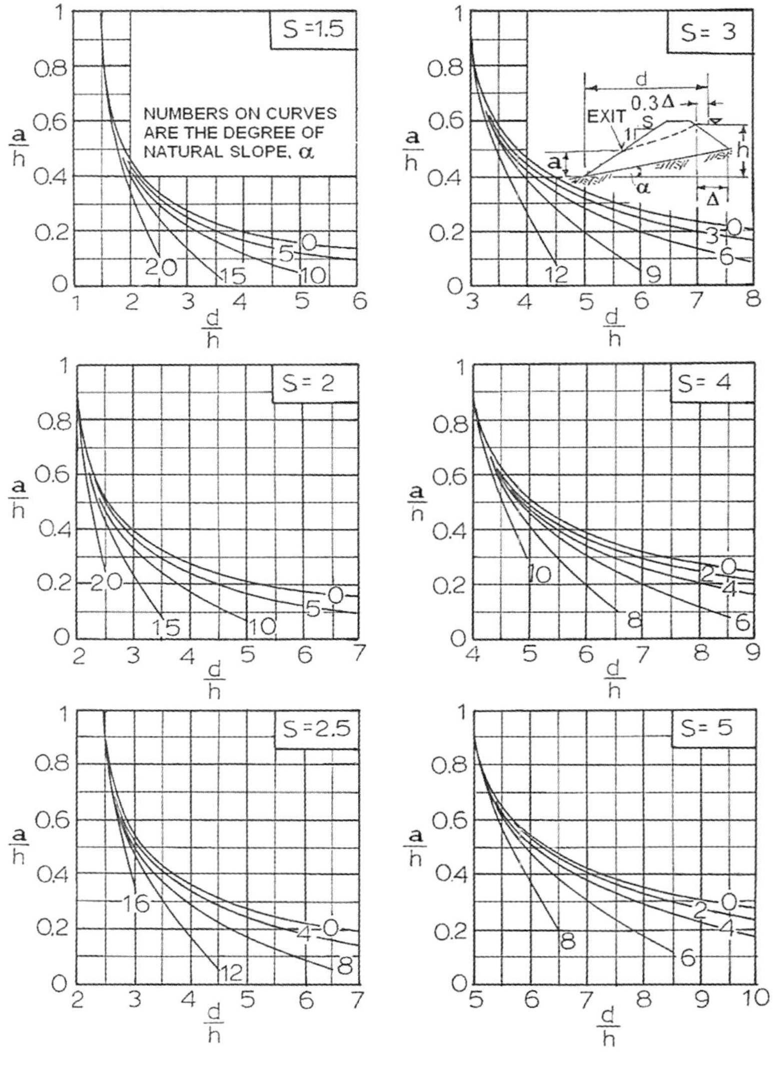

Fig. 4-5. Chart for determining point of exit (Huang 1981. Reproduced with permission)

When the dam is constructed on a horizontal base, i.e., $\alpha = 0$, the solution is the same as that developed by Schaffernak and Iterson in 1916, as reported by Casagrande (1937).

Example 4.2 Fig. 4-7(a) shows the cross section of a dam with all the necessary dimensions given. Determine the location of the phreatic surface.

Solution From the given coordinates, $\alpha = \tan^{-1}(60/530) = 6.5°$; $\Delta = 530 - 350 = 180$ ft.

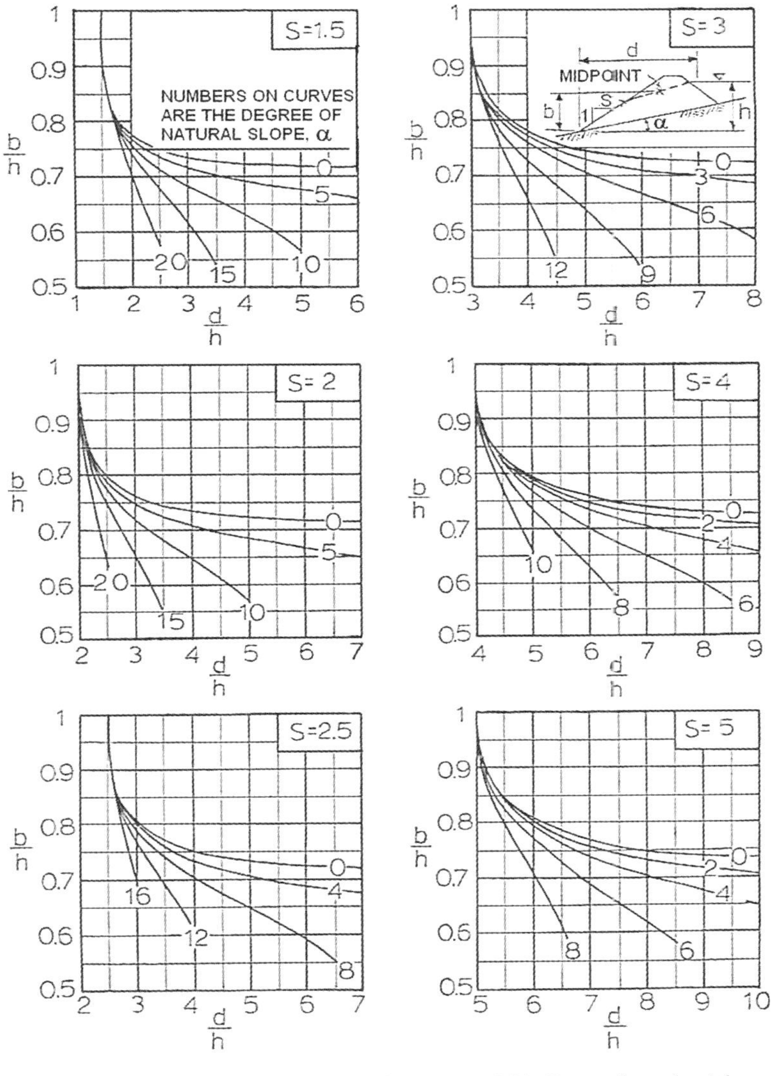

Fig. 4-6. Chart for determining midpoint (Huang 1981. Reproduced with permission)

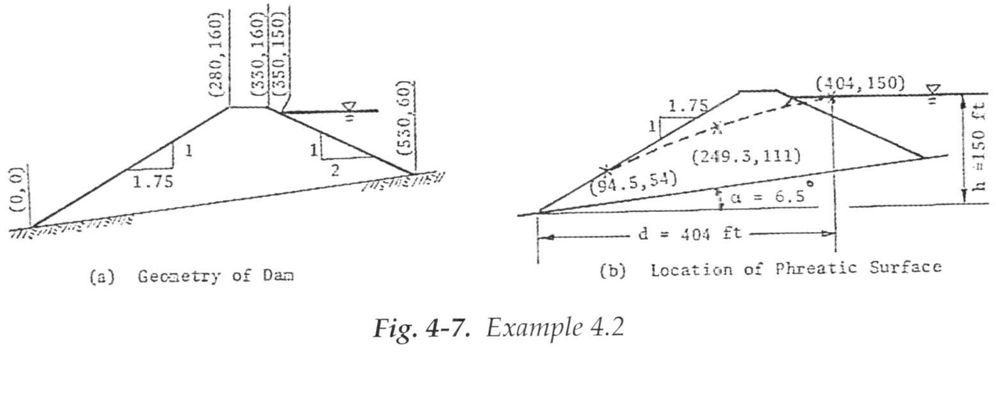

Fig. 4-7. Example 4.2

Location of starting point: $d = 350 + 0.3 \times 180 = 404$ ft, $h = 150$ ft, so the coordinates are (404, 150).

Location of exit point: From Fig. 4-5, with $\alpha = 6.5°$ and $d/h = 404/150 = 2.7$, when $S = 1.5$, $a/h = 0.29$; and when $S = 2$, $a/h = 0.43$. When $S = 1.75$, $a/h = (0.29 + 0.43)/2 = 0.36$, or $a = 0.36 \times 150 = 54$ ft. $aS = 1.75 \times 54 = 94.5$, so the coordinates are (94.5, 54).

Location of midpoint: From Fig. 4-6, when $S = 1.5$, $b/h = 0.72$; and when $S = 2$, $b/h = 0.76$. When $S = 1.75$, $b/h = (0.72 + 0.76)/2 = 0.74$, or $b = 0.74 \times 150 = 111$ ft. $x = (94.5 + 404)/2 = 249.3$ ft, so the coordinates are (249.3, 111).

Knowing the three points, the line of seepage can be drawn, as shown in Fig. 4-7(b).

4.3 Earth Dams with Filter Drains

If a drainage system is provided within the dam, such as the use of porous shells, toe drains, or underdrains, the line of seepage becomes steeper and Dupuit's assumption is no longer valid, so the method presented in Section 4.2 for determining the location of phreatic surface cannot be used. Depending on the angle β of the filter drain shown in Fig. 4-8, two methods can be used to determine the point of exit, one for $\beta \leq 60°$ and the other for $60° < \beta \leq 180°$.

4.3.1 When $\beta \leq 60°$

Based on the assumption that the hydraulic gradient is $dy/d\ell$, instead of dy/dx by Dupuit, where ℓ is the distance along the line of seepage, Gilboy (1933) developed a simple chart for a dam on a horizontal base, as shown in Fig. 4-9. The insert in the figure shows only the more impervious part of the dam; the porous shell, if any, is not shown. When β is less than 30°, the solutions check closely with Fig. 4-5 for $\alpha = 0$. The solutions by Fig. 4-9 are very satisfactory for slopes up to 60°. If deviations of 25% are permitted, it may even be used up to 90°, i.e., for a vertical discharge face.

Example 4.3 Fig. 4-10 shows a zoned dam with the porous shells outside and the clay core inside. The dimensions necessary to determine the exit point of the phreatic surface are indicated by the coordinates in parentheses. Determine the distance, a, of the exit point.

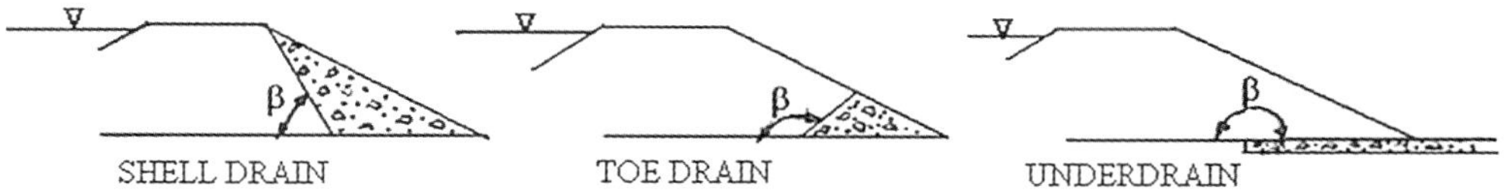

Fig. 4-8. Measurement of angle β

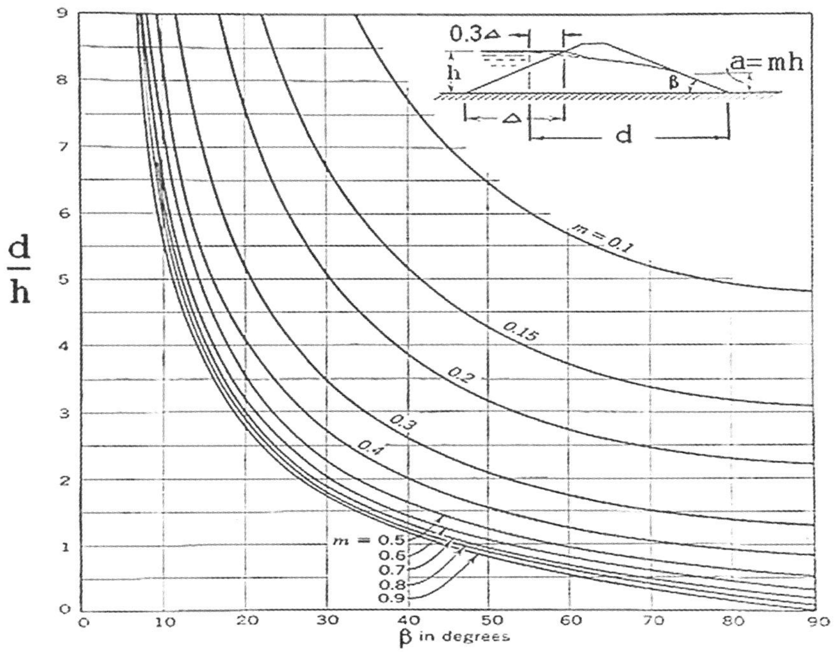

Fig. 4-9. Location of phreatic surface (Gilboy 1933. Reproduced with permission from the International Commission on Large Dams)

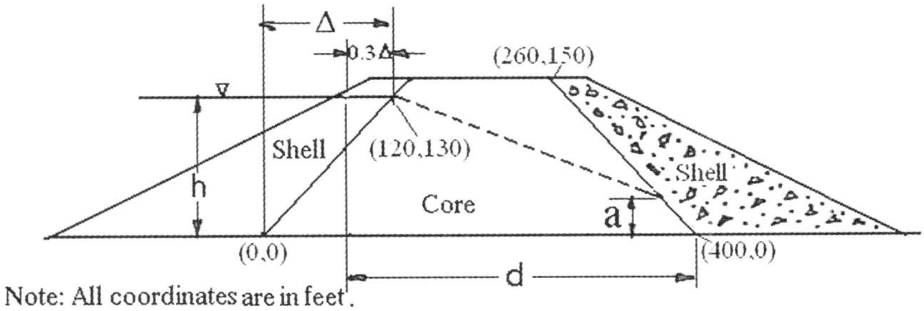

Fig. 4-10. Example 4.3

Solution Because of the porous nature of the shells, the starting point of the phreatic surface is at (120, 130). From the coordinates of the four given points, $h = 130$ ft, $\Delta = 120$ ft, $d = 0.3 \times 120 + (400 - 120) = 316$ ft, and $\beta = \tan^{-1}[150/(400 - 260)] = 47°$. With $d/h = 316/130 = 2.43$ and $\beta = 47°$, from Fig. 4-9, $m = 0.28$, or $a = 0.28 \times 130 = 36.4$ ft. The location of the phreatic surface can be approximated by connecting a straight line between the starting and exit points.

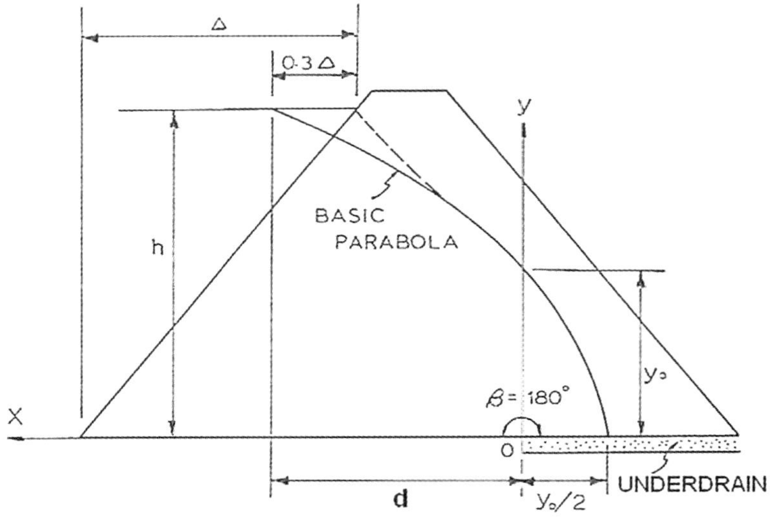

Fig. 4-11. Basic parabola for underdrain

4.3.2 When 60° < β ≤ 180°

Fig. 4-11 shows an earth dam with an underdrain, or β = 180°. The equation for the line of seepage was derived by Kozeny and can be expressed as a basic parabola (Harr 1962)

$$x = \frac{y^2 - y_o^2}{2y_o} \tag{4-11}$$

Note that when $x = 0$, $y = y_o$ and that when $y = 0$, $x = -0.5y_o$. The intercept y_o can be determined by letting $x = d$ and $y = h$, which are the coordinates of one known point on the line of seepage, and then solving y_o, or

$$y_o = \sqrt{d^2 + h^2} - d \tag{4-12}$$

Example 4.4 Fig. 4-12 shows a dam with an underdrain. Given the coordinates of the three points shown in parentheses, compute the x coordinates at every y interval of 20 ft (6.1 m) and sketch the line of seepage.

Solution Based on the given coordinates, $h = 140$ ft, $\Delta = 400 - 120 = 280$ ft, and $d = 120 + 84 = 204$ ft. From Eq. (4-12), $y_o = [(204)^2 + (140)^2]^{0.5} - 204 = 43.4$ ft. From Eq. (4-11), the equation of the basic parabola is

$$x = \frac{y^2 - 1{,}883}{86.8} \tag{4-13}$$

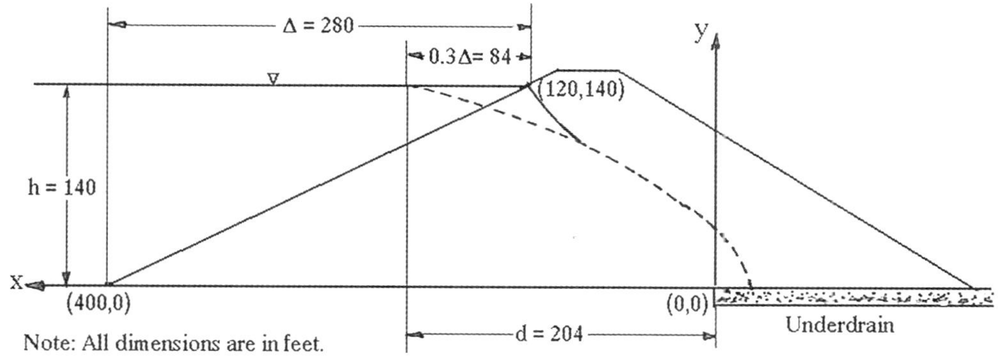

Fig. 4-12. Example 4.4

The x- and y-coordinates of eight points from $y = 0$ ft to $y = 140$ ft are tabulated as follows:

x (ft)	y (ft)	x (ft)	y (ft)
−21.7	0	52.0	80
−17.1	20	93.5	100
−3.3	40	144.2	120
19.8	60	204.0	140

The basic parabola is plotted in the figure as indicated by the dashed curve. Because the line of seepage is a flow line, which must be perpendicular to the equipotential line or the upstream slope, the basic parabola must be adjusted at the upstream side, as indicated by the solid curve.

When $\beta < 180°$, Casagrande (1937) developed a method for sketching the line of seepage. By comparing the line of seepage obtained from the flownets for various angles, β, he found the distance, $\Delta\lambda$, between point A on the basic parabola and point B on the line of seepage, as shown in Fig. 4-13 for a toe drain with $\beta = 135°$. To adjust the basic parabola, a correction factor, c_f, must be obtained from Fig. 4-14. The correction factor is defined as

$$c_f = \frac{\Delta\lambda}{\lambda + \Delta\lambda} \tag{4-14}$$

in which $\lambda + \Delta\lambda$ is the distance from the origin to the basic parabola along the slope surface. Eq. (4-14) can be used to determine $\Delta\lambda$. Having plotted the basic parabola and determined the discharge point by scaling a distance of $\Delta\lambda$ from the basic parabola, the entire line of seepage can be sketched in easily. The following rules are convenient for sketching the line of seepage:

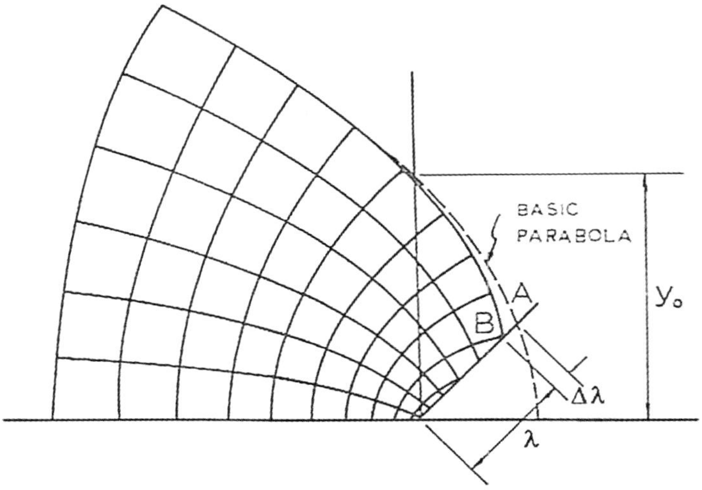

Fig. 4-13. Comparison of basic parabola and flownet

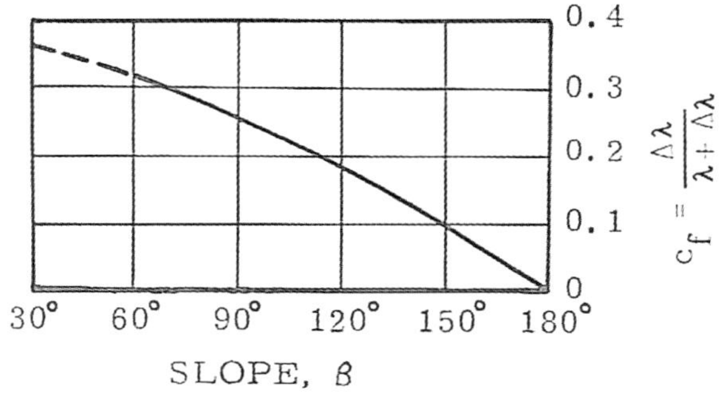

Fig. 4-14. Correction factor for phreatic surface (Casagrande 1937. Reproduced with permission of Journal of the New England Water Works Association*)*

1. When $\beta < 90°$, the line of seepage is tangent to the slope; when $\beta \geq 90°$, it is tangent to a vertical line.
2. When $y = 0$, $x = -y_o/2$.
3. When $x = 0$, $y = y_o$, and the line of seepage makes an angle of 45° with horizontal.

Example 4.5 Fig. 4-15(a) shows an earth dam with a toe drain. Sketch the line of seepage using Casagrande's procedure.

Solution Based on the given coordinates, it can be easily found that $d = 304$ ft and $h = 140$ ft. From Eq. (4.12), $y_o = [(304)^2 + (140)^2]^{0.5} - 304 = 30.7$ ft. From Eq. (4-11), the equation of the basic parabola is

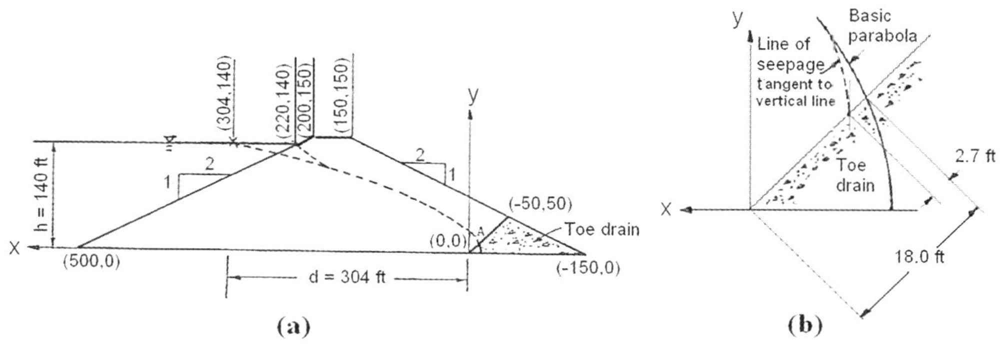

Fig. 4-15. Example 4.5

$$x = \frac{y^2 - 942}{61.4}$$

The basic parabola is plotted in Fig. 4-15(a). Some adjustments must be made at the entrance and exit ends. An exploded view of the line of seepage near the toe is shown in Fig. 4-15(b). The intersection of the toe with the basic parabola can be determined by substituting $y = -x$ into the equation, and a quadratic equation $y^2 - 61.4y - 942 = 0$ is obtained. The solution is $y = 12.7$ ft, or $\lambda + \Delta\lambda = 12.7/\sin 45° = 18.0$ ft. With $\beta = 135°$, from Fig. 4-14, $c_f = 0.15$. From Eq. (4-14), $\Delta\lambda = 0.15 \times 18 = 2.7$ ft. In Fig. 4-15(b), the basic parabola is shown in the solid curve and the adjusted line of seepage in the dashed curve. Because of the large β, there is not much adjustment at the exit end.

4.4 Unsteady-State Seepage

The phreatic surface discussed so far is concerned with the steady-state seepage. In most cases, it is assumed that, after a period of time, a steady-state seepage condition finally will develop, so the steady-state phreatic surface can be used for stability analysis. However, the assumption of steady-state seepage for temporary dams, such as those used for refuse disposal and sediment control, does not appear to be reasonable, because the steady-state condition might not be reached during the design life. In fact, many of the refuse dikes and silt dams constructed in the Eastern Kentucky Coal Field would be considered unsafe should a steady-state seepage condition be assumed. To achieve the required factor of safety, the designer has to assume an unsteady-state phreatic surface arbitrarily. Although unsteady-state seepage can be analyzed by numerical methods or transient flownets, they are too complex to be of general use. The uncertainty in determining the permeability and effective porosity of soils usually precludes the use of more refined methods.

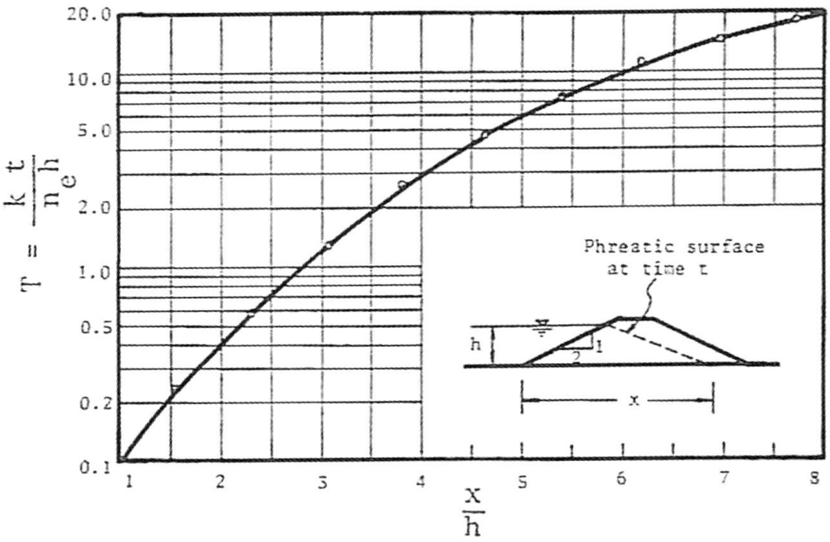

Fig. 4-16. Relationship between dimensionless time and dimensionless distance (Huang 1986. Reproduced with permission)

Based on the transient flownets presented by Cedergren (1977), Huang (1986) developed a simple chart for estimating the unsteady-state seepage in an earth dam, as shown in Fig. 4-16. The chart can be applied to an earth dam on a horizontal impervious base with an upstream slope of 2:1. Given the dimensionless time, T, the distance, x, traveled by the phreatic surface along the base of the dam can be determined.

The dimensionless time is expressed as

$$T = \frac{kt}{n_e h} \tag{4-15}$$

in which t = actual time for the phreatic surface to travel a distance x along the base of the dam, k = permeability, n_e = effective porosity of the soil, and h = depth of water in the pond.

It can be visualized that the phreatic surface progresses from upstream to downstream, with the upper end fixed at the pool elevation and the lower end moving along the impervious base. Although the phreatic surface is not a straight line between these two ends, the assumption of a straight line is quite reasonable, especially at the later stage. In the initial stage where the phreatic surface is curved, the assumption of a straight line is on the safe side, because it results in a higher phreatic surface.

Example 4.6 Fig. 4-17 shows a temporary dam with a horizontal impervious base. The soil in the dam has a permeability of 3×10^{-7} ft/s (2.7×10^{-5} cm/s) and an effective porosity of 0.2. If the dam is used for only five years, determine the location of the unsteady state phreatic surface at the end of the fifth year.

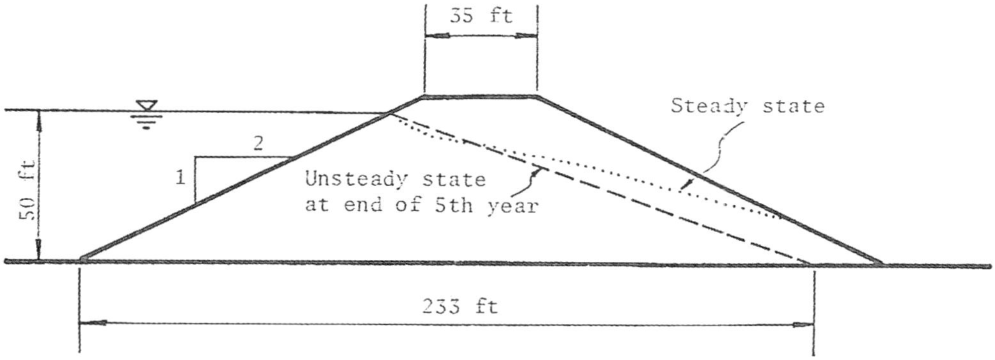

Fig. 4-17. Example 4.6

Solution Given t = 5 years, $k = 3 \times 10^{-7}$ ft/s = 9.5 ft/year, n_e = 0.2, and h = 50 ft, from Eq. (4-15), $T = 9.5 \times 5/(0.2 \times 50) = 4.75$. From Fig. 4-16, $x/h = 4.66$, or $x = 4.66 \times 50 = 233$ ft. The location of the phreatic surface at the end of the fifth year is shown in Fig. 4-17 by the dashed line.

The chart presented in Fig. 4-16 is based on the results obtained by Cedergren (1977) for dams with a horizontal impervious base and an upstream slope of 2:1. For dams with other configurations, as shown in Fig. 4-18, the chart should give a phreatic surface higher than reality, so using it for stability analysis is on the safe side.

In Fig. 4-18, if the impervious base is not horizontal, as shown in (a), or the dam is placed on a soil foundation, as shown in (b), an imaginary horizontal

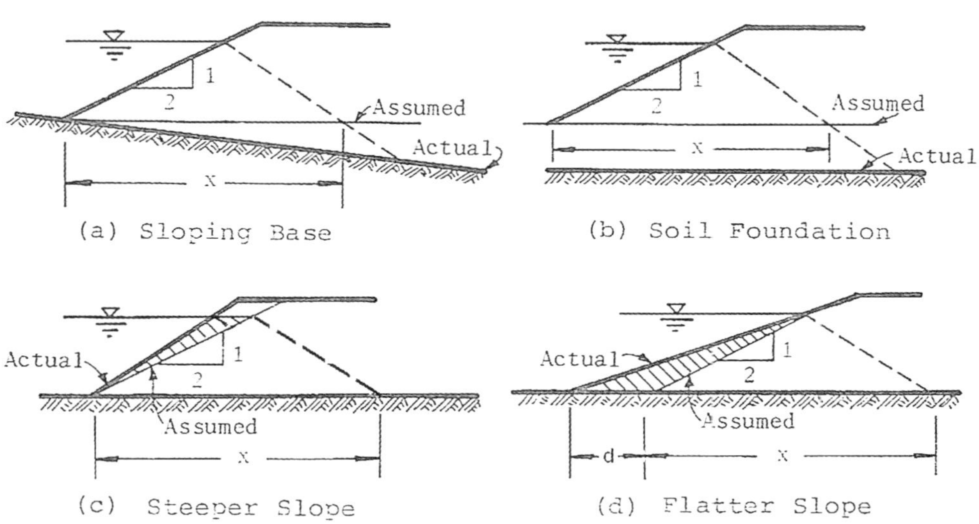

Fig. 4-18. Applications to special cases

base is assumed, and the method is applied as usual. The extra flow region below the imaginary base will lower the phreatic surface, so the assumption of a horizontal base is on the safe side. If the slope is steeper than 2:1, as shown in (c), a 2:1 slope can be drawn from the toe. Because the hatched portion of the dam is considered as water instead of soil, the value of x for the 2:1 slope should be greater than that for the steeper slope, so the use of the chart based on a 2:1 slope is on the safe side. If the slope is flatter than 2:1, as shown in (d), a 2:1 slope can be drawn from the pool elevation, and the consideration of the hatched portion of the dam as a body of water will give a larger and more conservative value of x. However, to measure the x distance from the toe of the actual slope, the distance d between the two toes must be added.

4.5 Pore Pressure Ratio

Pore pressure ratio can be used for two purposes: first, to replace the phreatic surface for steady-state seepage, and second, to simulate the excess water pressure due to consolidation during or immediately after construction. Both cases will be discussed in this section.

4.5.1 Pore Pressure Ratio for Steady-State Seepage

When the location of the phreatic surface is unknown or unpredictable, it is convenient to assume a pore pressure ratio so that the adverse effect of water can be included in the stability analysis. Even if the location of the phreatic surface is known a priori, the conversion of the phreatic surface to a pore pressure ratio can simplify the use of equations and charts for determining the factor of safety, as demonstrated in Chapters 6 and 7.

The pore pressure ratio is defined as a ratio between the water pressure and the overburden pressure, or

$$r_u = \frac{u}{\gamma h} = \frac{\gamma_w h_w}{\gamma h} \tag{4-16}$$

in which u = pore water pressure, γ_w = unit weight of water, h_w = depth of water between the phreatic surface and the failure surface, γ = total unit weight of soil, and h = depth of soil between the ground surface and the failure surface. Because the pore pressure ratio generally is not uniform throughout a slope, an average pore pressure ratio should be used in stability analysis.

Fig. 4-19 shows the conversion of a phreatic surface to a pore pressure ratio for both the plane and the cylindrical failure surfaces. As defined by Eq. (4-16), the pore pressure ratio is a ratio between the pore pressure and the overburden pressure, or between the total upward force due to water pressure and the total downward force due to the weight or overburden pressure. According to the Archimedes principle, the upward force is equal to the weight of water

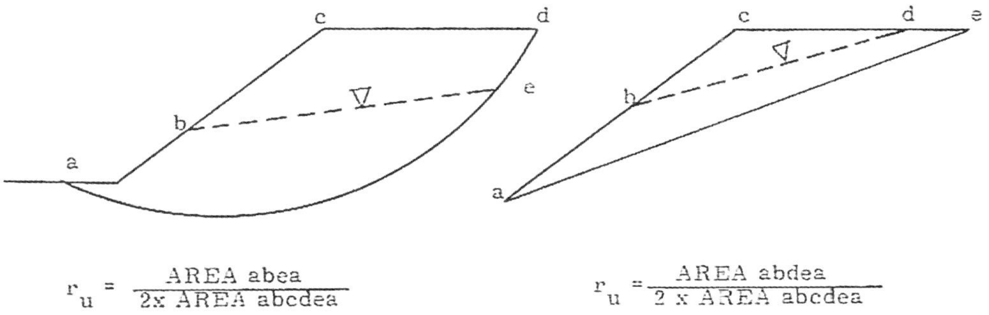

Fig. 4-19. Determination of pore pressure ratio

displaced, or the volume of sliding mass under water multiplied by the unit weight of water. The downward force is equal to the weight of the sliding mass. Therefore, the pore pressure ratio can be determined by

$$r_u = \frac{\text{Volume of sliding mass under water} \times \text{unit weight of water}}{\text{Volume of sliding mass} \times \text{unit weight of soil}} \tag{4-17}$$

Since the unit weight of water is approximately equal to one-half the unit weight of soil, the pore pressure ratio can be determined approximately by

$$r_u = \frac{\text{area of the sliding mass under water}}{2 \times \text{area of the sliding mass}} \tag{4-18}$$

If the location of the failure surface is known, the average pore pressure ratio can be determined by Eq. (4-17) or (4-18), as shown in the figure. If the location of the failure surface is not known, proper judgment or prior experience in estimating its location is needed in order to determine the pore pressure ratio.

In applying their stability charts, as presented in Section 7.5.1, Bishop and Morgenstern (1960) suggested the use of an average pore pressure ratio by weighing the pore pressure ratio over the area through the entire slope, or

$$r_u = \frac{\sum(\text{Pore pressure ratio} \times \text{area})}{\text{Total area}} \tag{4-19}$$

Eq. (4-19) is useful when the pore pressure is determined from field measurements or from the construction of flownets, because it is not necessary to know the location of the most critical circle a priori. However, the use of Eq. (4-17) based on the probable location of the most critical circle is more accurate and should be used whenever possible. If the total area is used, Eq. (4-17) is still valid except that the volume of the sliding mass is replaced by the volume of the entire slope, as illustrated in Example 4.7.

In an effective stress analysis, the pore pressure ratio is used to convert the total weight of soil, W, to the effective weight, W', or

$$W' = (1 - r_u)W \tag{4-20}$$

If there is no water table or no seepage in the slope, or $r_u = 0$, then $W' = W$. If the entire soil mass in under water, or $r_u = 0.5$, then $W' = 0.5W$. The effective weight is used to determine the shear resistance along the failure surface, and the total weight is used to determine the driving force.

Example 4.7 Fig. 4-20 is an earth dam on a horizontal ledge with the location of the phreatic surface shown by the dashed line. The most critical circle has a radius of 115 ft (35.1 m) and a central angle of 83°. The surface of the dam is along afbcd and the chord of the circle is along abd. If the total unit weight of the soil is 130 pcf (20.4 kN/m³), determine the average pore pressure ratio based on the area of the sliding mass and compare it to that based on the total area.

Solution

Based on area of sliding mass

Area of sliding mass = Area of the circular segment abde – Area abf + Area bcd. The area of the circular segment can be computed by

$$\text{Area of circular segment} = \frac{\pi R^2 \theta}{360} - \frac{R^2 \sin\theta}{2} = \frac{\pi(115)^2(83)}{360} - \frac{(115)^2 \sin(83°)}{2}$$
$$= 3{,}016 \text{ ft}^2$$

Some other areas are based on the measurements from the figure and should be considered as approximate only.

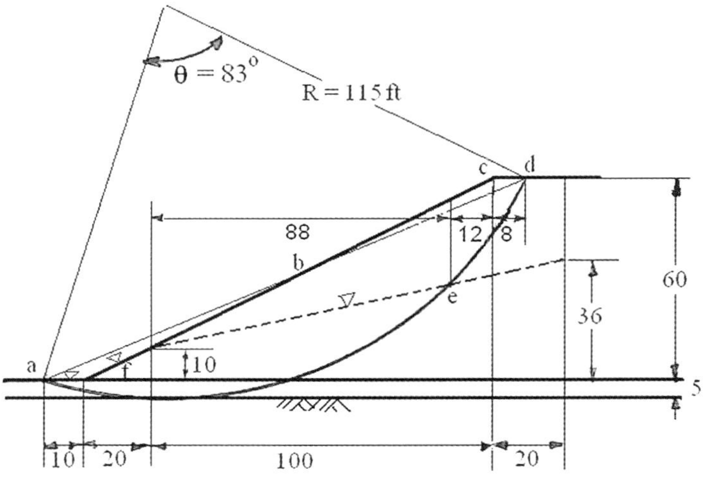

Fig. 4-20. Example 4.7

$$\text{Area of sliding mass} = 3{,}016 - \frac{1}{2} \times 81 \times 5 + \frac{1}{2} \times 72 \times 4 = 2{,}958 \text{ ft}^2$$

The area of the sliding mass above the water table can be determined by dividing the area into two triangles and one trapezoid, or

$$\text{Area above water table} = \frac{1}{2} \times 98 \times 22 + \frac{1}{2} \times (25 + 16) \times 12 + \frac{1}{2} \times 16 \times 8 = 1{,}388 \text{ ft}^2$$

Area of the sliding mass under water = 2,958 – 1,388 = 1,570 ft^2

From Eq. (4-17),

$$r_u = \frac{1{,}570 \times 62.4}{2{,}958 \times 130} = 0.255$$

Based on total area

It is estimated that the most critical circle will fall in a region within 20 ft beyond the top edge and 10 ft beyond the toe, as shown in Fig. 4-20. The total area consists of two rectangles and one trapezoid, or

$$\text{Total area} = 5 \times 10 + \frac{1}{2} \times (5 + 65) \times 120 + 65 \times 20 = 5{,}550 \text{ ft}^2$$

The area under water includes one rectangle and two trapezoids:

$$\text{Area under water} = 5 \times 10 + \frac{1}{2} \times (5 + 15) \times 20 + \frac{1}{2} \times (15 + 41) \times 120 = 3{,}610 \text{ ft}^2$$

$$r_u = 3{,}610 \times 62.4/(5{,}550 \times 130) = 0.312$$

If the width of the total area is reduced to 10 ft from the top edge, instead of the original 20 ft, the pore pressure ratio is 0.315, which is about the same as the original 0.312.

It can be seen that there is a significant difference between the two methods, which is expected. In most cases, the use of the average over the total area is more conservative and results in a higher pore pressure ratio and a lower factor of safety.

4.5.2 Pore Pressure Ratio due to Consolidation

When a new overburden is placed above a layer of clay located below the phreatic surface, an excess pore pressure, other than that due to steady-state seepage, will be developed in the clay layer. If the clay is completely saturated and the new overburden is applied instantaneously, an excess pore pressure equal to the new overburden pressure is developed instantaneously with a pore pressure ratio equal to 1, but this excess pore pressure will dissipate gradually and finally disappear after a long period of time. In the long-term stability analysis, it is assumed that the soil above the phreatic surface has no pore pressure, and the soil below the phreatic surface has a static pore pressure caused by the steady-state seepage. In the short-term stability analysis, the excess pore pressure in the soil below the phreatic surface caused by a new overburden also should be considered. The theory of consolidation so well known in soil mechanics can be applied to determine the excess pore pressure in any clay layer at any time after the load is applied. If a dam is built in stages, the short-term stability at the end of each stage must be analyzed by taking the excess pore pressure into account.

Fig. 4-21 shows the relationship between the consolidation ratio and the time factor based on Terzaghi's one-dimensional (1D) consolidation theory with a drainage layer at the top and an impervious boundary at the bottom. In the figure, z is the distance below the drainage layer, H is the length of drainage path, which is the same as the thickness of clay layer, U_z is the consolidation ratio, and T is the time factor defined as

$$T = \frac{c_v t}{H^2} \tag{4-21}$$

in which c_v = coefficient of consolidation, and t = time since consolidation started. The relationship between the consolidation ratio, U_z, and pore pressure ratio, r_u, is

$$r_u = 1 - U_z \tag{4-22}$$

Fig. 4-21 clearly shows that the pore pressure ratio varies significantly with depth. Because trial failure surfaces may cut through the clay layer at any depth, unless the clay layer is thin, it may be necessary to divide it into several layers, each represented by a pore pressure ratio at its midheight. Theoretically, a point somewhat above the midheight gives a better representation of the average pore pressure ratio. However, the use of midheight is recommended, because it gives a higher pore pressure ratio than average, so the design is on the safe side. The difference between the average pore pressure ratio and the pore pressure ratio at the midheight decreases as the time increases.

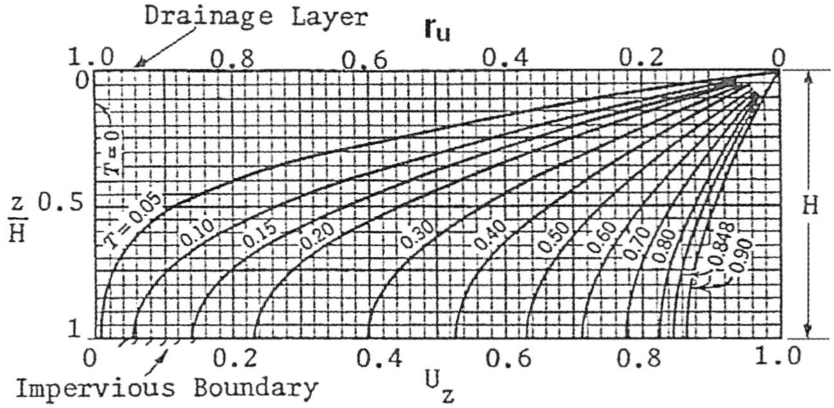

Fig. 4-21. Relationship between consolidation ratio and time factor (Taylor 1948)

It should be noted that Fig. 4-21 also could be applied to a clay layer with drainage at both the top and bottom. In this case, the thickness of layer is $2H$, and an impervious boundary can be assumed at the midheight to separate the clay layer into two halves. The water in the top half will drain to the top and that in the bottom half drain to the bottom. Because both parts are symmetrical with the same length of drainage path, H, Fig. 4-21 can be flipped down and applied to the lower part as well.

Example 4.8 A new dam is to be constructed over a clay layer 20 ft (12.2 m) thick, as shown in Fig. 4-22. The clay layer is located below the water table and is completely saturated. The soil above the clay layer is sand and can serve as a drainage layer. The clay layer on the downstream side is underlain by rock, which can be considered as an impervious boundary, and that on the upstream side by a thin layer of sand, which can provide adequate drainage. To obtain a better representation of the pore pressures, the clay is divided into four sublayers. From start to finish, it will take two years to complete the dam. If the clay has a coefficient of consolidation of 1.4×10^{-4} in.^{2}/s (5.4 mm^2/min), determine the excess pore pressure ratio at the end of construction in each of the sublayers on both sides.

Solution Given total construction time of two years, the average consolidation time $t = 1$ year.

Downstream side: With drainage on one side, $H = 20$ ft. Given $c_v = 1.4 \times 10^{-4}$ in.^{2}/s = 30.66 ft^2/year, from Eq. (4-21), $T = 30.66 \times 1/(20)^2 = 0.077$. From top to bottom, the values of z/H at the midheights of the sublayers are 0.125, 0.375, 0.625, and 0.875. From Fig. 4-21, the pore pressure ratios are 0.25, 0.68, 0.89, and 0.96.

Upstream side: With drainage on both sides, $H = 10$ ft. Since T is inversely proportional to H^2, $T = 4 \times 0.077 = 0.31$. From Fig. 4-21, the pore pressures for z/H at 0.25 and 0.75 are 0.22 and 0.55. The pore pressure ratio for each sublayer is shown in Fig. 4-22.

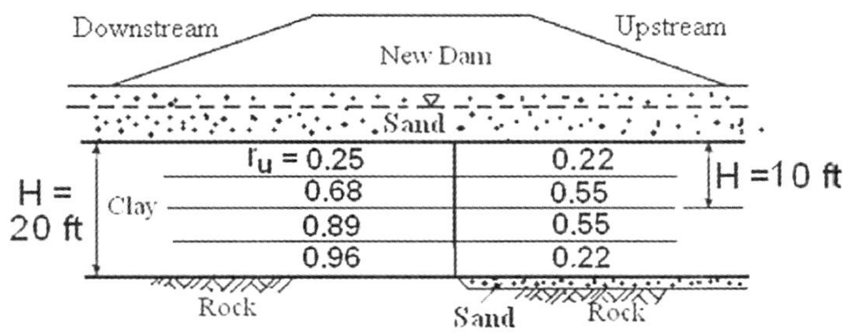

Fig. 4-22. Example 4.8

Summary

1. For homogeneous and isotropic cross sections, the location of the phreatic surface can be determined easily by drawing a flownet. If the cross section is anisotropic with permeability in the horizontal direction greater than that in the vertical direction, it can be transformed into an isotropic cross section by reducing the horizontal coordinates according to Eq. (4-2).

2. Charts are presented in Figs. 4-5 and 4-6 to determine the location of the phreatic surface in earth dams on an impervious base. These charts are based on Dupuit's theory and are applicable when the downstream slope is not greater than 30°. If the impervious base is horizontal and the downstream slope is >30° but <60°, Fig. 4-9 can be used. If the base is not impervious, the phreatic surface will be lower and the use of the charts for stability analysis is on the safe side.

3. For earth dams with filter drains, as shown in Fig. 4-8 for shell drains, toe drains, and underdrains, the exit point varies with the angle β. In the case of an underdrain with $\beta = 180°$, the equation of the phreatic surface is a basic parabola and can be plotted directly based on the distances h and d, as shown in Fig. 4-11. When $\beta > 60°$ but $< 180°$, the basic parabola must be adjusted by moving the exit point down the slope a distance of $\Delta\lambda$, using the correction chart developed by Casagrande (1937), as shown in Fig. 4-14.

4. A simple chart is presented to determine the location of the unsteady-state phreatic surface as a function of time. Although the chart is based on a dam with a horizontal impervious base and an upstream slope of 2:1, with slight modifications, it also can be applied to other configurations with more conservative results.

5. When the location of the steady-state phreatic surface is unknown or unpredictable, it is convenient to assume a pore pressure ratio, so the adverse effect of water can be included in the stability analysis. Even if the location of the phreatic surface is known a priori, the conversion of the phreatic surface to a pore pressure ratio can simplify the use of equations and charts for determining the factor of safety, as demonstrated in Chapters 6 and 7. Given the location of both the phreatic and failure surfaces, the pore pressure ratio can be computed by Eq. (4-17). If the location of the failure surface is unknown and cannot be roughly estimated, the same equation still can be applied by considering the entire slope as a failure mass, instead of the mass above the failure surface.

6. Another use of the pore pressure ratio is to evaluate the effect of excess pore pressure for short-term stability during or at the end of construction. When a new overburden is placed above a layer of clay located below the phreatic surface, an excess pore pressure, other than that due to

steady-state seepage, will be developed in the clay layer. If the clay is completely saturated and the new overburden is applied instantaneously, an excess pore pressure equal to the new overburden pressure is developed instantaneously with a pore pressure ratio equal to 1, but this excess pore pressure will dissipate gradually and finally disappear after a long period of time. Because trial failure surfaces may cut through the clay layer at any depth, unless the clay layer is thin, it may be necessary to divide it into several layers, each represented by a pore pressure ratio at its midheight. Methods for determining the excess pore pressure ratio during or at the end of construction are presented.

Problems

4.1 Fig. P4-1 shows an earth dam with an underdrain. Using the information provided, sketch the flownets for the following two cases: (a) the soil is isotropic with the same permeability in both the horizontal and vertical direction, and (b) the soil is anisotropic with the horizontal permeability four times greater than the vertical permeability.

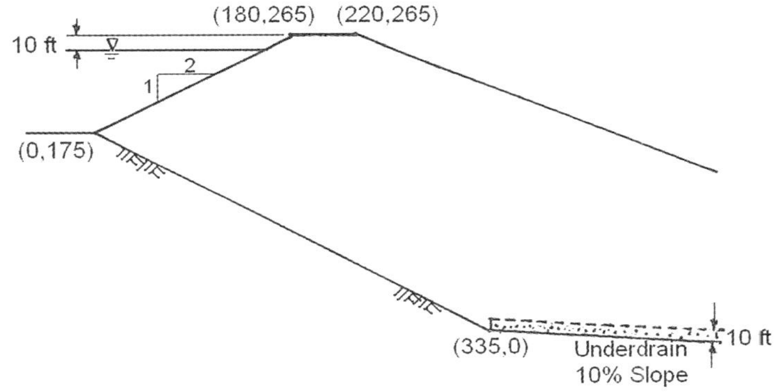

Fig. P4-1.

4.2 Same as Problem 4.1 except refer to Fig. P4-2.

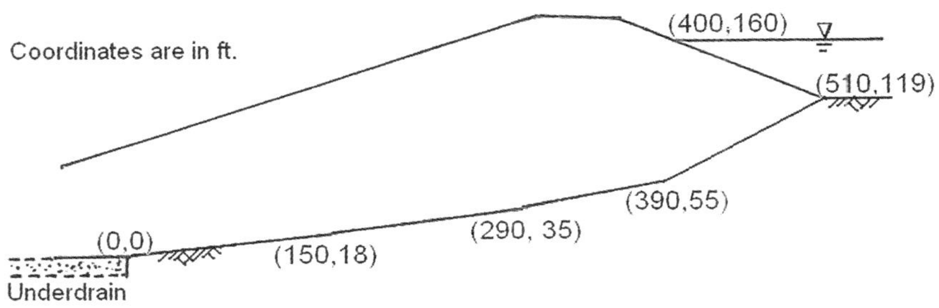

Fig. P4-2.

4.3 Fig. P4-3 shows an earth dam with an underdrain. Using the coordinates provided, sketch the flownet and determine the coordinates of the phreatic surface. If the dam has a permeability of 2×10^{-5} cm/s, compute the amount of seepage in cfs per ft of dam.
[Answer: 1.11×10^{-5} cfs per ft]

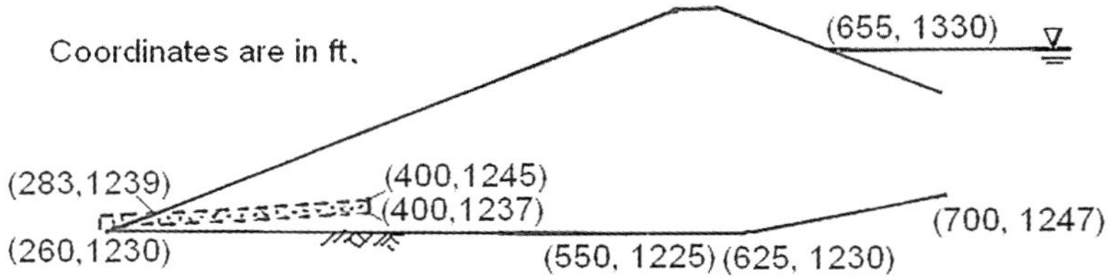

Fig. P4-3.

4.4 Based on the dimensions shown in Fig. P4-4, determine the values of "a" from Fig. 4-5 and "b" from Fig. 4-6 and sketch the phreatic surface.
[Answer: 72 ft, 144 ft]

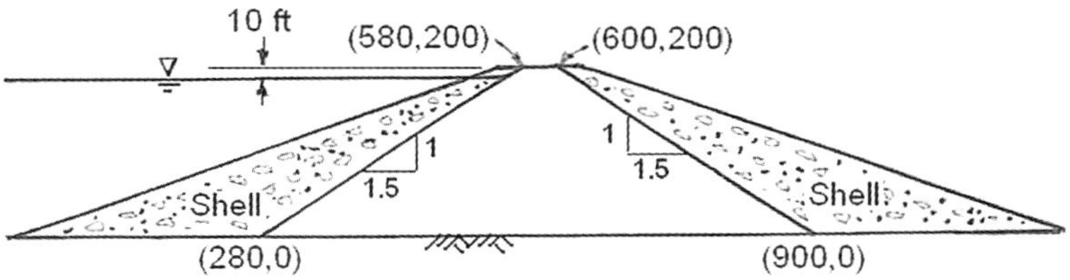

Fig. P4-4.

4.5 Fig. P4-5 shows an earth dam on an inclined ledge. Locate the midpoint and exit point and sketch the phreatic surface. If the dam has a permeability of 5×10^{-4} cm/s, what is the amount of seepage in cfs per foot of dam?
[Answer: 1.36×10^{-5} cfs per ft]

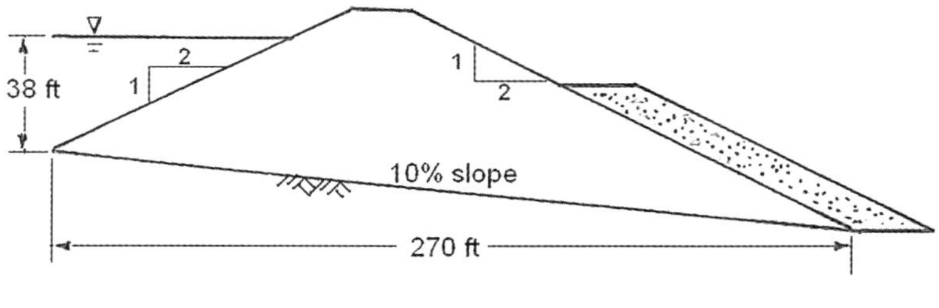

Fig. P4-5.

4.6 Same as Problem 4.5 except refer to Fig. P4-6.
[Answer: 7.87×10^{-5} cfs per ft]

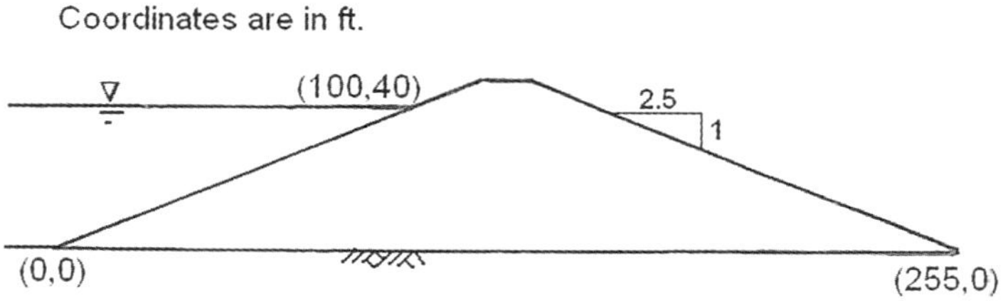

Fig. P4-6.

4.7 Fig. P4-7 shows an earth dam with a chimney drain. Locate the exit point of the phreatic surface by both Gilboy's and Casagrende's methods.
[Answer: 24.5 ft, 22.2 ft]

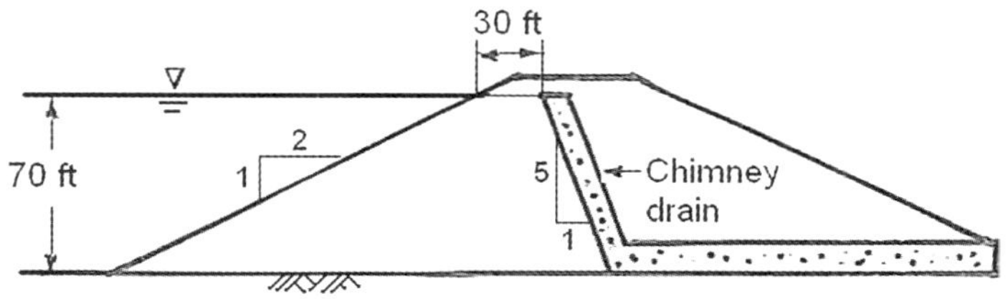

Fig. P4-7.

4.8 For the dam shown in Fig. P4-8, sketch the location of the phreatic surface.

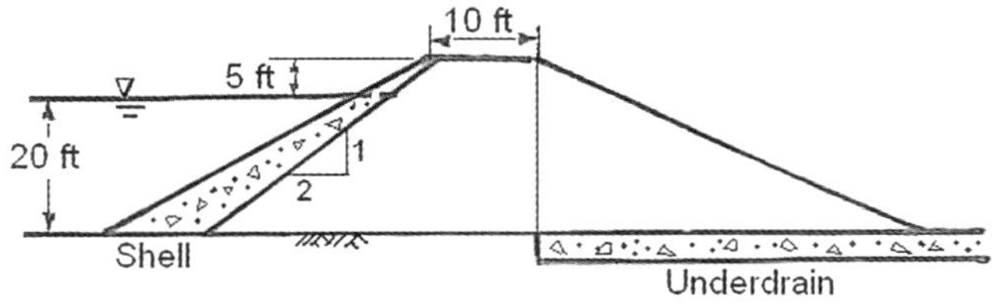

Fig. P4-8.

4.9 For the dam shown in Fig. P4-9, sketch the location of the phreatic surface.

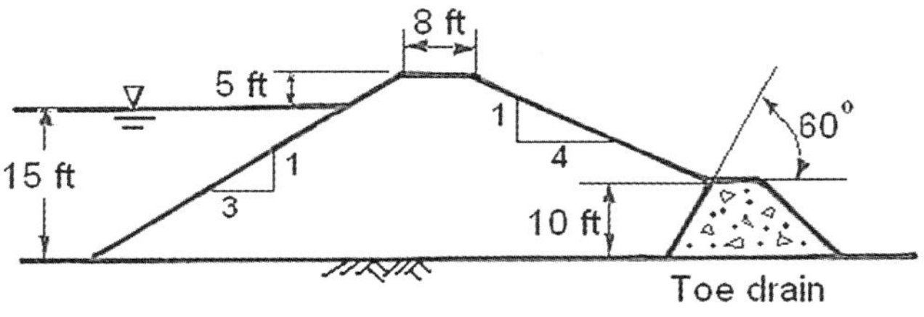

Fig. P4-9.

4.10 Fig. P4-10 shows a new earth dam on bedrock. The permeability of the dam is 5×10^{-5} cm/s and its effective porosity is 0.2. If the reservoir behind the dam is filled up instantaneously to a height of 35 ft, determine the location of the phreatic surface at the end of 120 days, as indicated by the distance x from the toe. If the face of the dam has a slope of 3:1, what should be the most conservative values of x?
[Answer: 133 ft, 168 ft from toe]

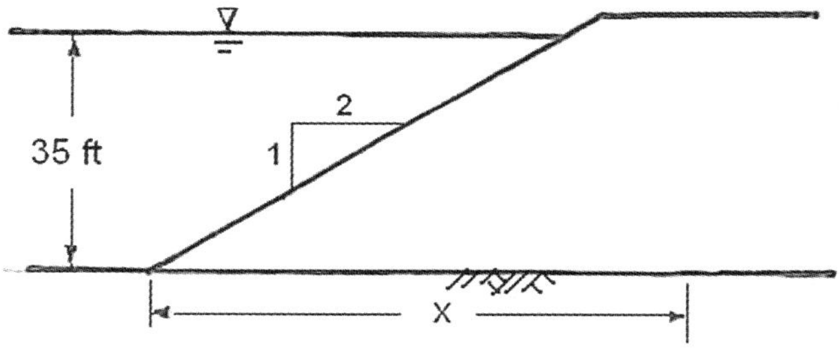

Fig. P4-10.

4.11 A coal waste embankment is placed on top of a slurry pond consisting of saturated fine refuse. The fine refuse is 30 ft thick with drainage layers on both the top and bottom, as shown in Fig. P4-11. The fine refuse has a coefficient of consolidation of 0.0035 cm^2/s and is divided into three layers for stability analysis. If the construction of the embankment takes one year to complete, or the average consolidation time is 183 days, estimate the pore pressure ratios at the end of construction for each layer. [Answer: 0.33, 0.66, 0.33]

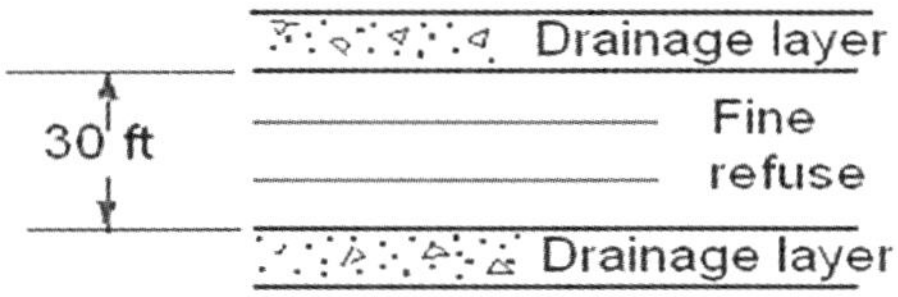

Fig. P4-11.

Chapter 5

Remedial Measures for Correcting Slides

Before planning and design of remedial measures, a field investigation must be undertaken. The scope of field investigation is described in this chapter. Also described are the various methods for correcting slides, including slope reduction or removal of weight, buttresses, surface drainage, subsurface drainage, retaining walls, anchor systems, bridging or tunneling, soil reinforcements, pile systems, vegetation and biotechnical stabilization, and hardening of soils. The selection of methods for stabilization also is discussed.

5.1 Field Investigation

The scope of a field investigation should include topography, geology, water, weather, and history of slope changes. If a slide has occurred, the shape of the sliding surface also should be determined. After the field investigation, some preliminary planning is needed before the start of the permanent remedial works.

5.1.1 Topography

The topography or geometry of the ground surface is an overt clue to past landslide activity and potential instability. More detail than that shown on existing topographic or project design maps usually is required for landslide studies.

Because the topography of a landslide is changing continually, the area must be mapped at different times, if possible, from several years before construction to several years after remedial measures are undertaken. Ultimately, the effectiveness of corrective measures is indicated by whether there is further change in the topography.

If a detailed survey of the area and the preparation of a contour map are not possible because of the lack of time, at least several cross sections must be surveyed from the accumulated masses at the toe of the slide to above the head scarp. The cross section must be long enough to cover part of the undisturbed area above and below the slide. The surface of the area should not be shown in a simplified form but with as many topographic features as possible, such as all marked edges, swells and depressions, scarps, cracks, and so on. The surveyed sections are supplemented by the logs of borings.

Aerial photos are most useful for the investigation of landslides, because they offer a perfect three-dimensional (3D) view of the area. From an aerial photo, one can determine precisely the boundaries of a landslide, because the slope surface below the scarp is irregularly undulated with ponded depressions. Also, the character of vegetation on the slope affected by the slide differs from that of the undisturbed adjacent slope. The amount of movement is determined easily from the offset of linear features, such as highways, railroads, and alleys, as soon as they continue to the undisturbed area.

5.1.2 Geology

Geologic structure is frequently a major factor in landslides. Although this topic includes major large-scale structural features, such as folds and faults, the minor structural details, including joints, small faults, and local shear zones, may be even more important. The landslide and the surrounding area should be mapped geologically in detail. On the map, the shape of the head scarp and the area of accumulation, outcrops of beds, offsets in strata, changes in joint orientation, dips, and strikes should be identified.

An important characteristic of the sliding slope is the shape of the cross section. If the slope was sculpted by erosion and covered with waste deposited by rainwash, the profile forms a gentle curve at its transition into the flood plain. Even a very ancient landslide is recognized from the curved, bulged shape of the toe.

5.1.3 Water

Water is a major factor in most slides. The plan for corrective measures requires a good knowledge of the hydrological conditions of the slide itself and of its surroundings. The first task is to determine the depth of the groundwater table and its fluctuation and to map all streams, springs, seeps, wet grounds, undrained depressions, aquiferous pressures, and permeable strata.

The changes of slope relief produced by sliding alter the drainage conditions of the surface water, as well as the regime of the groundwater. The seepage of groundwater has a significant effect on slope stability. Less pressure is built up when water is seeping out of the ground than when the exits for groundwater are blocked. For example, in one major slide area, landslide activity always was preceded by the stoppage of spring discharge near the toe; the cessation of movement was marked by an increase in spring discharge. Slip surfaces are generally impervious, retaining both surface water and groundwater. When slip surfaces approach the ground surface, new springs and wet grounds appear. In the boring logs, the depth and fluctuation of groundwater must be recorded. However, the pore pressure in clayey soils affected by sliding cannot be determined simply by observing the water table in a borehole, because by filling the borehole, the water loses the pressure in its vicinity. Therefore, the installation of piezometric instruments for pore pressure measurements is needed.

5.1.4 Weather

The climate of the area, including rainfall, temperature, evaporation, wind, snowfall, relative humidity, and barometric pressure, is the ultimate dynamic factor influencing most landslides. The effects of these factors seldom can be evaluated analytically, because the relations are too complex. Empirical correlations of one or more of these factors, particularly rainfall, snow, and melting temperatures, with episodes of movement or movement rates can point out their influences that must be controlled to minimize movements.

5.1.5 History of Slope Changes

Many clues can alert an investigator to past landslides and future risks. Some of these are hummocky ground, bulges, depressions, cracks, bowed and deformed trees, slumps, and changes in vegetation. Large features can be determined from large-scale maps and aerial photos; however, the evidence often is hidden by vegetation or is so small that it can only be determined by direct observation. Even then, only a person intimately familiar with the soils, geologic materials, and conditions in that particular area can recognize the potential hazards.

Among the most difficult kinds of slides to recognize and guard against are old slides that have been covered by glacial till or other, more recent sediments. Recent and active landslides can be recognized easily by their fresh appearance with steep and bared head scarps, open cracks, and strung tree roots. The state of tree growth is indicative of the age of the movements. Trees on unstable ground are tilted downslope but tend to return to a vertical position during the period of rest, so that the trunks become conspicuously bent. From the younger, vertically growing trunk segments, the date of the last sliding movement can be inferred.

5.1.6 Shapes of Failure Surfaces

As noted in Section 1.1, slides are divided into two types: translational and rotational. Translational slides are marked by lateral separation with very little vertical displacement, and by vertical cracks. Rotational slides are characterized by the rotation of the block or blocks of which they are composed, and by concave, rather than vertical, cracks. Fig. 5-1 is a schematic of these two types of slides, which took place during the 1964 Alaska earthquake (Hansen 1965).

In translational slides, the mass progresses down and out along a more or less planar or gently undulatory surface and has little of the rotational movement or backward tilting characteristics. The moving mass commonly slides out on the original ground surface.

The most common examples of rotational slides are slightly deformed slumps, which are slides along a surface of rupture that is concave upward. The exposed cracks are concentric in plan and concave toward the direction of movement. In many rotational slides, the underlying surface of rupture, together with the exposed scarps, is spoon-shaped. If the slide extends for a considerable distance along the slope perpendicular to the direction of movement, much of the rupture surface may approach the shape of a cylinder, the axis of which is parallel to the slope. In the head area, the movement may be almost wholly downward and have little apparent rotation. However, the top surface of each unit commonly tilts backward toward the slope, although some blocks may tilt forward. The classic purely rotational slide on a circular or cylindrical surface is relatively uncommon in natural slopes because of their internal inhomogeneities and discontinuities. Because rotational slides occur most frequently in fairly homogeneous materials, their incidence among constructed embankments and fills, and hence their interest to engineers, has been high relative to other types of failure.

The distinction between translational and rotational slides is useful in planning control measures. A translational slide may progress indefinitely if the surface on which it rests is inclined sufficiently, as long as the shear resistance along this surface remains lower than the more or less constant driving force. The movement of translational slides commonly is controlled structurally by surfaces of weakness, such as faults, joints, bedding planes, and variations in shear strength between layers of bedded deposits, or by the contact between firm bedrock and overlying detritus. The rotational slide, if the surface of rupture dips into the hill at the foot of the slide, tends to restore equilibrium in the unstable mass. The driving moment during movement decreases, and the slide may stop moving.

The location of the slip surface can be determined by an inclinometer. The inclinometer measures the change in inclination or tilt of a casing in a borehole, and thus allows the distribution of lateral movements to be determined as a function of depth below the ground surface and as a function of time. Inclinometers have undergone rapid development to improve reliability, provide accuracy, reduce weight and bulk of instruments, lessen data acquisition and reduction time, and improve versatility of operation under adverse conditions. Automatic data-recording devices, power cable reels, and other features are now available.

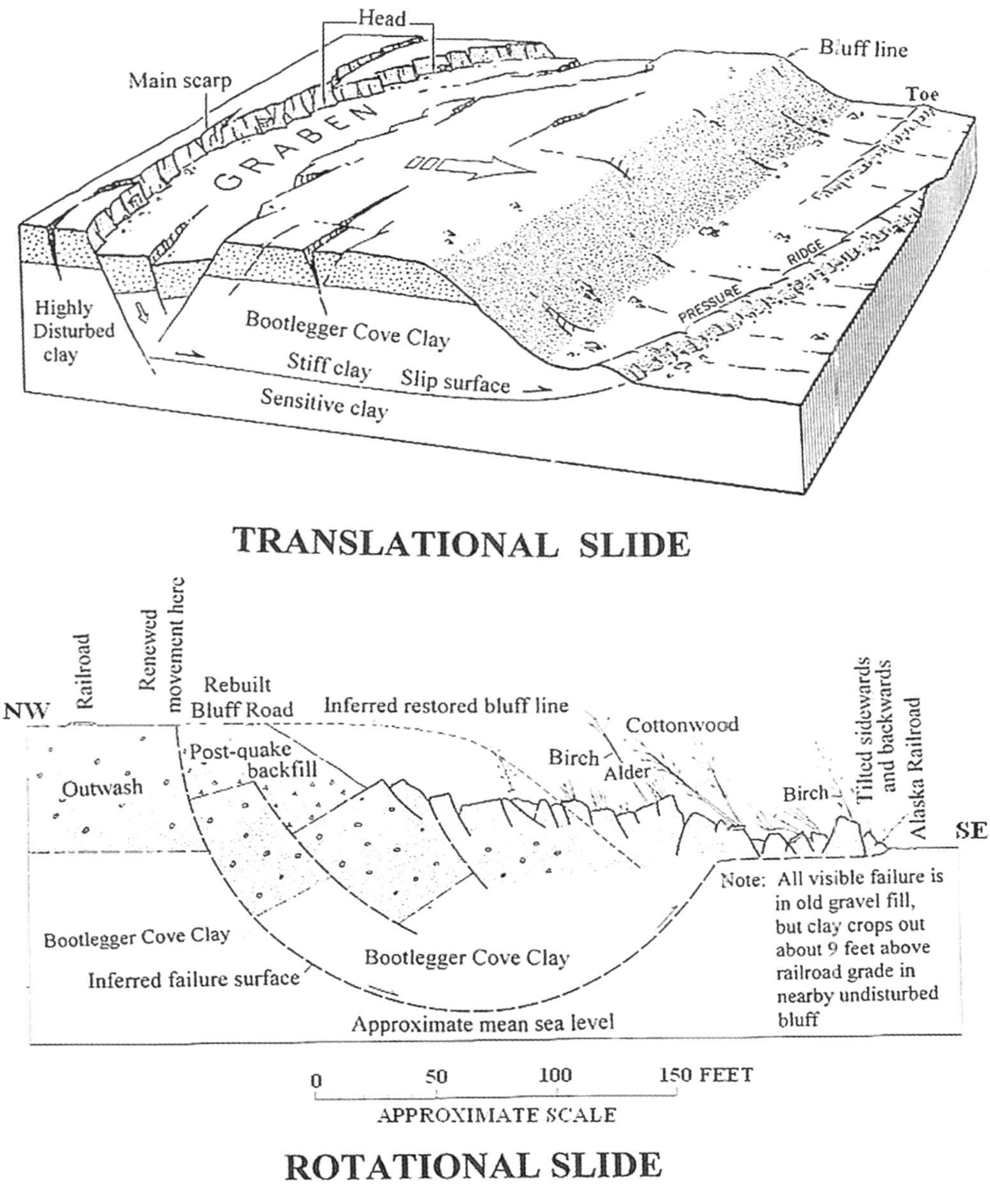

Fig. 5-1. Translational and rotational slides (Hansen 1965)

5.2 Preliminary Planning

When a slide takes place, it is necessary to determine the cause of the slide so that proper remedial measures can be taken to correct it. The processes involved in slides comprise a continuous series of events from cause to effect. Seldom, if ever, can a slide be attributed to a single definite cause. The detection of the causes may require continuous observations, and a final decision cannot be made within a short time. Because water is the major cause that may initiate a slide, or intensify a slide after it has occurred, the following initial remedial measures should be taken as soon as possible.

1. Capture and drain all the surface water that flows into the slide area;
2. Pump the groundwater out from all wells in the slide area and dewater all the undrained depressions; and
3. Fill and tamp all open cracks to prevent the infiltration of surface water.

Only after the completion of the initial measures should other permanent and more expensive measures based on a detailed investigation be undertaken.

Peck (1967) described the catastrophic slide in 1966 on the Baker River north of Seattle, Washington, which may be titled, "The Death of a Power Plant." He claimed that the state of the art at that time was unable to make reliable assessments of the stability of many, if not all, natural slopes under circumstances of practical importance. After the destruction of the power plant by the slide, he asked the following questions: "Was spending the time necessary to get information about subsurface conditions and movements a tactical error? Could the unfavorable developments have been prevented by providing extensive resloping, deep drainage, and other means rather than the use of the observational methods?" Although these questions cannot be answered satisfactorily, they clearly indicate the importance of prompt action in the correction of slides.

After the geometry of a slide, the location of the water table, and the soil parameters of various layers are determined, the factor of safety can be calculated. The factor of safety at the time of failure should be close to 1. If not, some of the parameters used in the analysis must be adjusted. If the soil is homogeneous and there is only one soil, the shear strength of the soil can be back-calculated, as described in Section 3.6. This shear strength then can be used for the redesign of the slope.

Based on the results of the investigation, a new slope is designed, and stability analysis can be used to determine the factor of safety. If a strong retaining structure is used, the stability or safety of the structure also should be separately considered by the principles of soil and structural mechanics.

5.3 Corrective Methods

Corrective methods can be used either to decrease the driving forces or increase the resisting forces. Because the factor of safety is a ratio between the resisting forces and the driving forces, a reduction of the driving forces or an increase of resisting forces will increase the factor of safety. Many of the practical examples presented here to illustrate the different corrective methods were described in a special report titled, "Landslides Analysis and Control" published by the

Transportation Research Board (Schuster and Krizek 1978). A much-expanded version of a special report titled, "Landslides: Investigation and Mitigation," which contains 25 chapters written by 30 authors covering a diversity of subjects, also was issued by the Transportation Research Board and can serve as an excellent reference for the study of landslides (Turner and Schuster 1996).

5.3.1 Slope Reduction or Removal of Weight

Fig. 5-2 shows three methods for slope reduction: direct reduction, flattening by cutting berms, and flattening without hauling material away. Although the third method is the most economical, care must be taken to ensure that the material to be placed on the toe is of good quality. If necessary, a drainage blanket should be placed to minimize the effect of water.

Fig. 5-3 shows the stabilization of the Cameo slide above a railroad in the Colorado River Valley by partial removal of weight (Peck and Ireland 1953). Stability analyses determined that the removal of volume B was more effective than the removal of volume A, as expected.

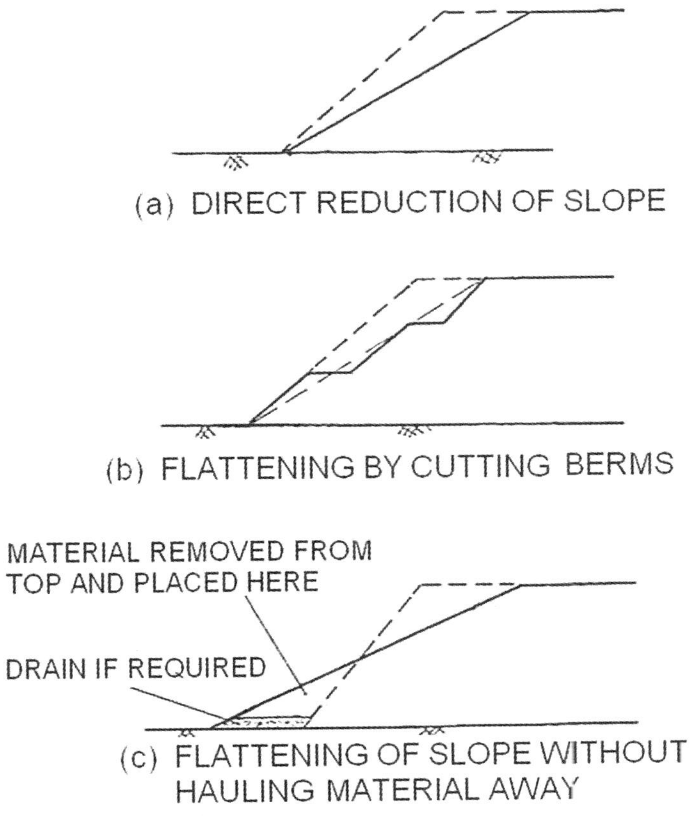

Fig. 5-2. Methods for slope reduction

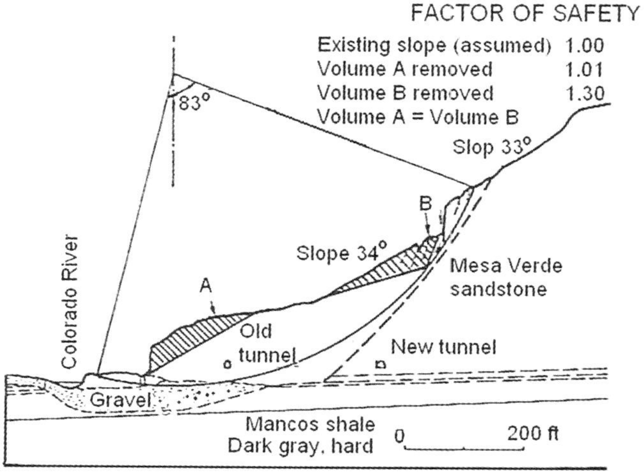

Note: $1 ft = 0.305 m$

Fig. 5-3. *Stabilization of the Cameo slide by partial removal of weight (Peck and Ireland 1953. Reproduced with permission from AREMA)*

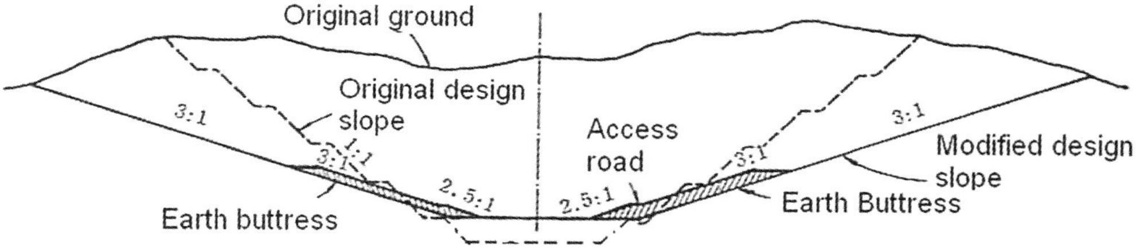

Note: $1 ft = 0.305 m$

Fig. 5-4. *Typical section of Mulholland cut showing original and modified design (Smith and Cedergren 1962, © ASTM International. Reproduced with permission)*

Fig. 5-4 shows a slope flattening, which was used effectively on a 320-ft (98-m) cut for a southern California freeway (Smith and Cedergren 1962). A failure took place during construction on a 1:1 benched cut slope composed predominately of sandstone and interbedded shale. After considerable study and analysis, the slope was modified to 3h:1v, and the final roadway grade was raised some 60 ft (18 m) above the original elevation. Moreover, to provide additional stability, earth buttresses were placed from roadway levels to a height of 70 ft (21 m).

5.3.2 Buttresses

There are two types of buttresses. The earth buttress, as shown in Fig. 5-4, is an additional earth berm placed near the toe of the slope to reduce the overturning

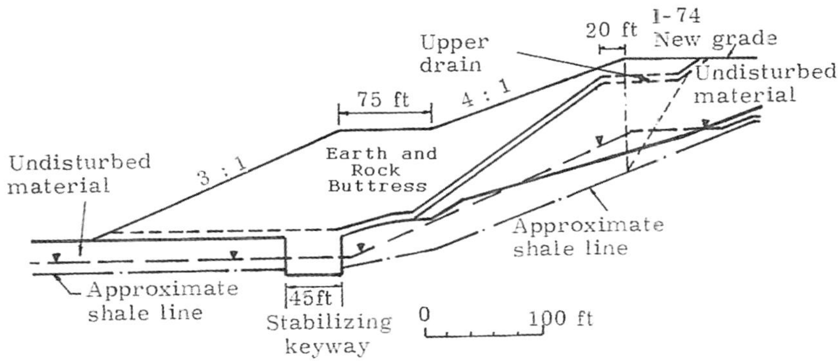

Note: 1 ft = 0.305 m

Fig. 5-5. *Stabilization berm to correct landslide in shale on I-74 in Indiana (Haugen and DiMillio 1974)*

moment, thus increasing the factor of safety. Because the purpose of an earth buttress is to serve as a counterweight to reduce the overturning moment, any earthy materials can be used.

Another buttress of high-strength, well-compacted material is shown in Fig. 5-5 for correcting a slide in a shale embankment on I-74 in southern Indiana (Haugen and DiMillio 1974). The borrow material used in the embankment was predominately local shale materials that were interbedded with limestone and sandstone. These shale materials, after being placed in the embankment, deteriorated with time and finally caused the embankment to fail. Careful studies of the in situ shear strength versus the original strength used in the design showed an approximate reduction of one-half in shear strength. After considering various alternatives, an earth and rock buttress design finally was selected. The use of a high-strength buttress will reduce not only the overturning moment and increase the resisting moment, thus enhancing the safety of the design, but also will protect the slope surface from erosion.

5.3.3 Surface Drainage

Of all possible methods for correcting slides, proper drainage of water is probably the most important. Good surface drainage is strongly recommended as part of the treatment for any slide. Every effort should be made to ensure that surface waters are carried away from a slope by interception trenches and diversion ditches. The surface of the area affected by sliding is generally uneven, hummocky, and transversed by unnoticed cracks and deep fissures. Therefore, reshaping the surface of a slide mass can be extremely beneficial in that cracks and fissures are sealed and water-collecting surface depressions are eliminated. This is particularly true for the cracks behind a scarp face where large volumes of water can seep into the failure zone and result in serious consequences.

Although surface drainage in itself is seldom sufficient for the stabilization of a slope in motion, it can contribute substantially to the drying of the material in the slope, thus controlling the slide.

5.3.4 Subsurface Drainage

Because groundwater is one of the major causes of slope instability, subsurface drainage is a very effective remedial measure. Methods frequently used are the installation of drainage layers, trench drains, horizontal drains, vertical drains, drainage tunnels or galleries, and wellpoints.

Drainage layers, sometimes also called underdrains, can be placed at the bottom of fills over the entire area to facilitate drainage. If the granular materials used for the drainage layer do not satisfy the filter criteria, a filter fabric or granular material that satisfies the filter criteria must be placed between the drainage layer and the adjoining soils to prevent clogging.

Trench drains are excavated trenches filled with granular materials that satisfy the filter criteria for the adjoining soils, or with rocks encased in filter fabric. An underdrain pipe may be required to facilitate drainage. Trenches usually are excavated at the steepest side slopes and should extend below the water-bearing layer. For large areas, an extensive system of trenches in the form of finger drains or a herringbone pattern may be needed.

A horizontal drain is a small-diameter well drilled into a slope on an approximately 5 to 10% grade and fitted with a perforated pipe. Pipes should be provided to carry the collected water to a safe point of disposal to prevent surface erosion. Fig. 5-6 shows the use of both surface and subsurface drainage for

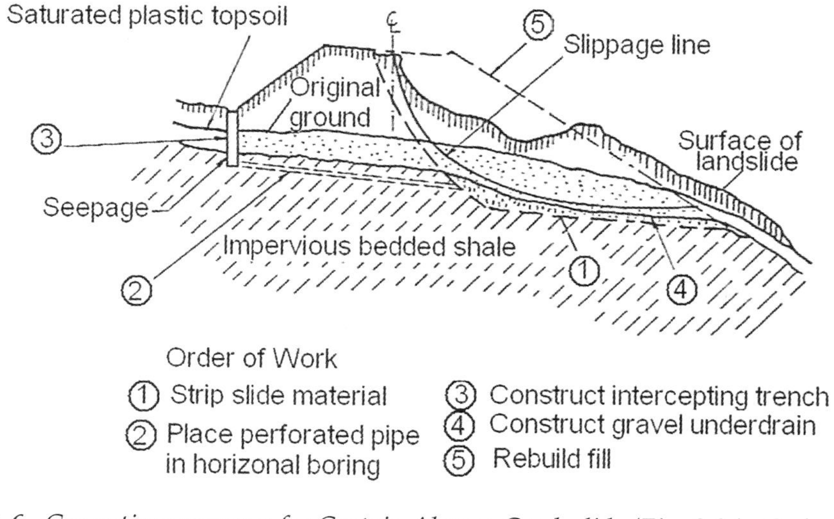

Fig. 5-6. Corrective measures for Castaic-Alamos Creek slide (Fig. 8.8 in Gedney and Weber 1978, © National Academy of Sciences, Washington, DC. Reproduced with permission of the Transportation Research Board, Washington, DC)

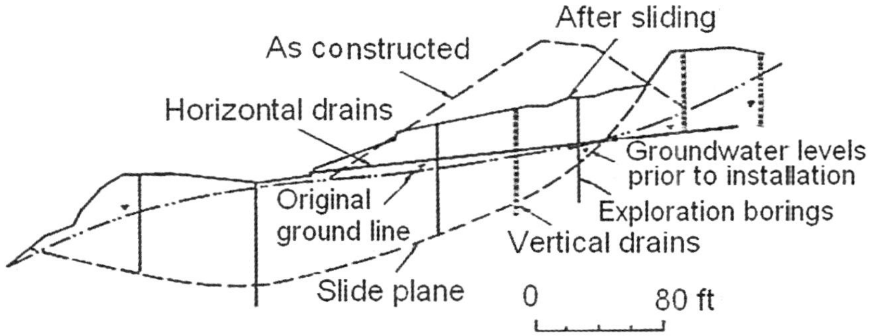

Fig. 5-7. Slide treatment consisting of horizontal and vertical drains (Root 1958, © National Academy of Sciences, Washington, DC. Reproduced with permission of the Transportation Research Board, Washington, DC)

correcting the slide on the Castaic-Alamos Creek in California (Dennis and Allan 1941). The surface water is collected by the intercepting trench connected to the perforated pipe and gravel underdrain, which are used also for subdrainage.

A vertical drainage well can be either a gravity drain or a pumped well, depending on whether there is an outlet for the water to drain by gravity. In many cases, a horizontal drain can be drilled to intercept the vertical drain at the bottom. Fig. 5-7 shows the use of both vertical and horizontal drains for correcting an active landslide that occurred at San Marcos Pass near Santa Barbara, California (Root 1958). The vertical wells were about 3 ft (1 m) in diameter, 40 ft (12 m) long, and belled at the bottom so that they interconnected to form a somewhat continuous curtain. The drains had 8-in. (20-cm) perforated pipes in the center for the full depth of the vertical drains and were backfilled with pervious material. The horizontal drains then were drilled to intersect the vertical drains at the belled portion of the vertical wells.

A drainage tunnel or gallery is a deep and large structure, usually about 3 ft (1 m) wide by 6 ft (2 m) high in cross section, constructed for the purpose of discharging a large amount of water. The effectiveness of a drainage tunnel may be increased by drilling short or long drainage borings in the wall, floor, or roof of the tunnel to collect water from various locations.

Wellpoints are small-diameter wells that are driven or jetted into place. Vacuum is applied to the top of the wellpoints through a header, or a horizontal pipe that applies vacuum to suck water up the wellpoints. They work very well in clean sands but not as well in fine-grained soils. Because the water is drawn up the riser pipe by vacuum, their maximum effective lift is limited to 20 to 25 ft (6.1–7.6 m). For greater lifts, pumped wells, each with its own pump to push water to the top of the well, can be used.

5.3.5 Retaining Walls

Fig. 5-8 shows the use of a retaining wall to correct a cut slope failure on I-94 in Minneapolis–St. Paul, Minnesota (Shannon and Wilson, Inc. 1968). The use of a

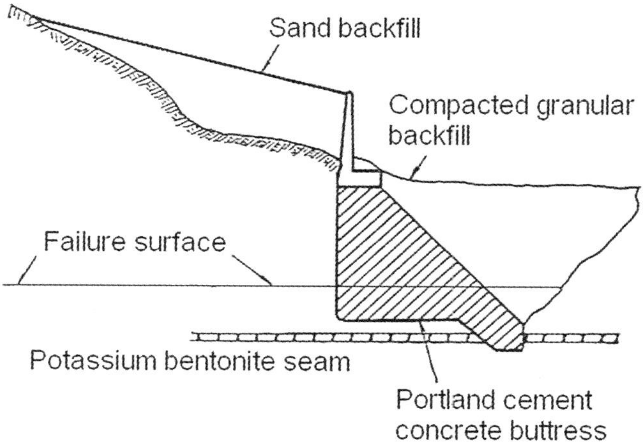

Fig. 5-8. Retaining wall to correct slide on I-94 in Minnesota (Adapted from Shannon and Wilson 1968. Reproduced with permission from Shannon and Wilson, Inc.)

retaining wall is often occasioned by the lack of space necessary for the development of the slope to a full length. Because retaining walls are subject to an unfavorable system of loading, a large wall width is necessary to increase stability. Although the methods of stability analysis can be applied to determine the factor of safety of a slope with failure surfaces below the wall, the design of the wall will require special considerations to ensure that the wall itself is stable against sliding, overturning, and bearing capacity failure.

5.3.6 Anchor Systems

One type of anchor system is the tieback wall, which carries the backfill forces on the wall by a "tie" system to transfer the imposed load to an area behind the slide mass where satisfactory resistance can be established. The ties may consist of pre- or post-tensioned cables, rods, or wires, and some form of deadmen or other methods to develop adequate passive earth pressure. Fig. 5-9 shows a section of tieback wall to correct the slide condition on New York Avenue in Washington, DC (O'Colman and Trigo 1970).

5.3.7 Bridging or Tunneling

In particularly serious cases, the removal of steep, long, and narrow unstable slopes may be too costly or dangerous, so a bridge can be constructed to span the unstable area. The bridge must be supported by driven piles or drilled shafts placed well below the foundation materials. Bridges are commonly applicable only to small landslides or unstable areas. For large slides with bridges more

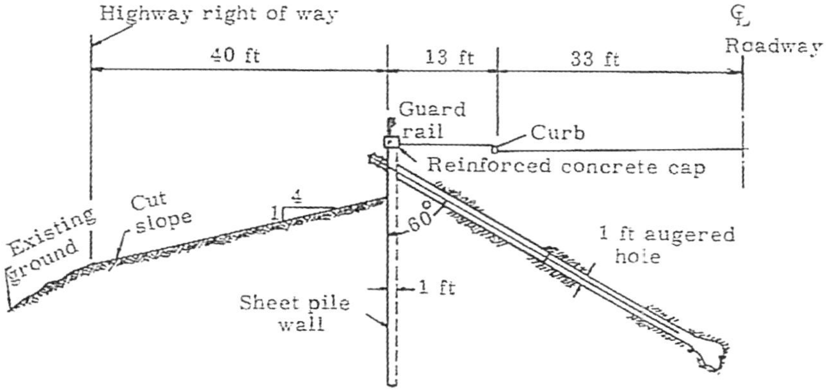

Note: *1 ft = 0.305 m*

Fig. 5-9. *Tieback wall to correct slide on New York Avenue in Washington, DC (O'Colman and Trigo 1970)*

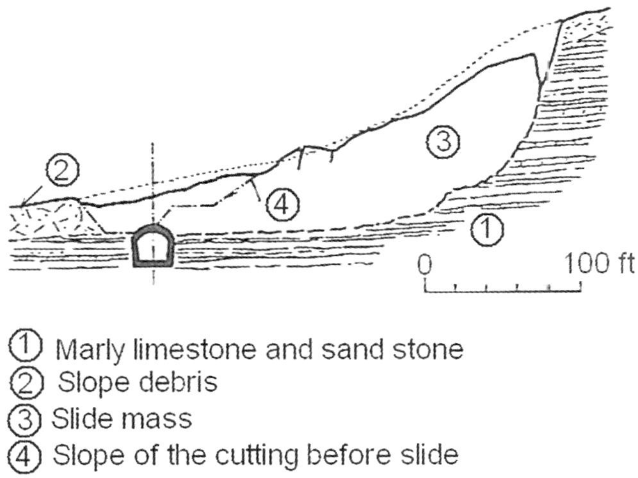

Note: *1 ft = 0.305 m*

Fig. 5-10. *Tunnel to correct slide on Spaichingen-Nusplingen Railway line in Germany (Zaruba and Mencl 1969, © Academia Publishing of the Czech Academy of Sciences)*

than 100 to 300 ft (30 to 90 m) long, this method usually is not economically justified when compared with other methods.

In mountainous areas with large slides, it may be necessary to construct a tunnel, as shown in Fig. 5-10 for the slide on the Spaichingen-Nusplingen Railway line in Germany (Zaruba and Mencl 1969).

5.3.8 Soil Reinforcements

Soil reinforcements include mechanically stabilized earth (MSE) walls, reinforced soil slopes (RSS), and soil nailing walls. One of the greatest advantages of using soil reinforcements is their flexibility and capacity to absorb deformation due to poor subsoil conditions in the foundation. Also, based on observations in seismically active zones, these structures have demonstrated a higher resistance to seismic loading than do rigid concrete structures. In addition to global stability to check the factors of safety over the entire slope and the adequacy of the reinforcements provided, these reinforced soil structures also should be treated as retaining walls by considering the external stability of the wall, including sliding, overturning, bearing capacity, and the position of the resultant within the middle third of the base, as is usually done for gravity retaining walls.

The majority of the MSE walls for permanent applications use a segmental precast concrete facing and galvanized steel reinforcement, whereas geotextile-faced MSE walls are used more frequently for temporary construction. Recently, modular block facings with geosynthetic reinforcements, principally geogrids, have gained acceptance because of their lower cost and nationwide availability. Fig. 5-11 shows the use of a MSE wall to correct a large landslide on a section of I-40 near Redwood, Tennessee (Royster 1974). The slope-forming materials were essentially a thick surface deposit of colluvium underlain by residual clays and clay shales. The groundwater table was seasonally variable but generally was found to be above the colluvium and residuum interface. This particular slide occurred within an embankment placed as a sidehill fill directly on a colluvium-filled drainage ravine. Because of blocked subsurface drainage and weakened foundation soils, the fill failed some four years after construction. Final design

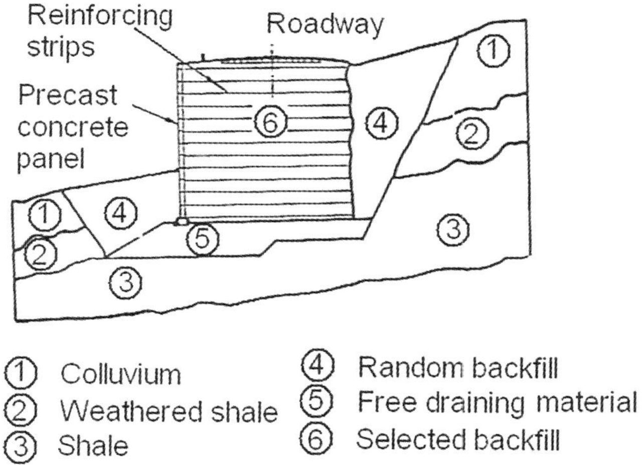

Fig. 5-11. Reinforced earth wall to correct slide on I-40 in Tennessee (Royster 1974. Reproduced with permission from the Highway Geology Symposium)

plans called for careful excavation of the failed portion of the fill to a firm, unweathered shale base, installation of a highly permeable drainage course below the wall area, placement of a MSE wall, and final backfill operations.

A major difference between a reinforced soil slope and a MSE wall is that the former has a surface slope less than 70° and does not require a precast or block facing, whereas the latter has a steep surface and requires such a facing to confine the selected backfill and facilitate the compaction. Fig. 5-12 shows the widening of a major highway to accommodate an additional 10-ft- (3-m-)wide paved shoulder. A 45° cut was made into the existing slope, which allowed for the construction of the reinforced soil slope. Primary geogrids usually are interspersed with secondary geogrids, which aid in compaction at the face of the slope and also tend to reduce surface erosion.

Soil nailing is an in situ soil reinforcement technique that has been used during the past four decades. The main components of a soil nailed retaining system are the in situ ground, the tension-resistant nails, and the facing element. The nails are usually corrosion-resistant steel bars or other metallic elements that can resist tensile stresses, shear stresses, and bending moments. They generally are placed in drilled boreholes and grouted along their total length or driven into the ground. The facing is not a major structural load-carrying element, but, rather, ensures local stability of the soil between the nails and protects the ground from surface erosion and weathering effects. It generally consists of a thin layer of shotcrete about 4 to 6 in. (100–150 mm) thick with wire or steel mesh between the nails. Prefabricated or cast-in-place concrete panels have increasingly been used in the construction of permanent structures to satisfy

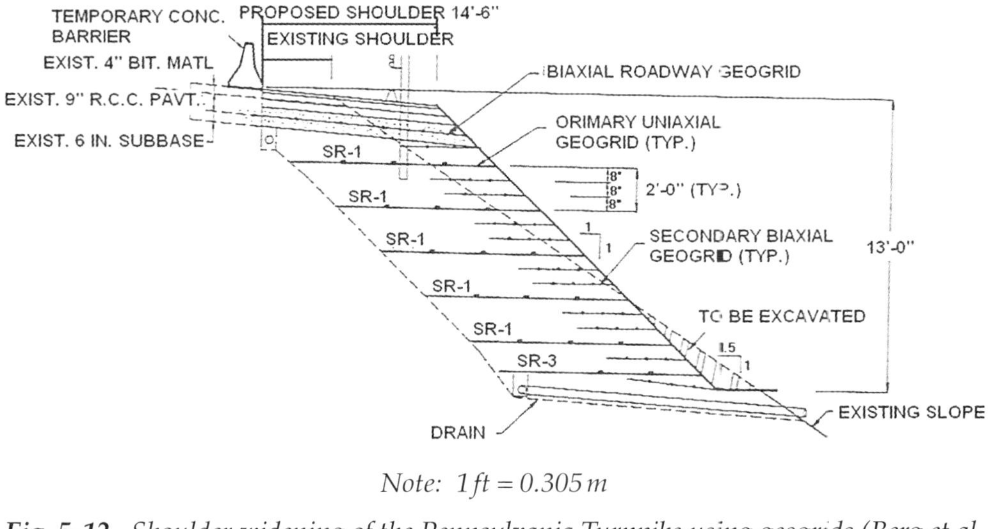

Note: 1 ft = 0.305 m

***Fig. 5-12.** Shoulder widening of the Pennsylvania Turnpike using geogrids (Berg et al. 1990, © National Academy of Sciences, Washington, DC. Reproduced with permission of the Transportation Research Board, Washington, DC)*

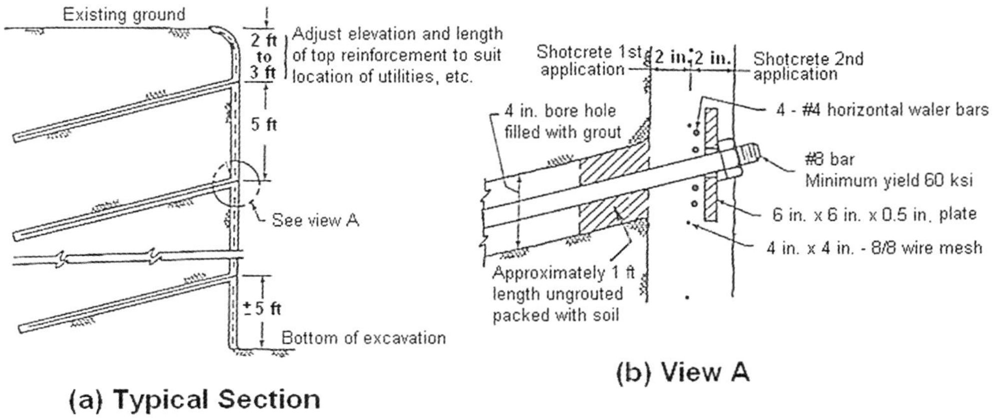

Note: 1 ft = 0.305 m; 1 in. = 25.4 mm; 1 ksi = 6.9 MN/m^2

Fig. 5-13. *Construction details of a lateral earth support system (Shen et al. 1978. Reproduced with permission)*

specific needs and accommodate adequate drainage. Successive incremental excavations with a height of 3 to 6 ft (0.9–1.8 m) are first made on the ground, the nails are then installed, and the shotcrete is applied. The next excavation is made and the process repeated until completion.

Soil nailing has most frequently been used in Europe for construction of temporary retaining structures in excavation. In North America, the behavior of a lateral support system similar to the European system was investigated by Shen et al. (1978, 1981a, 1981b). Fig. 5-13 shows a typical section of this soil nailed retaining system.

5.3.9 Pile Systems

Recorded attempts to use driven steel or wooden piles of nominal diameter to retard or prevent landslides seldom have been successful. Unless the slide is shallow, such piles are incapable of providing adequate shear resistance. Shallow slides can be controlled by piling, because the piles can be driven to an adequate depth. Otherwise, they may tilt from the vertical position, thus disturbing the adjacent material and the material underneath the piles and causing the development of a slip surface below the pile tip.

Fig. 5-14 shows the use of piles to correct a shallow slide in a railway cutting at East Slovakie, Czechoslovakia (Zaruba and Mencl 1969). The cutting has a slope of 4h:1v and was made in a fissured marly clay subjected to slaking. During the rainy spring of 1965, a small sheet slide developed at the toe of the slope, which extended to a length of 165 ft (50 m) and reached up to the top of the slope. Because the site was not accessible and the removal of a large volume of soil was difficult, piles were employed to prevent the further spreading of the slide.

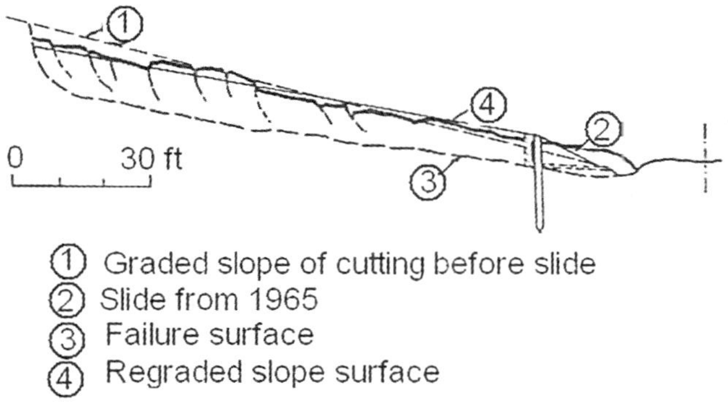

Note: 1 ft = 0.305 m

Fig. 5-14. *Stabilization of a slide by piles in Czechoslovakia (Zaruba and Mencl 1969, © Academia Publishing of the Czech Academy of Sciences)*

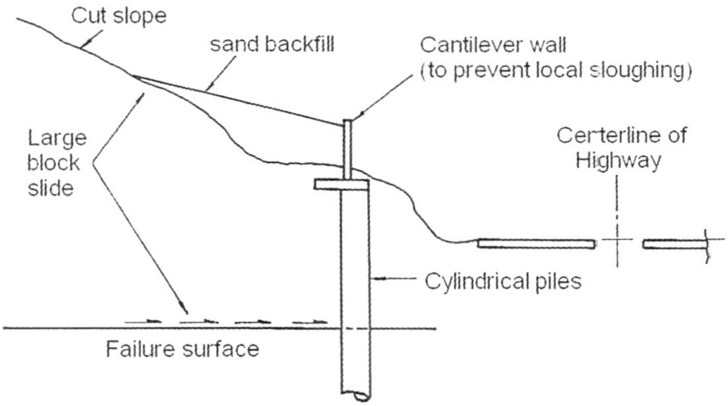

Fig. 5-15. *Cylinder pile to stabilize deep-seated slide on I-94 in Minnesota (Adapted from Shannon and Wilson Inc. 1968. Reproduced with permission from Shannon and Wilson, Inc.)*

Forty-two piles, 20 ft (6 m) long, were driven into prepared bore holes to a depth of 13 ft (4 m). Reinforced concrete slabs were supported against the piles to prevent movement of the soil between and around the piles. The pile spacing was 3 to 5 ft (1–1.5 m). A sand drain was constructed along the slab, discharging water to a ditch. After the treatment, the slope was flattened to 5h:1v.

Fig. 5-15 shows a cylinder pile wall system for stabilizing a deep-seated slope failure in I-94 in Minneapolis-St. Paul, Minnesota (Shannon and Wilson, Inc. 1968). The pile wall was placed as a restraining system, in which the forces tending to cause movement were predicted carefully. The cast-in-place piles were designed as cantilevers to resist the full earth thrust imposed by the soil.

5.3.10 Vegetation and Biotechnical Stabilization

Slope movements generally disturb the vegetative cover, including both trees and grass. The reforestation of the slope is an important task in corrective treatment. This is done during the final stage, invariably after at least partial stabilization of the slide. Forestation is most beneficial for shallow slides. Slides with deep-lying failure surfaces cannot be detained by vegetation, although in this case, too, vegetation can lessen the infiltration of surface water into the slope and thus contribute indirectly to the stabilization of the slide.

It generally is accepted that forest growth has two functions: drying out of the surface layers and consolidating them by a network of roots. Trees draw the water necessary for their growth from the slope surface, so the most suitable species will be those that have the largest consumption of water and the highest evapotranspiration rate. Therefore, it is more advantageous to plant deciduous trees than conifers.

Biotechnical stabilization combines the use of vegetation and other mechanical reinforcement, such as live-cut stems and branches, to prevent surfacial erosion and arrest shallow mass movement (Gray and Leiser 1982). In addition to the immediate effect on stabilization, secondary stabilization will occur as the result of rooting along the length of the buried stems. Gray and Leiser (1992, 1995) also found that alternate layers of earth and live brush can intercept water and divert it to the slope surface, thus reducing pore pressures in the process.

5.3.11 Hardening of Soils

If the water in the slope cannot be drained by subsurface drainage methods, foundation engineers may consider several methods of hardening of soils. These methods can be divided into chemical treatments, cement grouting, electro-osmosis, and thermal treatments.

Chemical treatments, which have been used with varying degrees of success, include lime or lime-soil mixtures, and ion exchange. One successful treatment, in which quicklime was placed in predrilled 0.5-ft- (0.2-m-)diameter holes on 5-ft (1.5-m) centers throughout an extensive slide area, was reported by Handy and Williams (1967). The ion exchange technique, which consists of treating the clay minerals along the plane of potential movement with a concentrated chemical, was reported by Smith and Forsyth (1971).

Cement grouting has been used in England for the stabilization of embankments and cuttings (Zaruba and Mencl 1969). Experience shows that this method yields fine results with rather shallow landslides in stiff materials such as clay-shales, claystones, and stiff clays, which break into blocks separated by distinct fissures. The method is actually a mechanical stabilization of the slope by filling the fissures with cement grout rather than changing the consistency of soil mass, because the cement mortar cannot enter into the soil mass. Cement grout was also used for a 300-ft (90-m) benched cut slope on I-40 along the Pigeon River in North Carolina (Schuster and Krizek 1978). Large volumes of cement grout were injected into the voids of broken rubble and talus debris to stabilize the slope.

The electro-osmosis technique has the same final effect as subsurface drainage but differs in that water is drained by an electric field rather than by gravity. The loss of pore water causes consolidation of the soil and a subsequent increase in shear strength. Casagrande et al. (1961) described the use of this method to stabilize a cut slope during the construction of a bridge foundation on the Trans-Canada Highway in Ontario. Casagrande et al. (1981) also reported on the use of electro-osmosis to stabilize an excavation in British Columbia.

The use of thermal treatments for preventing slides was first reported by Hill (1934). Since 1955, the Russians have experimented with and reported on the success of thermal treatment on plastic loess soils. High temperatures cause a permanent drying of the embankments and cut slopes. Beles and Stanculescu (1958) described the use of thermal methods to reduce the in situ water contents of heavy clay soils in Romania. Applications to highway landslides and unstable railroad fills also were cited.

5.4 Selection of Methods for Stabilization

Not all stabilization methods as described are appropriate for every type of slope failure. For instance, slope flattening and berms, in conjunction with surface drainage, are often the first methods to be considered. However, if there is no space to allow the use of a flatter slope, tieback walls or slope reinforcements by soil nails or geosynthetics are often the solution to the stability problem. Although retaining walls also can be used for small slides, they are not effective for large slides. If the water table is high, subsurface drainage is a very effective and rapid method for mitigating slides. There are no hard and fast rules for prescribing the treatment of slides. Many slides result from the combination of several causes, so the most economical and effective means of treating slopes consist of a combination of several methods. In choosing the methods that are technically possible, the following factors need to be considered.

1. When a slide occurs, it is necessary to determine the purpose of stabilizing the slope: whether to prevent further movements or to restore the load-carrying capacity. For example, an embankment may fail because of poor compaction and the infiltration of surface water. If diversion ditches are constructed to intercept surface water, the slope surface is graded and compacted to facilitate drainage, and all the cracks and fissures are sealed properly, the movements and the slope may become stable. Conversely, the restoration of an embankment to support a pavement or structure will be much more difficult, especially when the ground has been disrupted by large movements already.

2. The time required to complete the repair work may dictate the stabilizing methods to be used. If the work must be completed within a short time, an expeditious method that can be undertaken without delay may be considered as the most appropriate. If the repair work is not urgent and

can be delayed, more studies should be made so that a less expensive method may be devised. However, these long and laborious studies should not delay the requisite measures that obviously must be taken, as the treatment of active landslides is always a contest with time. In scheduling stability work, weather conditions always should be taken into account. In the northern part of the United States, extensive operations are very difficult, or even impossible, in winter when the slide areas may be frozen or covered by snow. It may be better to postpone some operations until spring or more favorable weather conditions.

3. Site accessibility and conditions may limit the methods of stabilization. In mountainous areas with only small roads, methods involving the use of heavy equipment might not be employed. Limitation of the right of way and adjacent facilities may require the use of retaining structures, soil reinforcements, or tieback walls.

4. The total cost of stabilization must be reasonable and within the limit of financial resources. The total cost should include the stabilizing system itself, right of way, temporary and permanent easements, disposal of unsuitable materials, and drainage. If the total cost exceeds the benefits, less expensive methods should be used.

5. Other factors to be considered include safety, availability of materials and equipment, aesthetics, environmental impact, and political issues.

Summary

1. Before taking any remedial measures, a field investigation of the slide area must be made. The scope of field investigation should include topography, geology, water, weather, history of slope changes, and shape of the sliding surface.

2. The distinction between translational and rotational slides is useful in planning control measures. A translational slide may progress indefinitely if the surface on which it rests is sufficiently inclined, as long as the shear resistance along this surface remains lower than the more or less constant driving force. A rotational slide, if the surface of rupture dips into the hill at the foot of the slide, tends to restore equilibrium in the unstable mass, so the driving moment during movement may decrease, and the slide may stop moving.

3. When a slide takes place, it is necessary to determine the cause of the slide so that proper remedial measures can be taken to correct it. The detection of the cause may require continuous observations, and a final decision cannot be made within a short time. Because water is the major

cause that may initiate a slide or intensify a slide after it has occurred, some initial remedial measures should be taken as soon as possible, such as capturing and draining all the surface water that flows into the slide area, pumping the groundwater out from all wells in the slide area, dewatering all the undrained depressions, and filling and tamping all open cracks to prevent the infiltration of surface water.

4. One of the simplest ways to stabilize a slope is to flatten it by removing the weight from the top and placing it at the bottom. The placement of a buttress as a counterweight near the toe will reduce the overturning moment and increase the factor of safety. The use of a high-strength buttress will not only reduce the overturning moment and increase the resisting moment but also protect the slope surface from erosion.

5. Of all the possible methods for correcting slides, proper drainage of water is probably the most important. Good surface drainage is recommended strongly as part of treatment for any slide. Every effort should be made to ensure that surface waters are carried away from a slope by intercepting trenches or division ditches. Because groundwater is one of the major causes of slope instability, subsurface drainage is a very effective remedial measure. Methods frequently used are the installation of drainage layers, trench drains, horizontal drains, vertical drains, drainage tunnels or galleries, and wellpoints.

6. The use of retaining walls to correct slides often is occasioned by the lack of space necessary for the development of the slope to a full length. Although the methods of stability analysis can be applied to determine the global factor of safety for a slope with failure surfaces below and behind a wall, the design of a wall requires special considerations to ensure that the wall itself is stable against sliding, overturning, and bearing capacity failures. A special type of retaining wall is the tieback wall, which carries the backfill forces on the wall by a "tie" system to transfer the imposed load to an area behind the slide mass where satisfactory resistance can be established. In particularly serious cases, a retaining wall might not be sufficient, making it necessary to construct a bridge or a tunnel to avoid slides.

7. Soil reinforcements, which include mechanically stabilized earth (MSE) walls, reinforced soil slope (RSS), and soil nailing walls, also can be used as remedial measures. One of the greatest advantages of using soil reinforcements is their flexibility and capability to absorb deformation due to poor subsoil condition in the foundation. Also, based on observations in seismically active zones, these structures have demonstrated a higher resistance to seismic loading than do rigid concrete structures. In addition to global stability to check the factors of safety over the entire slope and the adequacy of the reinforcements provided, these reinforced soil structures also should be treated as gravity retaining walls and designed by considering the external stability of the wall, including

sliding, overturning, bearing capacity, and position of the resultant within the middle third of the base.

8. Recorded attempts to use driven steel or wooden piles of nominal diameter to retard or prevent landslides seldom have been successful. Unless the slide is shallow, such piles are incapable of providing adequate shear resistance. For large and deep-seated slides, a pile wall system consisting of a retaining wall supported by large-cylinder piles can be used. The cast-in-place cylinder piles are designed as cantilevers to resist the full earth thrust imposed by the soil.

9. The reforestation of a slope is an important task in corrective treatment. Forestation is most beneficial for shallow slides. Slides with deep-lying failure surfaces cannot be detained by vegetation, although vegetation can lessen the infiltration of surface water into the slope and thus contribute indirectly to the stabilization of the slide. Biotechnical stabilization combines the use of vegetation and other mechanical reinforcements, such as live-cut stems and branches, to prevent surficial erosion and arrest shallow mass movement. In addition to the immediate effect on stabilization, secondary stabilization will occur as the result of rooting along the length of the buried stems. It also was found that alternate layers of earth and live brush could intercept water and divert it to the slope surface, thus reducing pore pressures in the process.

10. If the water in the slope cannot be drained by subsurface drainage, methods used for hardening of soils may be considered. These methods include chemical treatments such as placing quicklime in predrilled holes, using the ion exchange technique to treat the clay minerals along the plane of potential movement with a concentrated chemical, cement grouting by filling the fissures of stiff materials with cement grout, electro-osmosis by applying an electric field to drain out the subsurface water, and thermal treatment by heating in situ soils to reduce water content.

11. Not all stabilization methods are appropriate for every type of slope failure. There are no hard and fast rules for prescribing the treatment of slides. Many slides result from a combination of several causes, so the most economical and effective means of treating slopes consists of a combination of several methods. In choosing the methods that are technically possible, the major factors to be considered include the purpose of stabilization, the time required to complete the work, site accessibility and conditions, and total cost. Other factors to be considered include safety, availability of materials and equipment, aesthetics, environmental impact, and political issues.

Part 2

Methods of Stability Analysis

Chapter 6

Simplified Methods for Plane Failure Surfaces

This chapter presents several simplified methods for the stability analysis of plane failure surfaces. These methods are called simplified because of the following limitations: (1) only force equilibrium is satisfied and the requirement for moment equilibrium is ignored; (2) the effect of seepage can be considered only approximately by using a pore pressure ratio rather than the more precise phreatic surface; and (3) the number of failure planes is limited to three so that solutions can be obtained without the service of a computer. To remove these stated limitations, the method of slices described in Chapter 8 should be used instead.

6.1 Infinite Slopes

Fig. 6-1 shows an infinite slope underlain by a rock surface with an angle of inclination, β. The slope is considered as infinite, because it has a length much greater than the depth, d. If a free-body of width b is taken, the forces on the two vertical sides are the same, because every plane, which has the same infinite distances to both top and bottom, should be considered the same with the same side force. Because the side forces neutralize each other, the only forces to be considered are the weight, W, the seismic force, C_sW (C_s is the seismic coefficient),

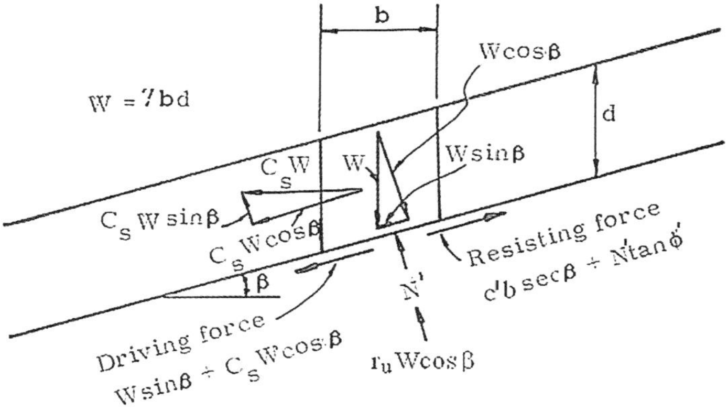

Fig. 6-1. Analysis of infinite slope

the effective normal force, N', and the neutral force, $r_u W\cos\beta$, where r_u is the pore pressure ratio. Note that the use of $\cos\beta$ to compute the neutral force, instead of $\sec\beta$, is based on the normal method explained in Section 2.4.1. For an infinite slope with both phreatic and failure surfaces inclined at the same angle β, the use of $r_u W\cos\beta$ as the neutral force theoretically is correct, as explained in Section 4.1.3.

In all the derivations that follow, only the effective stress analysis will be presented. The equations also can be applied to a total stress analysis by simply replacing the effective strength parameters by the total strength parameters. It will be shown that for a soil with a cohesion and a friction angle, the factor of safety decreases with the increase in d, so the most critical failure surface is parallel to the slope along the bottom of the soil overburden. The factor of safety is defined as a ratio of the resistant force due to the shear strength of soil along the failure surface to the driving force due to the weight of the sliding mass. The resisting force is composed of two parts: one due to cohesion and equal to $c'b\sec\beta$, and the other due to friction and equal to $N'\tan\phi'$, where N' is the effective force normal to the failure plane. Consider force equilibrium normal to the failure surface,

$$N' = W[(1-r_u)\cos\beta - C_s\sin\beta]\tan\phi' \tag{6-1}$$

The driving force is always equal to the components of weight and seismic force parallel to the failure surface, or $W\sin\beta + C_s W\cos\beta$. Therefore, the factor of safety, F, can be written as

$$F = \frac{c'b\sec\beta + W[(1-r_u)\cos\beta - C_s\sin\beta]\tan\phi'}{W(\sin\beta + C_s\cos\beta)} \tag{6-2}$$

Replacing W by γbd,

$$F = \frac{\dfrac{c'}{\gamma d}\sec\beta + [(1-r_u)\cos\beta - C_s \sin\beta]\tan\phi'}{(\sin\beta + C_s \cos\beta)} \tag{6-3}$$

in which γ = total unit weight of the soil and d = depth of the failure surface below the slope surface.

Eq. (6-3) is applicable to an infinite slope possessing both a cohesion and an angle of internal friction. Note that when there is no seepage, $r_u = 0$, and when water flows along the surface of the slope, $r_u = \gamma_w/\gamma \cong 0.5$. For the most general case where water flows parallel to and at a depth of d_w below the slope surface, from Eq. (4-17),

$$r_u = \frac{\gamma_w(d-d_w)}{\gamma d} = \frac{\gamma_w}{\gamma}\left(1 - \frac{d_w}{d}\right) \tag{6-4}$$

It can be seen from Eqs. (6-3) and (6-4) that F decreases with the increase in d. For a cohesionless material not subjected to earthquake, or $c' = 0$ and $C_s = 0$, Eq. (6-3) can be simplified to

$$F = (1-r_u)\frac{\tan\phi'}{\tan\beta} \tag{6-5}$$

Eq. (6-5) shows that the factor of safety for a cohesionless material is independent of d, so every plane parallel to the slope is a critical plane and has the same factor of safety. However, this statement is true only when $r_u = 0$ or 0.5, or when a pore pressure ratio is assumed a priori, regardless of the location of the failure surface. If there is a phreatic surface within the soil overburden, r_u will increase with the increase in d, as indicated by Eq. (6-4), so the most critical failure plane is still at the bottom above the rock surface.

It is interesting to note that the failure of a cohesionless soil in an infinite slope is similar to the dumping of sand from a dump truck. When sand is dumped to form a pile, the particles on the surface will roll down the slope, indicating a plane failure. This type of failure also can be considered as a cylindrical failure with a center at infinity. In an effective stress analysis, the effective cohesion, c', is usually quite small but should not be assigned 0. If $c' = 0$, the soil is cohesionless and the most critical failure surface is a shallow circle with a very large radius, which barely scratches the surface of the slope. To design a homogeneous embankment with zero cohesion, it is not necessary to use any computer software to determine the minimum factor of safety, because the most critical failure surface is a shallow circle and the minimum factor of safety simply can be computed by Eq. (6-5).

Example 6.1 Fig. 6-2 is an example of an infinite slope with a slope angle of 16°, a soil overburden of 10 ft (3.1 m), and a phreatic surface as shown. Assuming a soil unit weight of 125 pcf (19.7 kN/m^3) and a seismic coefficient of 0.1, determine both the static and the seismic factors of safety for the following two cases: (a) the soil has an effective cohesion of 200 psf (9.6 kPa) and an effective friction angle of 30°, and (b) the soil is cohesionless with an effective friction angle of 30°.

Solution Because 50% of the sliding mass is under water, from Eq. (4-16), $r_u = 0.25$.

(a) With $c' = 200$ pcf, $\phi' = 30°$, $\gamma = 125$ pcf, $\beta = 16°$, and $C_s = 0$, from Eq. (6-3), the static factor of safety with $C_s = 0$ is

$$F = \frac{\left(\frac{200}{125\times 10}\right)\sec 16° + (1-0.25)\cos 16° \tan 30°}{\sin 16°} = \frac{0.583}{0.276} = 2.112$$

If $C_s = 0.1$, a term involving C_s is added to both the numerator and denominator, so the seismic factor of safety is

$$F = \frac{0.583 - 0.1\times \sin 16° \tan 30°}{0.276 + 0.1\times \cos 16°} = \frac{0.567}{0.372} = 1.524$$

(b) With $c' = 0$ pcf, $\phi' = 30°$, $\gamma = 125$ pcf, $\beta = 16°$, and $C_s = 0$, from Eq. (6-3), the static factor of safety is

$$F = \frac{(1-0.25)\cos 16° \tan 30°}{\sin 16°} = \frac{0.416}{0.276} = 1.507$$

If $C_s = 0.1$, from Eq. (6-3), the seismic factor of safety is

$$F = \frac{0.416 - 0.1\times \sin 16° \tan 30°}{0.276 + 0.1\times \cos 16°} = \frac{0.400}{0.372} = 1.075$$

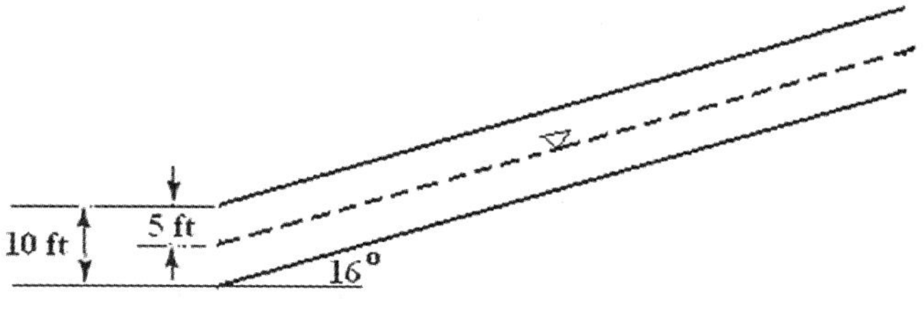

Fig. 6-2. Example 6.1

6.2 Triangular Cross Section

Fig. 6-3 shows a triangular fill on a sloping surface. It is assumed that the failure plane is along the bottom of the fill. Good examples for this type of failure are the spoil banks created by surface mining, where the original ground surface is not properly scalped, and a weak layer exists at the bottom of the fill. In addition to the vertical weight, a horizontal seismic force equal to C_sW also is applied.

Similar to Eq. (6-2) for an infinite slope, except for the change of notation from β to α and the replacement of $b\sec\beta$ by $H\csc\alpha$, the factor of safety can be written as

$$F = \frac{c'H\csc\alpha + W[(1-r_u)\cos\alpha - C_s\sin\alpha]\tan\phi'}{W(\sin\alpha + C_s\cos\alpha)} \tag{6-6}$$

$$\text{and } W = \frac{1}{2}\gamma H^2\csc\beta\csc\alpha\sin(\beta-\alpha) \tag{6-7}$$

in which γ = unit weight of soil and β = angle of the outslope. Substituting W from Eq. (6-7) into Eq. (6-6),

$$F = \frac{2\sin\beta\csc(\beta-\alpha)\left(\dfrac{c'}{\gamma H}\right) + [(1-r_u)\cos\alpha - C_s\sin\alpha]\tan\phi'}{\sin\alpha + C_s\cos\alpha} \tag{6-8}$$

If the fill width, W_f, is given instead of the height, H,

$$F = \frac{c'W_f\sin\beta\csc(\beta-\alpha) + W[(1-r_u)\cos\alpha - C_s\sin\alpha]\tan\phi'}{W(\sin\alpha + C_s\cos\alpha)} \tag{6-9}$$

in which

$$W = \frac{1}{2}\gamma W_f^2\sin\alpha\sin\beta\csc(\beta-\alpha) \tag{6-10}$$

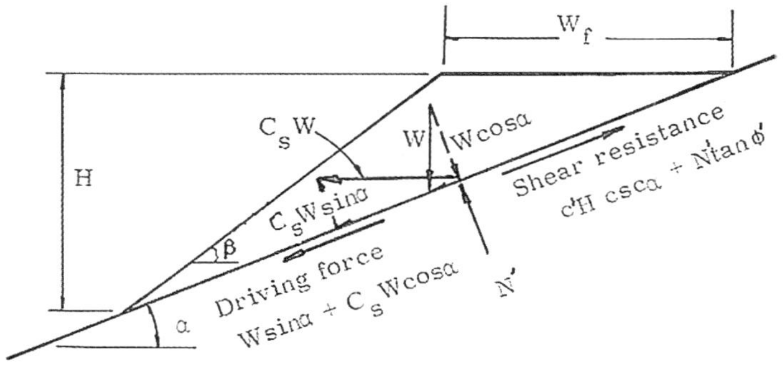

Fig. 6-3. Plane failure of triangular fill

Substituting W from Eq. (6-10) into Eq. (6-9),

$$F = \frac{2\csc\alpha\left(\dfrac{c'}{\gamma W_f}\right) + [(1-r_u)\cos\alpha - C_s\sin\alpha]\tan\phi'}{\sin\alpha + C_s\cos\alpha} \tag{6-11}$$

Eq. (6-11) shows that the factor of safety is independent of the angle of outslope, β, when the fill width, W_f, is given, because when β changes, both the resisting force and the driving force change in the same proportion.

Example 6.2 In Fig. 6-3, given H = 40 ft (12.2 m), α = 20°, β = 36°, γ = 125 pcf (19.7 kN/m^3), c' = 160 psf (7.7 kPa), ϕ' = 24°, and r_u = 0.05, determine both the static factor of safety with $C_s = 0$ and the seismic factor of safety with $C_s = 0.1$.

Solution From Eq. (6-8) with $C_s = 0$, the static factor of safety is

$$F = \frac{2\sin 36^\circ \csc(36^\circ - 20^\circ)\left(\dfrac{160}{125\times 40}\right) + (1-0.05)\cos 20^\circ \tan 24^\circ}{\sin 20^\circ} = \frac{0.534}{0.342} = 1.561$$

For $C_s = 0.1$, a term involving C_s is added to both the numerator and denominator, so the seismic factor of safety is

$$F = \frac{0.534 - 0.1\times\sin 20^\circ \tan 24^\circ}{0.342 + 0.1\times\cos 20^\circ} = \frac{0.519}{0.436} = 1.190$$

6.3 Trapezoidal Cross Section

Fig. 6-4(a) shows the forces acting on a trapezoidal fill. An example of this type is a hollow fill where the top part of the fill is placed on a natural slope and the

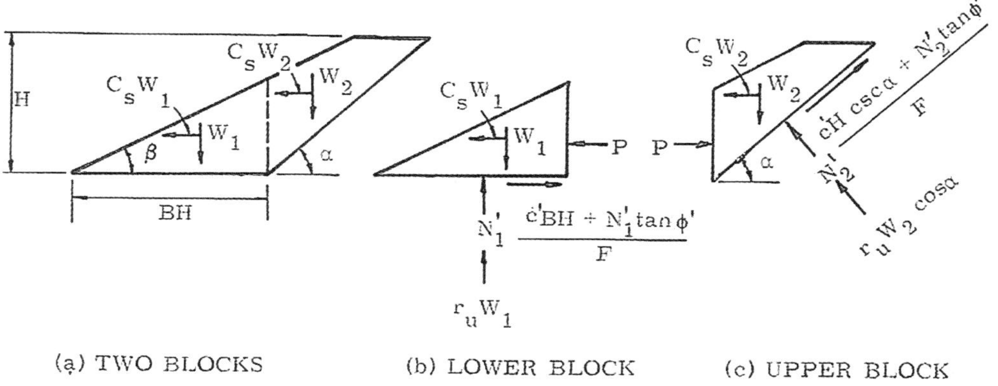

Fig. 6-4. Plane failure along bottom of trapezoidal fill

bottom part on a horizontal ground. If the natural slope and original ground surfaces are not properly scalped, plane failures will occur along the bottom of the fill.

The factor of safety with respect to failure along the bottom of the fill can be determined by dividing the fill into two sliding blocks. Assuming that the force, P, acting between the two blocks is horizontal, and that the shear force has a factor of safety, F, four equations (two from each block) can be determined from statics to solve the four unknowns, P, F, N_1', and N_2', where N_1', and N_2' are the effective forces normal to the failure planes. It has been found that the assumption of a horizontal P, or neglecting the friction between the two blocks, always results in a smaller factor of safety and is therefore on the safe side.

From the equilibrium of forces on the lower block in both the vertical and horizontal directions, as shown in Fig. 6-4(b),

$$N_1' = (1 - r_u)W_1 \tag{6-12}$$

$$P + C_s W_1 = \frac{c'BH + N_1' \tan\phi'}{F} \tag{6-13}$$

in which W_1 = weight of the lower block and B is the ratio between the base width and height. Substituting Eq. (6-12) into Eq. (6-13),

$$P = \frac{c'BH + (1 - r_u)W_1 \tan\phi'}{F} - C_s W_1 \tag{6-14}$$

The equilibrium of forces on the upper block in both normal and tangential to the failure plane gives

$$N_2' = P\sin\alpha + W_2[(1 - r_u)\cos\alpha - C_s \sin\alpha] \tag{6-15}$$

$$W_2[\sin\alpha + C_s\cos\alpha] = P\cos\alpha + \frac{c'H\csc\alpha + N_2'\tan\phi'}{F} \tag{6-16}$$

in which W_2 = weight of the upper block. Substituting Eqs. (6-14) and (6-15) into Eq. (6-16), a quadratic equation can be obtained to solve the factor of safety, F:

$$a_1 F^2 + a_2 F + a_3 = 0 \tag{6-17}$$

in which

$$a_1 = a_4 \sin\alpha + C_s(a_4 + a_5)\cos\alpha \tag{6-18}$$

$$a_2 = -\left\{\frac{c'}{\gamma H}\left(B\cos\alpha + \frac{1}{\sin\alpha}\right) + [(1 - r_u)\cos\alpha - C_s\sin\alpha](a_4 + a_5)\tan\phi'\right\} \tag{6-19}$$

$$a_3 = -B\sin\alpha\tan\phi'\left[\frac{c'}{\gamma H} + (1 - r_u)\frac{a_5}{B}\tan\phi'\right] \tag{6-20}$$

For an irregular slope with W_1 and W_2 given,

$$a_4 = \frac{W_2}{\gamma H^2} \tag{6-21}$$

$$a_5 = \frac{W_1}{\gamma H^2} \tag{6-22}$$

For type I fill, as shown in Fig. 6-5(a),

$$a_4 = \frac{1}{2}\left[\frac{1}{\tan\alpha} - \frac{(1-B\tan\beta)^2}{\tan\beta}\right] \tag{6-23}$$

$$a_5 = \frac{1}{2}B^2\tan\beta \tag{6-24}$$

For type II fill, as shown in Fig. 6-5(b),

$$a_4 = \frac{1}{2\tan\alpha} \tag{6-25}$$

$$a_5 = B - \frac{1}{2\tan\beta} \tag{6-26}$$

Note that in using Eqs. (6-21) and (6-22) for irregular slopes, the weights W_1 and W_2 must be calculated from measurements on the cross section. If the slope is uniform, as shown in Fig. 6-5, either Eqs. (6-23) and (6-24) or Eqs. (6-25) and (6-26) may be used, depending on whether the fill is type I or type II. The type I fill, in which $B \le \cot\beta$, has a triangular lower block, whereas the type II fill, in which $B > \cot\beta$, has a trapezoidal lower block.

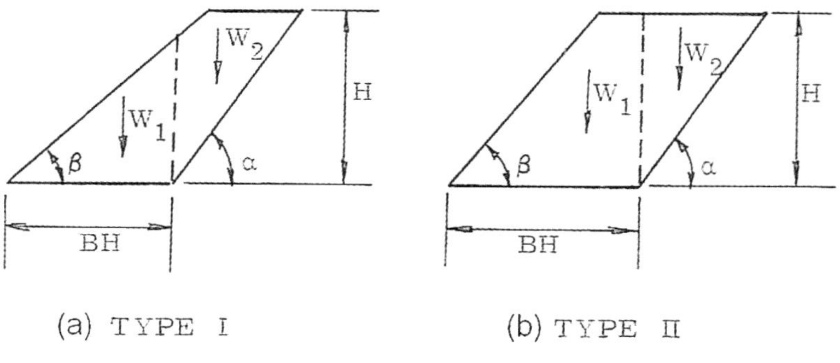

Fig. 6-5. Two types of trapezoidal fills

Example 6.3 A trapezoidal fill, with the dimensions shown in Fig. 6-6, has a total unit weight, γ, of 125 pcf (19.7 kN/m^3) and subjected to seepage with a pore pressure ratio, r_u, of 0.05, and earthquake with a seismic coefficient, C_s, of 0.1. Assuming plane failures along the bottom of the fill, where the soil has an effective cohesion, c', of 160 psf (7.7 kN/m^2) and an effective friction angle, ϕ', of 24°, determine the seismic factor of safety.

Solution The fill is type I, because it has a triangular cross section at the bottom. The base width is 460 ft and the height, H, is 230 ft, or $B = 460/230 = 2$. With $\beta = 24°$, $\alpha = 37°$, and $B = 2$, from Eqs. (6-23) and (6-24),

$$a_4 = \frac{1}{2}\left[\frac{1}{\tan 37°} - \frac{(1 - 2\tan 24°)^2}{\tan 24°}\right] = 0.650$$

$$a_5 = \frac{1}{2}(2)^2 \tan 24° = 0.890$$

From Eqs. (6-18) to (6-20),

$$a_1 = 0.65 \sin 37° + 0.1 \times (0.65 + 0.89)\cos 37° = 0.514$$

$$a_2 = -\left\{\frac{160}{125 \times 230}\left(2\cos 37° + \frac{1}{\sin 37°}\right) + [(1 - 0.05)\cos 37° - 0.1\sin 37°](0.65 + 0.89)\tan 24°\right\}$$

$$= -0.497$$

$$a_3 = -2\sin 37 \tan 24°\left[\frac{160}{120 \times 230} + (1 - 0.05)\frac{0.89}{2}\tan 24°\right] = -0.104$$

Thus, the quadratic equation becomes

$$0.514F^2 - 0.497F - 0.104 = 0$$

$$F = \frac{0.497 \pm \sqrt{(0.497)^2 + 4 \times 0.514 \times 0.104}}{2 \times 0.514} = 1.144 \text{ or } -0.182$$

Disregarding the negative value, the seismic factor of safety is 1.144.

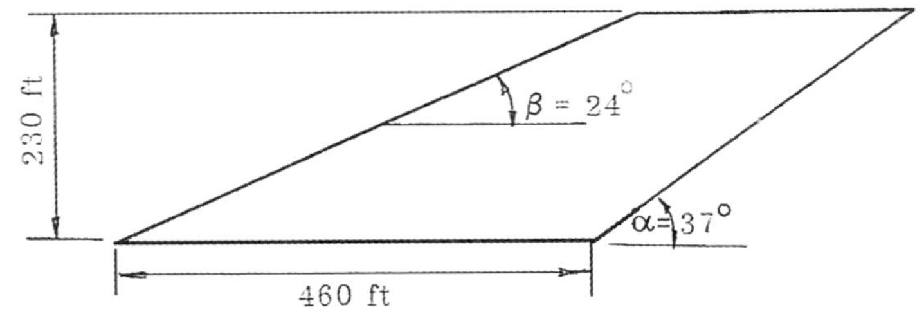

Fig. 6-6. Example 6.3

6.4 Two Sliding Blocks

Fig. 6-7 shows a plane failure surface consisting of two sliding blocks. This type of failure rarely occurs unless a weak layer exists at the bottom of each block. There are a total of four unknowns (P, N_1', N_2', and F), which can be solved by four equilibrium equations, two for each block. As in the normal method described previously, the total weight is used to compute the driving force and the submerged weight for the shear resistance. By resolving the forces in two directions, one parallel to and the other perpendicular to the failure plane at the bottom of each block, the following two equations are obtained for the lower block, or block 1:

$$\frac{c'L_1 + N_1' \tan\phi'}{F} = W_1[\sin\alpha_1 + C_s \cos\alpha_1] + P\cos(\phi_d - \alpha_1) \tag{6-27}$$

$$N_1' = W_1[(1-r_u)\cos\alpha_1 - C_s \sin\alpha_1] + P\sin(\phi_d - \alpha_1) \tag{6-28}$$

in which L = length of the failure plane, α = angle of inclination of the failure plane, and ϕ_d = developed angle of internal friction between the two blocks. The subscript 1 refers to the lower block. Substituting Eq. (6-28) into (6-27) and solving for P,

$$P = \frac{c'L_1 + W_1[(1-r_u)\cos\alpha_1 - C_s \sin\alpha_1]\tan\phi' - FW_1[\sin\alpha_1 + C_s\cos\alpha_1]}{F\cos(\phi_d - \alpha_1) - \sin(\phi_d - \alpha_1)\tan\phi'} \tag{6-29}$$

Similar to block 1, P for the upper block, or block 2, can be obtained by changing the subscript from 1 to 2 and changing the sign, because P in block 2 is opposite in direction to that in block 1, or

$$P = -\frac{c'L_2 + W_2[(1-r_u)\cos\alpha_2 - C_s \sin\alpha_2]\tan\phi' - FW_2[\sin\alpha_2 + C_s\cos\alpha_2]}{F\cos(\phi_d - \alpha_2) - \sin(\phi_d - \alpha_2)\tan\phi'} \tag{6-30}$$

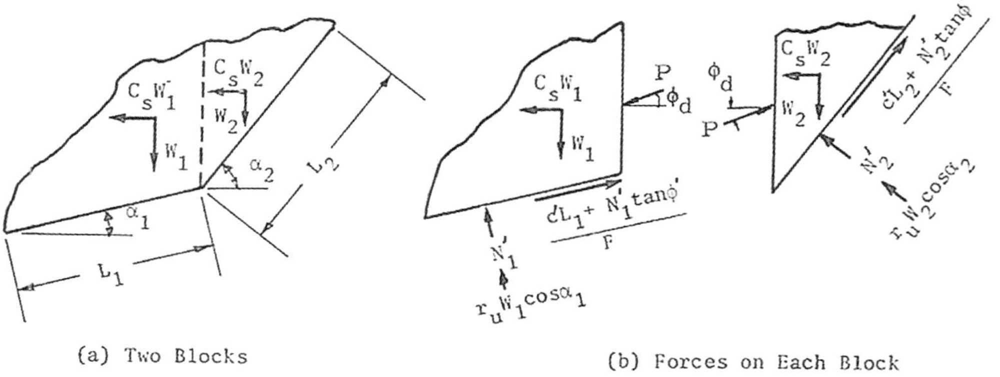

Fig. 6-7. Plane failure with two sliding blocks

Table 6-1. Coefficients of Quadratic Equation for Two Sliding Blocks

$a_1 = c'L_1 + W_1[(1 - r_u)\cos\alpha_1 - C_s \sin\alpha_1]\tan\phi'$ $a_2 = W_1(\sin\alpha_1 + C_s \cos\alpha_1)$	
$a_3 = \cos(\phi_d - \alpha_1)$ $a_4 = \sin(\phi_d - \alpha_1)\tan\phi'$	$a = a_2b_3 + b_2a_3$
$b_1 = c'L_2 + W_2[(1 - r_u)\cos\alpha_2 - C_s \sin\alpha_2]\tan\phi'$ $b_2 = W_2(\sin\alpha_2 + C_s \cos\alpha_2)$	$b = -(a_1b_3 + b_1a_3 + a_2b_4 + b_2a_4)$
$b_3 = \cos(\phi_d - \alpha_2)$ $b_4 = \sin(\phi_d - \alpha_2)\tan\phi'$	$c = a_1b_4 + b_1a_4$ $a F^2 + b F + c = 0$

Changing Eqs. (6-29) and (6-30) to simplified forms and setting them equal,

$$\frac{a_1 - a_2F}{a_3F - a_4} = -\frac{b_1 - b_2F}{b_3F - b_4} \tag{6-31}$$

Solving Eq. (6-31) for F, the following quadratic equation is obtained:

$$aF^2 + bF + c = 0 \tag{6-32}$$

For ease of reference, expressions leading to the coefficients a, b, and c are shown in Table 6-1.

Example 6.4 Fig. 6-8 shows a fill divided into two sliding blocks. The soil at the bottom of the fill has an effective cohesion of 100 psf (4.8 kPa), an effective friction angle of 25°, and a unit weight of 125 pcf (19.7 kN/m^3). Given a pore pressure ratio of 0.05 and a seismic coefficient of 0.1, determine the seismic factors of safety for ϕ_d of 0°, 17.2° (0.3 rad), and 34.4° (0.6 rad), respectively.

Solution From the coordinates in Fig. 6-8,

$\beta = \tan^{-1}(250/150) = 31°$
$\alpha_1 = \tan^{-1}(50/300) = 9.5°$
$\alpha_2 = \tan^{-1}[(150 - 50)/(470 - 300)] = 30.5°$
$L_1 = [(300)^2 + (50)^2]^{0.5} = 304.1\text{ ft}$
$L_2 = [(470 - 300)^2 + (150 - 50)^2]^{0.5} = 197.2\text{ ft}$

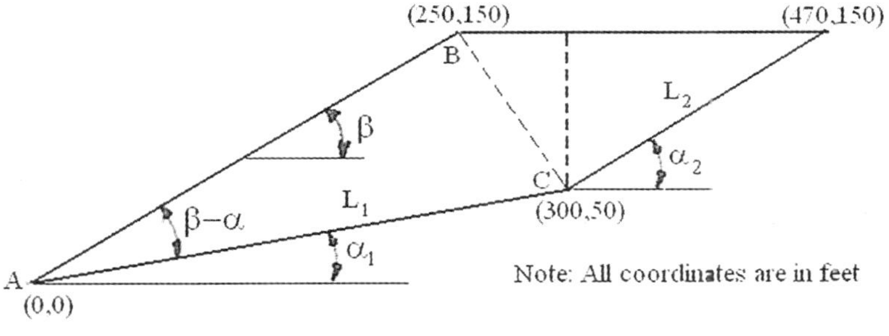

Fig. 6-8. Example 6.4

Block 1 is divided into two triangles:

$$\overline{AB} = [(150)^2 + (250)^2]^{0.5} = 291.5 \text{ ft}$$

$$\text{Area of } \Delta ABC = 0.5 \times \overline{AB} \times L_1 \times \sin(\beta - \alpha) = 0.5 \times 291.5 \times 304.1 \times \sin(31° - 9.5°) = 16{,}244 \text{ ft}^2$$

$$W_1 = 125 \times (16{,}244 + 0.5 \times 100 \times 50) = 2{,}343{,}000 \text{ lb}$$

Block 2 is a right triangle:

$$W_2 = 0.5 \times 100 \times 170 \times 125 = 1{,}062{,}500 \text{ lb}$$

From Table 6-1,

$$a_1 = 100 \times 304.1 + 2{,}343{,}000[(1 - 0.05)\cos 9.5° - 0.1 \times \sin 9.5°]\tan 25° = 1{,}036{,}074$$
$$a_2 = 2{,}343{,}000[\sin 9.5° + 0.1 \times \cos 9.5°] = 617{,}793$$
$$b_1 = 100 \times 197.2 + 1{,}062{,}500[(1 - 0.05)\cos 30.5° - 0.1 \times \sin 30.5°]\tan 25° = 400{,}125$$
$$b_2 = 1{,}062{,}500[\sin 30.5° + 0.1 \times \cos 30.5°] = 630{,}808$$

When $\phi_d = 0$,

$$a_3 = \cos(-9.5°) = 0.986$$
$$a_4 = \sin(-9.5°)\tan 25° = -0.077$$
$$b_3 = \cos(-30.5°) = 0.862$$
$$b_4 = \sin(-30.5°)\tan 25° = -0.237$$
$$a = 617{,}793 \times 0.862 + 630{,}808 \times 0.986 = 1{,}154{,}506$$
$$b = -[1{,}036{,}074 \times 0.862 + 400{,}125 \times 0.986 + 617{,}793 \times (-0.237) + 630{,}808 \times (-0.077)] = -1{,}092{,}630$$
$$c = 1{,}036{,}074 \times (-0.237) + 400{,}125 \times (-0.077) = -276{,}359$$

The quadratic equation is $1{,}154{,}506\, F^2 - 1{,}092{,}630\, F - 276{,}359 = 0$, or $F^2 - 0.946\, F - 0.239 = 0$, so the solution is $F = 1.153$.

When $\phi_d = 17.2°$,

$$a_3 = \cos(17.2° - 9.5°) = 0.991$$
$$a_4 = \sin(17.2° - 9.5°)\tan 25° = 0.0625$$
$$b_3 = \cos(17.2° - 30.5°) = 0.973$$
$$b_4 = \sin(17.2° - 30.5°)\tan 25° = -0.107$$
$$a = 617{,}793 \times 0.973 + 630{,}808 \times 0.991 = 1{,}226{,}243$$
$$b = -[1{,}036{,}074 \times 0.973 + 400{,}125 \times 0.991 + 617{,}793 \times (-0.107) + 630{,}808 \times 0.0625] = -1{,}377{,}946$$
$$c = 1{,}036{,}074 \times (-0.107) + 400{,}125 \times 0.0625 = -85{,}852$$

The quadratic equation is $1{,}226{,}243\, F^2 - 1{,}377{,}946\, F - 85{,}852 = 0$, or $F^2 - 1.124\, F - 0.070 = 0$, so the solution is $F = 1.183$.

When $\phi_d = 34.4°$,

$$a_3 = \cos(34.4° - 9.5°) = 0.907$$
$$a_4 = \sin(34.4° - 9.5°)\tan 25° = 0.196$$
$$b_3 = \cos(34.4° - 30.5°) = 0.998$$
$$b_4 = \sin(34.4° - 30.5°)\tan 25° = 0.0317$$

$a = 617{,}793 \times 0.998 + 630{,}808 \times 0.907 = 1{,}188{,}700$
$b = -[1{,}036{,}074 \times 0.998 + 400{,}125 \times 0.907 + 617{,}793 \times 0.0317 + 630{,}808 \times 0.196]$
$= -1{,}540{,}138$
$c = 1{,}036{,}074 \times 0.0317 + 400{,}125 \times 0.196 = 111{,}268$

The quadratic equation is $1{,}188{,}700\ F^2 - 1{,}540{,}138\ F + 111{,}268 = 0$, or $F^2 - 1.296\ F + 0.094 = 0$, so the solution is $F = 1.219$.

It can be seen that the factor of safety increases with the increase in ϕ_d, which is as expected. These three ϕ_d angles of 0°, 17.2°, and 34.4° were selected because they are recommended by LEAME for use in the original Spencer method, as described in Section 8.4.

Example 6.5 If the fill in Example 6.4 has a friction angle of 32°, which is much greater than the 25° at the bottom, determine the factor of safety when $\tan\phi_d = \tan 32°/F$, that is, the same factor of safety is applied to the vertical interface between the two blocks as to the failure planes at the bottom.

Solution The results of Example 6.5 are plotted in Fig. 6-9, as indicated by the curve with three small circles. Based on the given relationship $\tan\phi_a = \tan 32°/F$, the values of $\tan\phi_d$ for F values of 1.1, 1.2, and 1.3 are computed and plotted as the curve with three crosses. The intersection of these two curves gives a safety factor of 1.203.

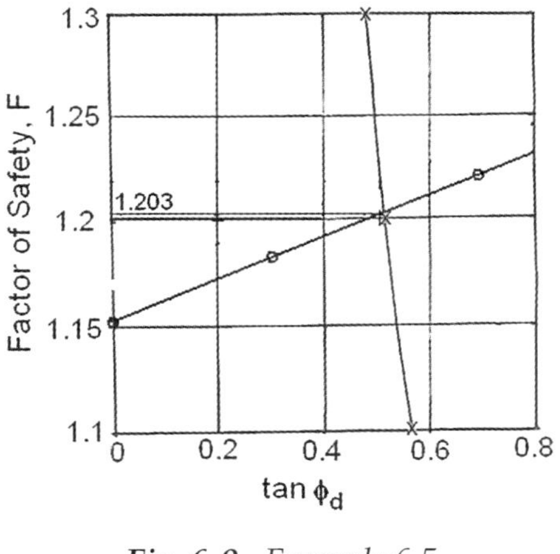

Fig. 6-9. Example 6.5

It can be seen that the assumption of $\phi_d = 0$ is more conservative and results in a safety factor of 1.153, which is smaller than the more theoretically correct value of 1.203. In view of the fact that the sliding block analysis only considers the force equilibrium while neglecting the moment equilibrium completely, it is better to be more conservative by using a lower factor of safety. Furthermore, the use of $\phi_d = 0$ eliminates the necessity of plotting the two curves shown in Fig. 6-9 and is more amenable to hand calculation.

Example 6.4 is repetitive and requires the computation of safety factors three times, one for each ϕ_d. After these tedious computations, Example 6.5 uses a graphical method, which is neither convenient nor accurate. It is much more efficient to solve the same problems by a spreadsheet, as illustrated by the following example.

Example 6.6 Solve both Examples 6.4 and 6.5 by using a spreadsheet.

Solution Table 6-2 is the spreadsheet for both Examples 6.4 and 6.5. All the equations used for the spreadsheet are presented in Table 6-1. Details about the spreadsheet are as follows:

1. The first three rows include all the input parameters. For convenience, all angles in degrees are converted to radians in row 4 before calling any of the trigonometric functions. For example, ϕ in cell B4 is converted from degrees to radians by the expression RADIANS(D1). Row 5 contains the four parameters not affected by ϕ_d or the factor of safety.
2. Rows 6 to 18 give the solutions to Example 6.4, and rows 19 to 23 to Example 6.5. The most important part, which requires much time to work out, is rows 7 to 10. Once these rows are completed, they can be copied and pasted repeatedly three times with only slight changes, if needed.
3. Cell D7 is the angle of internal friction between the two blocks, which may not be equal to that along the failure surface at the bottom of the fill. The expression for ϕ_d in cell A9 is ATAN(TAN(RADIANS(D7))/H7) so, if the assumed factor of safety in cell H7 is 1, D7 is the given developed angle of internal friction between the two blocks, such as the given 0, 17.2°, and 34.4° in Example 6.4, but expressed in terms of radians. After A9 is determined, a_3 in cell B9 can be determined by COS(A9-D$4). Because cell B9 will be copied to different rows but not to different columns, a $ sign must be placed before row 4 so that row 4 will not be changed when copied. The same should be applied to other variables, such as variable a in cell F9 can be expressed as H$2*D9+H$3*B9. Any variables with a row number of less than 6 must have a $ sign on the row number. The computed factor of safety in cell H10 is the solution for Example 6.4 with $\phi_d = 0$.
4. To find the factors of safety in Example 6.4 for other values of ϕ_d, simply change the value in cell D11 to 17.2 and that in cell D15 to 34.4, and the computed factor of safety will appear automatically.
5. To find the factor of safety for Example 6.5, change the value in cell D20 to 32, which is the given angle of internal friction of the fill. By trial and error, an assumed safety factor of 1.203, exactly equal to the computed factor of safety, can be easily obtained.

Table 6-2. Spreadsheet for Solving Examples 6.4 and 6.5

	A	B	C	D	E	F	G	H
1	c in psf =	100	ϕ in deg =	25	Cs =	0.1	ru =	0.05
2	W1 in lb =	2343000	L1 in ft =	304.1	α1 in deg =	9.5		
3	W2 in ib =	1062500	L2 in ft =	197.2	α2 in deg =	30.5		
4	ϕ in rad =	0.436332	α1 in rad =	0.165806	α2 in rad =	0.532325		
5	a_1 =	1036074	a_2 =	617793	b_1 =	400125	b_2 =	630808
6	*Example 6.4*							
7	**ϕ between blocks in deg =**			0	Assumed factor of safety, F =			1.000
8	ϕ_d	a_3	a_4	b_3	b_4	a	b	c
9	0	0.986286	−0.07696	0.861629	−0.236669	1154465	−1092588	−276001
10					**Computed factor of safety, F =**			1.154
11	**ϕ between blocks in deg =**			17.2	Assumed factor of safety, F =			1.000
12	ϕ_d	a_3	a_4	b_3	b_4	a	b	c
13	0.300197	0.990983	0.062479	0.973179	−0.107274	1226343	−1377941	−86144
14					**Computed factor of safety, F =**			1.183
15	**ϕ between blocks in deg =**			34.4	Assumed factor of safety, F =			1.000
16	ϕ_d	a_3	a_4	b_3	b_4	a	b	c
17	0.600393	0.907044	0.196332	0.997684	0.031716	1188533	−1540047	111418
18					**Computed factor of safety, F =**			1.219
19	*Example 6.5*							
20	**ϕ between blocks in deg =**			32	Assumed factor of safety, F =			1.203
21	ϕ_d	a_3	a_4	b_3	b_4	a	b	c
22	0.479067	0.951334	0.143699	0.998582	−0.024823	1217026	−1490568	31779
23					**Computed factor of safety, F =**			1.203

6.5 Three Sliding Blocks

Fig. 6-10 shows a plane failure surface consisting of three sliding blocks. There are a total of six unknowns (P_1, P_2, N_1', N_2', N_3', and factor of safety, F), which can be solved by six equilibrium equations, two for each block. Blocks 1 and 2 are the same as the two sliding blocks shown in Fig. 6-7, except that P is replaced by P_1 or P_2, so Eq. (6-29) can still be applied and written in the following simplified forms:

$$P_1 = \frac{a_1 - a_2 F}{a_3 F - a_4} \tag{6-33}$$

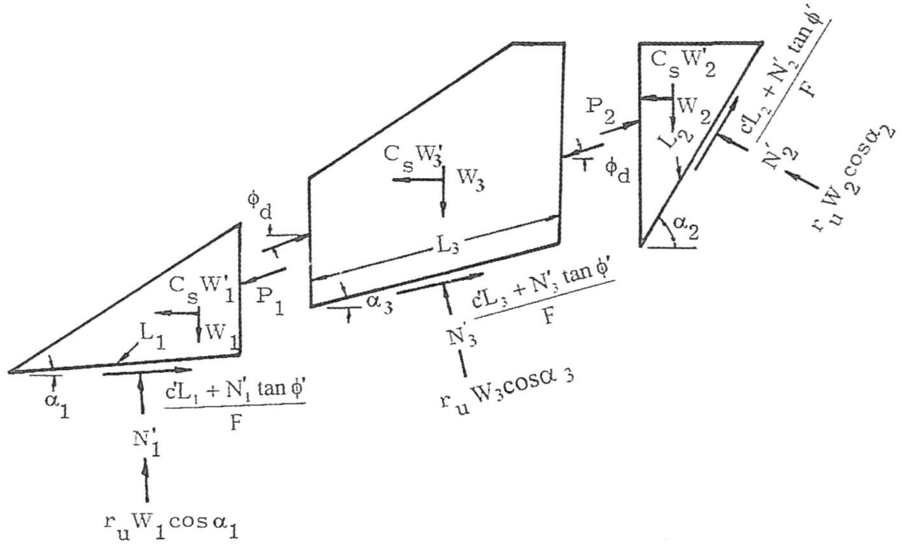

Fig. 6-10. Plane failure with three sliding blocks

$$P_2 = \frac{b_1 - b_2F}{b_3F - b_4} \tag{6-34}$$

The resultant of the forces on both sides of block 3 is P_3 and can be obtained from Eq. (6-29) by simply changing all the subscripts to 3:

$$P_3 = \frac{c'L_3 + W_3[(1 - r_u)\cos\alpha_3 - C_s \sin\alpha_3]\tan\phi' - FW_3(\sin\alpha_3 + C_s \cos\alpha_3)}{F\cos(\phi_d - \alpha_3) - \sin(\phi_d - \alpha_3)\tan\phi'} \tag{6-35}$$

To satisfy force equilibrium $P_2 + P_1 + P_3 = 0$, or in simplified forms,

$$\frac{b_1 - b_2F}{b_3F - b_4} + \frac{a_1 - a_2F}{a_3F - a_4} + \frac{c_1 - c_2F}{c_3F - c_4} = 0 \tag{6-36}$$

After simplification, the following cubic equation is obtained:

$$aF^3 + bF^2 + cF + d = 0 \tag{6-37}$$

Expressions for these coefficients are shown in Table 6-3. Eq. (6-37) can be solved by trial and error, as illustrated by the following example.

Example 6.7 Fig. 6-11 shows a fill consisting of three sliding blocks. The soil at the bottom of the fill has a cohesion of 160 psf (7.7 kPa) and a friction angle of 24°. Assuming that the unit weight of the fill is 125 pcf (19.7 kN/m³), the forces between blocks are horizontal, and there is no seepage, determine the static factor of safety for plane failure.

Table 6-3. Coefficients of Cubic Equation for Three Sliding Blocks

$a_1 = c'L_1 + W_1[(1 - r_u)\cos\alpha_1 - C_s\sin\alpha_1]\tan\phi'$
$a_2 = W_1(\sin\alpha_1 + C_s\cos\alpha_1)$
$a_3 = \cos(\phi_d - \alpha_1)$
$a_4 = \sin(\phi_d - \alpha_1)\tan\phi'$
$b_1 = c'L_2 + W_2[(1 - r_u)\cos\alpha_2 - C_s\sin\alpha_2]\tan\phi'$
$b_2 = W_2(\sin\alpha_2 + C_s\cos\alpha_2)$
$b_3 = \cos(\phi_d - \alpha_2)$
$b_4 = \sin(\phi_d - \alpha_2)\tan\phi'$
$c_1 = c'L_3 + W_3[(1 - r_u)\cos\alpha_3 - C_s\sin\alpha_3]\tan\phi'$
$c_2 = W_3(\sin\alpha_3 + C_s\cos\alpha_3)$
$c_3 = \cos(\phi_d - \alpha_3)$
$c_4 = \sin(\phi_d - \alpha_3)\tan\phi'$
$a = a_2b_3c_3 + a_3b_2c_3 + a_3b_3c_2$
$b = -(a_1b_3c_3 + a_2b_3c_4 + a_2b_4c_3 + a_3b_1c_3 + a_3b_2c_4 + a_3b_3c_1 + a_3b_4c_2 + a_4b_2c_3 + a_4b_3c_2)$
$c = a_1b_3c_4 + a_1b_4c_3 + a_2b_4c_4 + a_3b_1c_4 + a_3b_4c_1 + a_4b_1c_3 + a_4b_2c_4 + a_4b_3c_1 + a_4b_4c_2$
$d = -(a_1b_4c_4 + a_4b_1c_4 + a_4b_4c_1)$
$a\,F^3 + b\,F^2 + c\,F + d = 0$

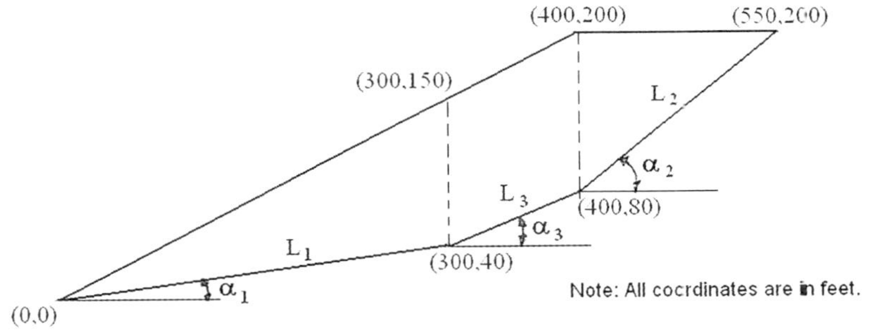

Fig. 6-11. Example 6.7

Solution From the coordinates shown in the figure,

$$\alpha_1 = \tan^{-1}(40/300) = 7.6^\circ \qquad L_1 = [(40)^2 + (300)^2]^{0.5} = 302.7\,\text{ft}$$
$$\alpha_2 = \tan^{-1}(120/150) = 38.7^\circ \qquad L_2 = [(120)^2 + (150)^2]^{0.5} = 192.1\,\text{ft}$$
$$\alpha_3 = \tan^{-1}(40/100) = 21.8^\circ \qquad L_3 = [(40)^2 + (100)^2]^{0.5} = 107.7\,\text{ft}$$
$$W_1 = 0.5 \times 300 \times 110 \times 125 = 2{,}062{,}500\,\text{lb}$$
$$W_2 = 0.5 \times 150 \times 120 \times 125 = 1{,}125{,}000\,\text{lb}$$
$$W_3 = 0.5 \times 100 \times (110 + 120) \times 125 = 1{,}437{,}500\,\text{lb}$$

From Table 6-3,

$a_1 = 160 \times 302.7 + 2{,}062{,}500 \times \cos 7.6° \times \tan 24° = 958{,}650$
$b_1 = 160 \times 192.1 + 1{,}125{,}000 \times \cos 38.7° \times \tan 24° = 421{,}640$
$c_1 = 160 \times 107.7 + 1{,}437{,}500 \times \cos 21.8° \times \tan 24° = 611{,}478$
$a_2 = 2{,}062{,}500 \times \sin 7.6° = 272{,}779$
$b_2 = 1{,}125{,}000 \times \sin 38.7° = 703{,}398$
$c_2 = 1{,}437{,}500 \times \sin 21.8° = 533{,}841$
$a_3 = \cos(0 - 7.6°) = 0.991$
$b_3 = \cos(0 - 38.7°) = 0.780$
$c_3 = \cos(0 - 21.8°) = 0.928$
$a_4 = \sin(0 - 7.6°) \times \tan 24° = -0.059$
$b_4 = \sin(0 - 38.7°) \times \tan 24° = -0.278$
$c_4 = \sin(0 - 21.8°) \times \tan 24° = -0.165$
$a = 272{,}779 \times 0.780 \times 0.928 + 0.991 \times 703{,}398 \times 0.928 + 0.991 \times 0.780 \times 533{,}841 = 1{,}256{,}975$
$b = -(958{,}650 \times 0.780 \times 0.928 - 272{,}779 \times 0.780 \times 0.165 - 272{,}779 \times 0.278 \times 0.928 + 0.991 \times 421{,}640 \times 0.928 - 0.991 \times 703{,}398 \times 0.165 + 0.991 \times 0.780 \times 611{,}478 - 0.991 \times 0.278 \times 533{,}841 - 0.059 \times 703{,}398 \times 0.928 - 0.059 \times 0.780 \times 533{,}841) = -1{,}123{,}683$
$c = -958{,}650 \times 0.780 \times 0.165 - 958{,}650 \times 0.278 \times 0.928 + 272{,}779 \times 0.278 \times 0.165 - 0.991 \times 421{,}640 \times 0.165 - 0.991 \times 0.278 \times 611{,}478 - 0.059 \times 421{,}640 \times 0.928 + 0.059 \times 703{,}398 \times 0.165 - 0.059 \times 0.780 \times 611{,}478 + 0.059 \times 0.278 \times 533{,}841 = -631{,}210$
$d = -(958{,}650 \times 0.278 \times 0.165 + 0.059 \times 421{,}640 \times 0.165 + 0.059 \times 0.278 \times 611{,}478) = -58{,}107$

The cubic equation is $1{,}256{,}975\, F^3 - 1{,}123{,}683\, F^2 - 631{,}210\, F - 58{,}107 = 0$, or

$$\text{Function}(F) = F^3 - 0.894F^2 - 0.502F - 0.0462 = 0 \tag{6-38}$$

Eq. (6-38) can be solved by trial and error:

Assume $F = 1.25$, Function$(F) = -0.1175$
Assume $F = 1.30$, Function$(F) = -0.0127$
Assume $F = 1.305$, Function$(F) = -0.00137$
Assume $F = 1.306$, Function$(F) = 0.00091$

The factor of safety is between 1.305 and 1.306 and Function(1.306) is closer to 0, so the factor of safety accurate to three decimal points is $F = 1.306$.

The same graphical method can be used to determine the factor of safety for $\tan\phi_d = \tan\phi/F$. However, to do this by hand is very hectic and prone to error. It is much easier to solve the problem using a spreadsheet.

Example 6.8 Check the results of Example 6.7 using a spreadsheet. If the angle of internal friction between the two blocks is 30° and the developed friction angle, ϕ_d, between the two blocks is $\phi_d = \tan^{-1}(\tan 30°/F)$, determine the factor of safety.

Solution The top part of Table 6-4 is the spreadsheet for checking Example 6.6 with ϕ between blocks = 0°, or $\phi_d = 0°$. All the computed values in the spreadsheet check very well with those in Example 6.7. The factor of safety can be obtained by solving the following cubic equation, Function(F) = $aF^3 + bF^2 + cF + d = 0$. By trial and error, it was found that when F is changed from 1.305 to 1.306, Function(F) changes from negative to positive. Because Function(F) is closer to 0 when F = 1.306, the factor of safety is 1.306, which checks exactly with Example 6.6.

The bottom part of Table 6-4 is the spreadsheet for this example with ϕ between blocks = 30° and $\phi_d = \tan^{-1}(\tan 30°/F)$. When the assumed F is 1.369, Function(F) is negative and, when the assumed F is 1.370, Function(F) is positive. Because the positive value is closer to 0 than the negative, the factor of safety is 1.370.

Table 6-4. *Spreadsheet Solution for Examples 6.6 and 6.7*

c in psf =	160	ϕ in deg =	24	Cs =	0	ru =	0
W1 in lb =	2062500	L1 in ft =	302.7	α1 in deg =	7.6		
W2 in ib =	1125000	L2 in ft =	192.1	α2 in deg =	38.7		
W3in lb =	1437500	L3 in ft =	107.7	α3 in deg =	21.8		
ϕ in rad =	0.418879	α1 in rad =	0.132645	α2 in rad =	0.675442	α3 in rad =	0.380482
a_1 =	958649.5	b_1 =	421639.8	c_1 =	611478		
a_2 =	272778.8	b_2 =	703398	c_2 =	533841.3		
ϕ between blocks in deg = 0				Assumed factor of safety, F =			1.000
ϕ_d	a_3	a_4	b_3	b_4	c_3	c_4	
0	0.991216	−0.05888	0.78043	−0.278376	0.928486	−0.165344	
a	b	c	d	F	Function(F)	F	Function(F)
1257985	−1124449	−632310.7	-58252.9	1.305	−2578.578	1.306	285.2384
				When ϕ between blocks = 0, F =			**1.306**
ϕ between blocks in deg = 30				Assumed factor of safety, F =			1.369
ϕ_d	a_3	a_4	b_3	b_4	c_3	c_4	
0.399099	0.964711	0.085635	0.962059	−0.088733	0.999827	0.006054	
a	b	c	d				
1436305	−1936331	−46707.36	4942.825	Function(F) = aF3 + bF2 +cF + d =			−2830.108
ϕ between blocks in deg = 30				Assumed factor of safety, F =			1.370
ϕ_d	a_3	a_4	b_3	b_4	c_3	c_4	
0.398837	0.96478	0.08549	0.961988	−0.08875	0.999832	0.005965	
a	b	c	d				
1436337	−1936017	−47003.95	4931.959	Function(F) = aF3 + bF2 +cF + d =			155.583
				When ϕ between blocks = 30, F =			**1.370**

The factor of safety for any slopes with three sliding blocks can be found by simply changing the input parameters on the first four lines. In Fig. 6-11, the failure surface at the bottom of block 1 is uphill, so α_1 is positive. If the failure surface is downhill, α_1 should be entered as negative.

6.6 Earth Pressure Method

The application of earth pressure theory can be illustrated by the simple example shown in Fig 6-12. By assuming the active force, P_A, and the passive force, P_P, as horizontal, it can be proved easily by Rankine's or Coulomb's theory that the failure plane inclines at an angle of 45° + $\phi/2$ for the active wedge and 45° – $\phi/2$ for the passive wedge. From basic soil mechanics,

$$P_A = \frac{1}{2}\gamma H_A^2 \tan^2\left(45° - \frac{\phi}{2}\right) - 2cH_A \tan\left(45° - \frac{\phi}{2}\right) \tag{6-39}$$

$$P_P = \frac{1}{2}\gamma H_P^2 \tan^2\left(45° + \frac{\phi}{2}\right) + 2cH_P \tan\left(45° + \frac{\phi}{2}\right) \tag{6-40}$$

in which H_A and H_P are the heights of the active and passive wedges, respectively. The factor of safety can be determined by

$$F = \frac{cL + W \tan\phi}{P_A - P_P} \tag{6-41}$$

in which L = length of failure surface at the middle wedge, and W = weight of the middle wedge.

Although this case is very similar to the case of three sliding blocks, the factor of safety defined by Eq. (6-41) applies only to the middle wedge, whereas in the three-block analysis the factor of safety applies to all three blocks. Without the

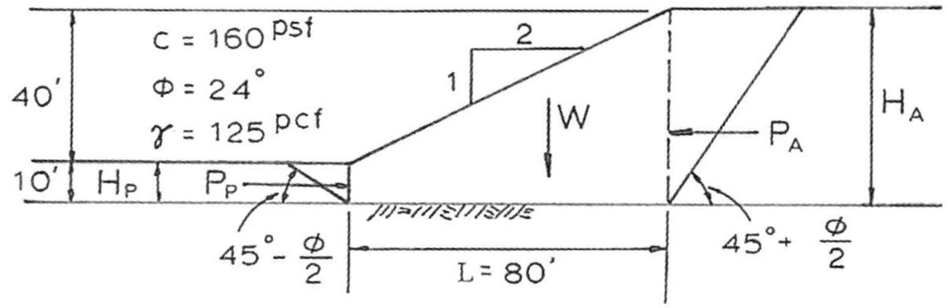

Fig. 6-12. Active and passive earth pressures, simple case
Note: 1 ft = 0.305 m; 1 psf = 47.9 kN/m²; 1 pcf = 157.1 N/m³

reduction of shear strength by a factor of safety, the shear resistance is greater and so is the factor of safety. Consequently, the factors of safety computed by Eq. (6-41) are much larger than those by the limit equilibrium method.

Example 6.9 For the case shown in Fig. 6-12 with $c = 160$ psf (7.7 kPa), $\phi = 24°$, $\gamma = 125$ pcf (19.6 kN/m^3), $H_A = 50$ ft (15.2 m), $H_P = 10$ ft (3.0 m), and $L = 80$ ft (24.4 m), determine the factor of safety by Eq. (6-41).

Solution From the dimensions shown in Fig. 6-12, $W = 0.5 \times 80 \times (10 + 50) \times 125 = 300{,}000$ lb. From Eq. (6-39), $P_A = 0.5 \times 125 \times (50)^2 \times (\tan 33°)^2 - 2 \times 160 \times 50 \times \tan 33° = 55{,}505$ lb. From Eq. (6-40), $P_P = 0.5 \times 125 \times (10)^2 \times (\tan 57°)^2 + 2 \times 160 \times 10 \times \tan 57° = 19{,}747$ lb. From Eq. (6-41),

$$F = \frac{160 \times 80 + 300{,}000 \tan 24°}{55{,}505 - 19{,}747} = \frac{146{,}369}{35{,}758} = 4.093$$

The factor of safety obtained from the three-block analysis based on $\phi_d = 0$ is 1.880, which is much smaller than the 4.093 obtained in this example.

This concept was employed also by Mendez (1971) for solving a more complex case, as shown in Fig. 6-13. A computer program was developed in which the inclination of the failure planes, θ_A and θ_P, and the inclination of earth pressures, α_A and α_P, can be varied. The wedge ABC is used to determine the active force, P_{A1}, in soil 1. The wedge CDF, together with the weight of BCDE and the active force P_{A1}, is used to determine the active force P_{A2} in soil 2. The same procedure is applied to the passive wedge, as shown by P_{P1}, P_{P2}, and α_P in the figure. The factor of safety is determined by

$$F = \frac{cL + [W\cos\theta + P_A \sin(\alpha_A - \theta) + P_P \sin(\theta - \alpha_P)]\tan\phi}{W\sin\theta + P_A \cos(\alpha_A - \theta) - P_P \cos(\theta - \alpha_P)} \tag{6-42}$$

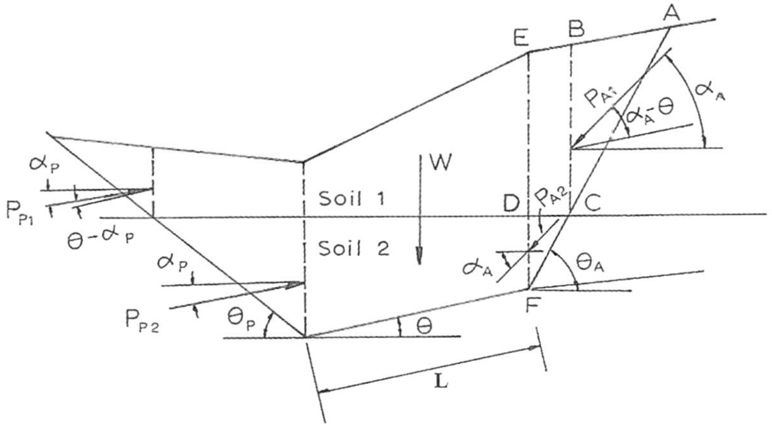

Fig. 6-13. Active and passive earth pressures, complex case

in which $P_A = P_{A1} + P_{A2}$, $P_P = P_{P1} + P_{P2}$, and θ is the angle of inclination of failure surface at middle block. It was found that the factor of safety is a minimum when $\alpha_A = \alpha_P = 0$.

The U.S. Navy (1971) also suggests the use of active and passive earth pressures for stability analysis but in a different way. Fig. 6-14(a) shows a slope composed of three different soils. It is assumed that the earth pressures are horizontal and the inclinations of failure planes in the active wedges are $45° + \phi/2$ and those in the passive wedges are $45° - \phi/2$.

In this method, P_A and P_P are not considered as the driving force, because they include the shear resistance along the failure planes in the active and passive wedges. The factor of safety is defined as the ratio between the resisting force and the driving force. The driving force is caused only by the soil weight, excluding any shear resistance, which should be included in the resisting force. As shown by the force diagram in Fig. 6-14(b), the active earth pressure applied to the central wedge is equal to P_A when the shear resistance along the failure plane is considered, and is equal to D_A when only the weight, W_A, and the normal force, R, are considered without the shear resistance. Therefore, D_A should be taken as the driving force and the shear resistance, R_A, as $D_A - P_A$. For each passive wedge, as shown in Fig. 6-14(c), $R_P = P_P - D_P$. The factor of safety is determined by

$$F = \frac{\sum R_A + \sum R_P + c_3 L + W \tan \phi_3}{\sum D_A - \sum D_P} \tag{6-43}$$

in which Σ is the summation over all wedges. For the case shown in Fig. 6-14(a), there are three active wedges and two passive wedges.

Example 6.10 Solve Example 6.9 by the U.S. Navy's method using Eq. (6-43).

$W_A = 0.5 \times 50 \times 50 \times \tan(45° - 12°) \times 125 = 101{,}470\,\text{lb}$
$D_A = 101{,}470 \times \tan(45° + 12°) = 156{,}250\,\text{lb}$
$W_P = 0.5 \times 10 \times 10 \times \tan(45° + 12°) \times 125 = 9{,}624\,\text{lb}$
$D_P = 9624 \times \tan(45° - 12°) = 6{,}250\,\text{lb}$

As determined in Example 6.9, $P_A = 55{,}505$ lb and $P_P = 19{,}747$ lb, so $R_A = 156{,}250 - 55{,}505 = 100{,}745$, and $R_P = 19{,}747 - 6250 = 13{,}497$ lb. From Eq. (6-43),

$$F = \frac{100{,}745 + 13{,}497 + 160 \times 80 + 300{,}000 \times \tan 24°}{156{,}250 - 6{,}250} = \frac{260{,}611}{150{,}000} = 1.737$$

A safety factor of 1.737 by the U.S. Navy's method checks better with the 1.880 by the three-block analysis.

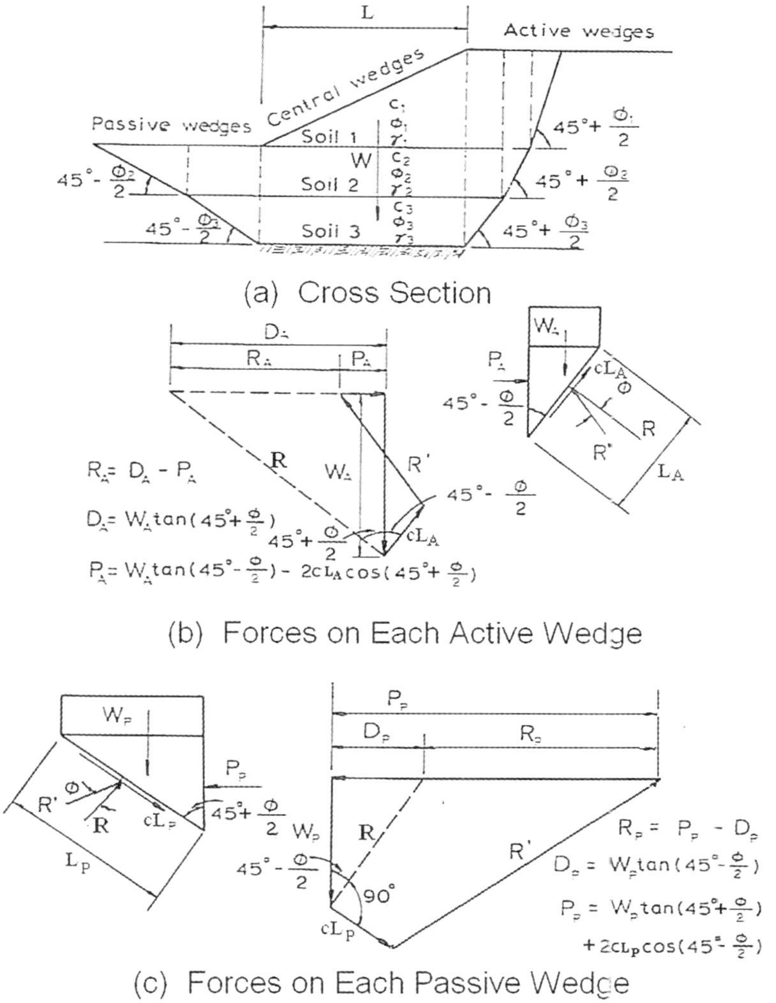

Fig. 6-14. U.S. Navy's method for plane failure (U.S. Navy 1971)

Summary

1. A case of practical interest, which occurs frequently in nature, is the infinite slope where a thin soil overburden is laid above a rock slope of considerable length. The factor of safety for such a slope can be determined by Eq. (6-3) with the pore pressure ratio computed by Eq. (6-4). The most critical failure surface is located at the bottom of the overburden along the rock surface. For a homogeneous embankment with zero cohesion, it is not necessary to use any computer software to determine the minimum

factor of safety, because the most critical failure surface is a shallow circle with a center at infinity, which is similar to an infinite slope, so the minimum factor of safety simply can be computed by Eq. (6-5).

2. When soils or rocks are pushed and placed over a hillside, the new fill will form a triangular cross section. If the natural slope surface is not properly scalped and a layer of weak materials exists at the bottom of the new fill, the most critical failure surface will be a plane through the bottom of the weak layer. The factor of safety for such a fill can be determined by Eq. (6-8) or (6-11), depending on whether the fill height, H, or the fill width, W_f, is given.

3. If a fill is placed partially on a natural slope and partially on horizontal ground, a trapezoidal cross section will be formed. This is a special case of the two-block analysis discussed in Section 6.4. By assuming that the lower block has a horizontal base and the force between the two blocks is horizontal, a simple quadratic equation amenable to hand calculations can be obtained.

4. Sliding-block analysis may be used when a layer of weak material exists at the bottom of each block. In the case of two sliding blocks, a quadratic equation is available to determine the factor of safety for a given ϕ_d, which is the developed friction angle between the two blocks. The coefficients of the quadratic equation can be found in Table 6-1. It is shown that the factor of safety increases with the increase in ϕ_d. The case of $\phi_d = 0$ implies that there is no friction between the two blocks, so the factor of safety is minimum. When $\phi_d = \phi$, where ϕ is the friction angle of the material in the blocks, the factor of safety is maximum. The most correct solution is to find a ϕ_d between 0 and ϕ such that $\tan\phi_d = \tan\phi/F$, where F is the factor of safety. However, this solution requires the use of a graphical method by determining the factor of safety at three different values of ϕ_d. This problem can be solved easily by trial and error using a spreadsheet. In view of the fact that the sliding-block analysis only considers the force equilibrium while neglecting the moment equilibrium completely, it is better to be more conservative by assuming $\phi_d = 0$.

5. For three sliding blocks, a cubic equation is presented to determine the factor of safety for any given ϕ_d. The coefficients of the cubic equation can be found in Table 6-3. The summary presented here for the two sliding blocks also applies to the three sliding blocks. Although manual solutions of three sliding blocks with $\phi_d = 0$ are possible, as illustrated by Example 6.7, it is more convenient to use a spreadsheet, especially when $\tan\phi_d = \tan\phi/F$ is assumed, as demonstrated by Example 6.8.

6. The active and passive earth pressure theory also has been used for the stability analysis of earth slopes. One method uses the central block as a free body with an active force, P_A, on one side and a passive force, P_B, on the other. Due to the difference in definition, the factor of safety ob-

tained by this method is unreasonably high when compared to the three-block analysis. Another method, as proposed by the U.S. Navy, also uses the active and passive earth pressures but in a different way and results in a factor of safety more comparable to that based on the three-block analysis.

Problems

6.1 Consider an infinite slope consisting of 20 ft of soil underlain by bedrock, oriented at 10° from horizontal. The soil has a cohesion of 500 psf, an angle of internal friction of 18°, and a unit weight of 110 pcf. Determine the factor of safety.
[Answer: 3.167]

6.2 If there is seepage in Problem 6.1 and the line of seepage is parallel to the bedrock at a distance of 8 ft below the surface, determine the factor of safety. If there is an earthquake with a seismic coefficient of 0.1 in addition to the seepage, what is the seismic factor of safety?
[Answer: 2.540, 1.603]

6.3 Fig. P6-3 shows a sidehill bench having a total unit weight of 125 pcf and subjected to a pore pressure ratio of 0.05. If the failure surface is a plane along the bottom of the fill where the soil has an effective cohesion of 160 psf and an effective friction angle of 24°, determine the factor of safety.
[Answer: 1.708]

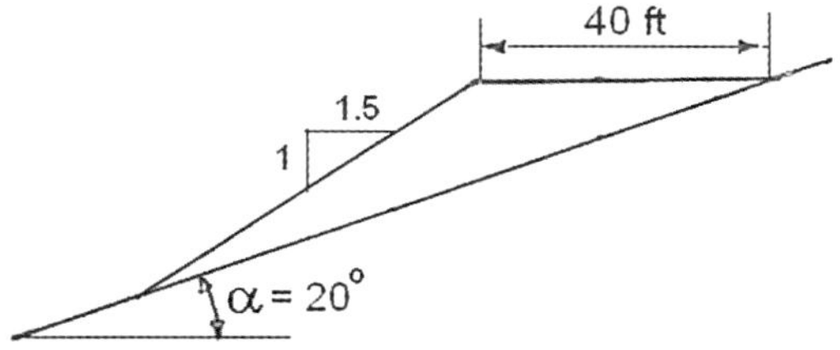

Fig. P6-3.

6.4 Same as Problem 6.3 except that that the degree of the natural slope, α, is steeper than 20°. What is the degree of the natural slope, α, when failure is imminent or the factor of safety is reduced to 1?
[Answer: 29.7°]

6.5 Fig. P6-5 is a trapezoidal fill on a stiff ground. The soil in the fill has a cohesion of 400 psf, an internal friction of 24°, and a total unit weight of 120 pcf. If the fill is divided into two blocks and the force between them is horizontal, determine the factor of safety.
[Answer: 2.784]

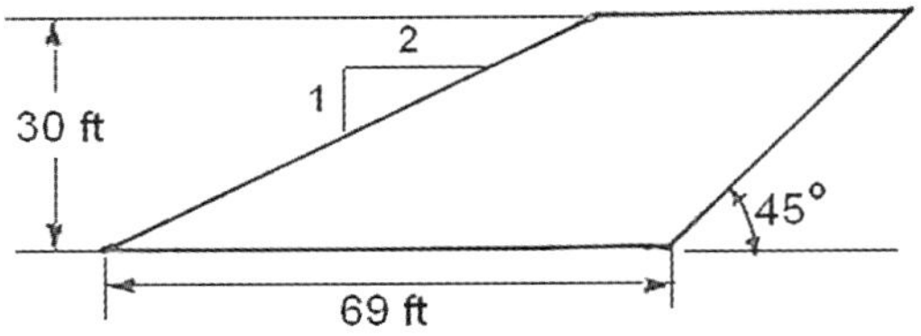

Fig. P6-5.

6.6 Same as Problem 6.5 except that the force between the two blocks inclines at an angle of 30° with the horizontal.
[Answer: 3.562]

6.7 Fig. P6-7 shows a sidehill fill. The soil in the fill has an effective cohesion of 160 psf, an effective friction angle of 24°, and a total unit weight of 125 pcf. It is assumed that the force between the two blocks inclined at an angle, ϕ_d, of 20° with the horizontal and that the fill is subjected to a pore pressure ratio of 0.05 and a seismic coefficient of 0.1. Determine the seismic factor of safety.
[Answer: 1.579]

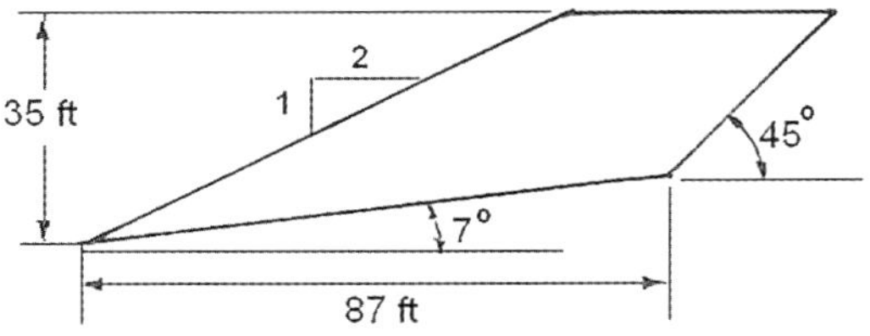

Fig. P6-7.

6.8 Same as Problem 6.7 except that the soil parameters are changed as follows: $c' = 300$ psf, $\phi' = 25°$, and $\gamma = 120$ pcf.
[Answer: 1.918]

6.9 Fig. P6-9 shows the dimensions of the fill. The soil at the bottom of the fill has a cohesion of 500 psf, a friction angle of 20°, and a total unit weight of 125 pcf. By dividing the fill into three blocks and assuming no friction between the blocks, determine the factor of safety.
[Answer: 2.354]

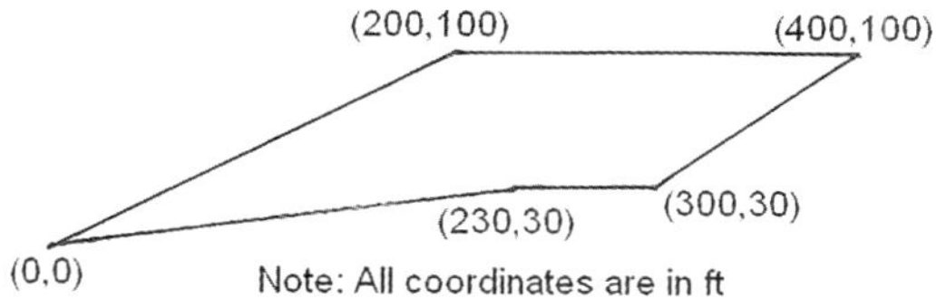

Fig. P6-9.

6.10 Same as Problem 6.9 except that the friction angle developed between two blocks is ϕ_d, where $\phi_d = \tan^{-1}(\tan 20°/F)$ and F is the factor of safety to be determined.
[Answer: 2.390]

6.11 Fig. P6-11 shows a slope divided into three blocks. The soil has a cohesion of 500 psf, a friction angle of 20°, and a total unit weight of 125 pcf. By assuming the forces between two blocks as horizontal, determine the factor of safety by the conventional three-block analysis.
[Answer: 1.767]

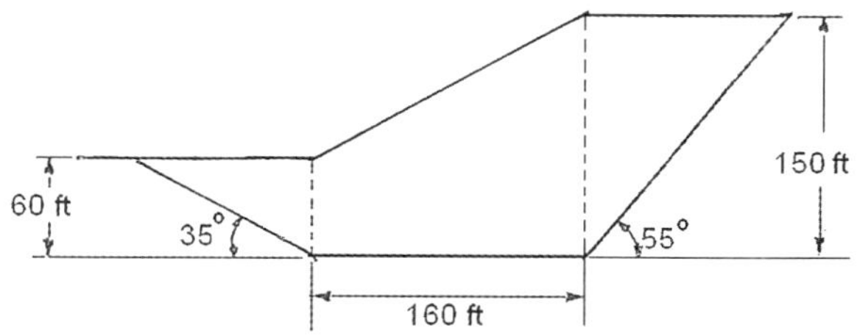

Fig. P6-11.

6.12 Same as Problem 6.11, except that the friction angle developed between two blocks is ϕ_d, where $\phi_d = \tan^{-1}(\tan 20°/F)$.
[Answer: 1.987]

6.13 Same as Problem 6.11, but determine the factor of safety by using the earth pressure method and considering the central block as a free body.
[Answer: 21.193]

6.14 Same as Problem 6.11, but determine the factor of safety by the U.S. Navy's method.
[Answer: 1.682]

6.15 Fig. P6-15 shows a slope divided into three blocks. The soil has a cohesion of 500 psf, a friction angle of 20°, and a total unit weight of 125 pcf. By assuming the forces between two blocks as horizontal, determine the factor of safety by the conventional three-block analysis.
[Answer: 1.651]

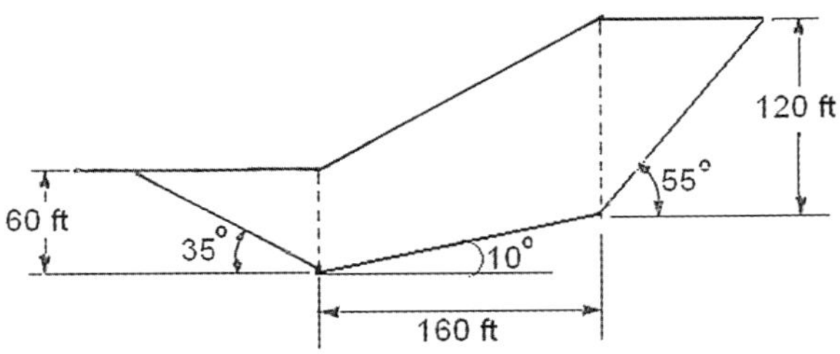

Fig. P6-15.

6.16 Same as Problem 6.15 except that the friction angle developed between two blocks is ϕ_d, where $\phi_d = \tan^{-1}(\tan 20°/F)$.
[Answer: 1.848]

6.17 Based on Rankine's theory and Mohr's envelope, derive Eqs. (6-39) and (6-40).

6.18 Based on Coulomb's theory and the force diagrams presented in Fig. 6-14, prove the following equations:

$$P_A = W_A \tan(45° - \phi/2) - 2cL_A \cos(45° + \phi/2)$$

$$P_P = W_P \tan(45° + \phi/2) + 2cL_P \cos(45° - \phi/2)$$

Chapter 7

Stability Charts and Other Solutions

Since Taylor (1937) first published his stability charts, various charts have been presented by Bishop and Morgenstern (1960), Morgenstern (1963), and Spencer (1967). These charts are applicable only for cylindrical failure surfaces and will be discussed in this chapter. Also included are the charts developed by Huang for triangular and trapezoidal fills on rock or stiff slopes (1977a), triangular fills on soil slopes (1978b), and earth dams and embankments (1975), and some new charts for the effective stress analysis of nonhomogeneous dams. With simple calculations, these charts can provide a quick answer to the safety factor of a proposed slope and are therefore particularly useful for preliminary design and estimating purposes. The well-known friction circle and logarithmic-spiral methods originally developed by Taylor (1937) also are presented.

7.1 Homogeneous Slopes with $\phi = 0$

Fig. 7-1 shows the stability chart for $\phi = 0$ analysis of a simple slope (Taylor, 1937, 1948). The slope has an angle, β, a height, H, and a ledge at a depth of DH below the toe, where D is a depth ratio, the depth to bedrock divided by the height of the slope.

The chart can be used to determine not only the developed cohesion, c_d, as shown by the solid curves but also nH, which is the distance from the toe to the

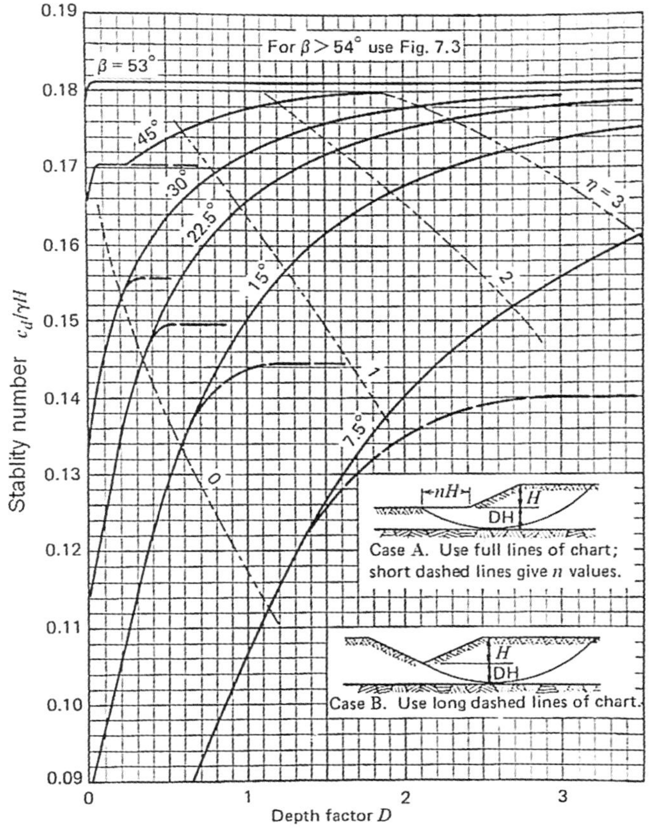

Fig. 7-1. Stability chart for soils with zero friction angle (Taylor 1937, 1948)

failure circle, as indicated by the short dashed curves. When $\phi = 0$, the most critical circle is always tangent to the rock. If there is no loading outside the toe, the most critical circle is a midpoint circle with its center on a vertical line through the midpoint of the slope, as indicated by case A in the figure. If there are loadings outside the toe, the most critical circle is a toe circle, as shown by case B, and the long dashed curves should be used. If the curve falls on the left side of the $n = 0$ line, the most critical circle is a slope circle, which intersects the slope surface and does not pass below the toe, so the loading outside the toe has no effect on the developed cohesion. A slope circle occurs only when the ledge is at the same elevation as or closely below the toe.

Given β and D, a stability number defined as $c_d/\gamma H$, where γ is the total unit weight and H is the height of the slope, can be found from the chart, so the amount of cohesion actually developed can be determined. The factor of safety can be obtained by dividing the allowable cohesion, or the shear strength, with the developed cohesion. Note that the greater the stability number, the smaller the factor of safety.

Example 7.1 Fig. 7-2 shows a simple slope with a height, H, of 40 ft (12.2 m), a slope angle, β, of 22.5°, and a ledge 60 ft (18.3 m) below the toe. If the soil has a cohesion, c, of 1,200 psf (57.5 kPa), and a total unit weight, γ, of 120 pcf (18.9 kN/m^3), determine the factor of safety and the distance from the toe to the point where the most critical circle appears on the ground surface. What is the factor of safety if there are heavy loadings outside the toe?

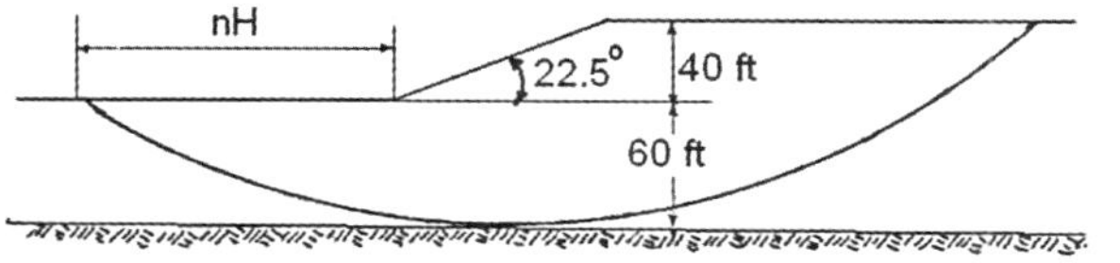

Fig. 7-2. Example 7.1

Solution For $D = 60/40 = 1.5$ and $\beta = 22.5°$, from the solid curve in Fig. 7-1, $c_d/\gamma H = 0.1715$, or $c_d = 0.1715 \times 120 \times 40 = 823.2$ psf. The factor of safety is $F = c/c_d = 1{,}200/823.2 = 1.46$. From the short dashed curve, $n = 1.85$, or the distance between the toe and the failure circle $nH = 1.85 \times 40 = 74$ ft.

When there are loadings outside the toe, the point falls on the horizontal portion of the long dashed curve with $c_d/\gamma H = 0.1495$, or $c_d = 0.1495 \times 120 \times 40 = 717.6$ psf. The factor of safety is $F = 1{,}200/717.6 = 1.67$.

7.2 Homogeneous Slopes with Both c and ϕ

By the use of the friction circle method, as described in Section 7.9, Taylor (1937, 1948) determined the developed cohesion, c_d, for a given developed friction angle, ϕ_d, and plotted a series of curves shown in Fig. 7-3. When the friction angle is not zero, the most critical circle is a shallow circle. If the ledge lies at a considerable depth below the toe, the location of the ledge, as indicated by the depth ratio, D, should have no effect on the developed cohesion.

In Fig. 7-3, the most critical circle may pass through the toe, designated as case 1 and shown by the solid curves, or pass below the toe, designated as case 2 and shown by the long dashed curves. However, if $D = 0$, the most critical circle will lie above the toe, designated as case 3 and shown by the short dashed curves. The figure can be used to determine the factor of safety with respect to cohesion, F_c, by assuming that the angle of internal friction is developed fully, or to determine the factor of safety with respect to internal friction, F_ϕ, by assuming that the cohesion is developed fully. To find the factor of safety with respect to shear strength, F, a trial-and-error or graphical method is required, as illustrated in Example 7.2.

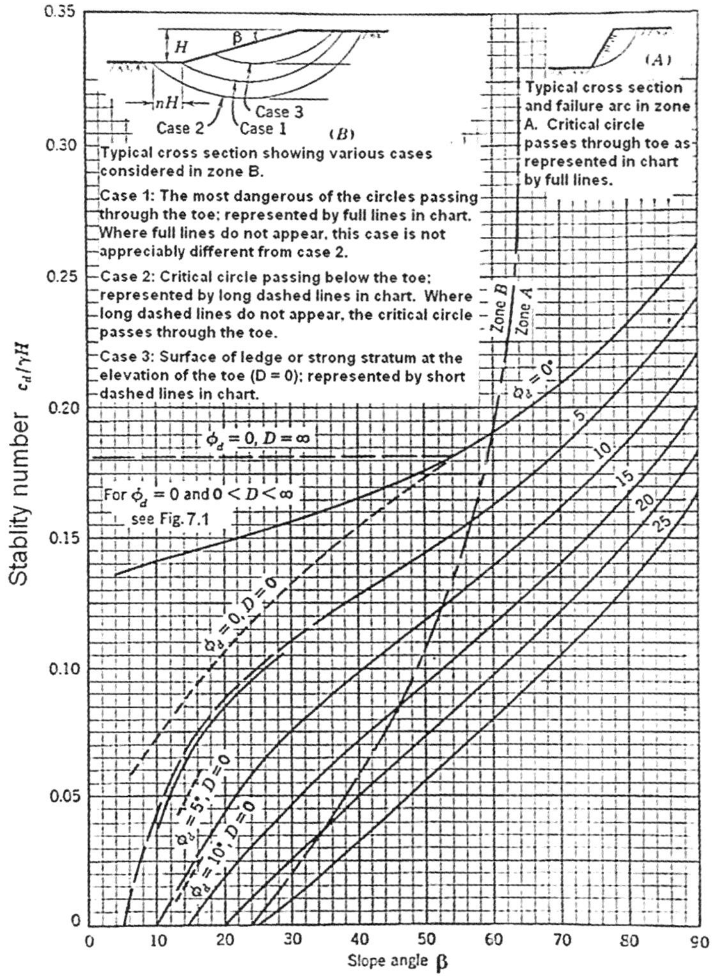

Fig. 7-3. Stability chart for soils with friction angle (Taylor 1937, 1948)

Example 7.2 Given $H = 40$ ft (12.2 m) and $\beta = 30°$ and the ledge is far away from the surface. If the soil has a cohesion, c, of 800 psf (38.3 kPa), a friction angle, ϕ, of 10°, and a total unit weight, γ, of 100 pcf (15.7 kN/m^3), determine F_c, F_ϕ, and F.

Solution Assume that the angle of internal friction is developed fully, or $\phi_d = 10°$. From Fig. 7-3, for $\beta = 30°$, $c_d/\gamma H = 0.075$, or $c_d = 0.075 \times 100 \times 40 = 300$ psf, so $F_c = c/c_d = 800/300 = 2.67$.

Next, assume that the cohesion is developed fully, or $c_d/\gamma H = 800/(100 \times 40) = 0.2$. It can be seen from Fig. 7-3 that when $c_d/\gamma H = 0.2$ and $\beta = 30°$, the developed friction angle is less than zero, or the factor of safety with respect to internal friction is infinity. This situation occurs when the resisting moment because of cohesion is greater than the driving moment.

To determine the factor of safety with respect to shear strength, the same factor of safety should be applied to both cohesion and internal friction. A value of F_c is assumed, and a value of F_ϕ, which is equal to $\tan\phi/\tan\phi_d$, is determined from the chart. By trial and error, the factor of safety with respect to shear strength is obtained when $F_c = F_\phi$. Instead of trial and error, a graphical method may be used by plotting F_c versus F_ϕ and finding its intersection with a 45° line, as shown in Fig. 7-4.

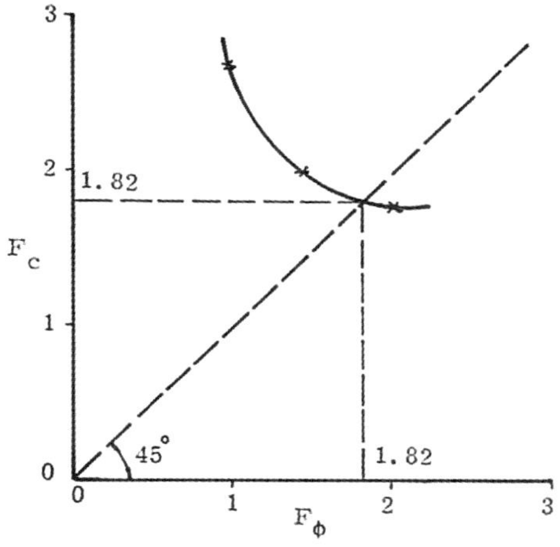

Fig. 7-4. *Factor of safety with respect to shear strength*

In the figure, one point on the F_c versus F_ϕ curve was determined previously as $F_c = 2.67$ and $F_\phi = 1$. It is necessary to have two more points in order to plot the curve. First, assume $F_c = c/c_d = 2$, or $c_d = 800/2 = 400$ psf. For $c_d/\gamma H = 400/(100 \times 40) = 0.1$, from Fig. 7-3, $\phi_d = 7°$, or $F_\phi = \tan 10°/\tan 7° = 1.44$. Next, assume $F_c = 1.8$, or $c_d = 800/1.8 = 444$ psf. For $c_d/\gamma H = 444/(100 \times 40) = 0.111$ from Fig. 7-3, $\phi_d = 5$ or $F_\phi = \tan 10°/\tan 5° = 2.02$. Fig. 7-4 shows the plot of the three points. The factor of safety with respect to shear strength is 1.82.

A variational limiting equilibrium approach was used by Leshchinski and San (1994) to develop seismic stability charts for a simple slope, as shown in Fig. 7-5. This case is similar to Taylor's but includes the seismic coefficient, C_s, as a variable. Because most of their charts are concerned with slopes steeper than 1:1, only two charts with slopes of 1:1 and 2:1 are of practical interest to earth slopes and therefore presented here.

A particular feature of the chart is the dimensionless parameter, λ, defined as

$$\lambda = \frac{1}{\gamma H}\frac{c}{\tan\phi} \tag{7-1}$$

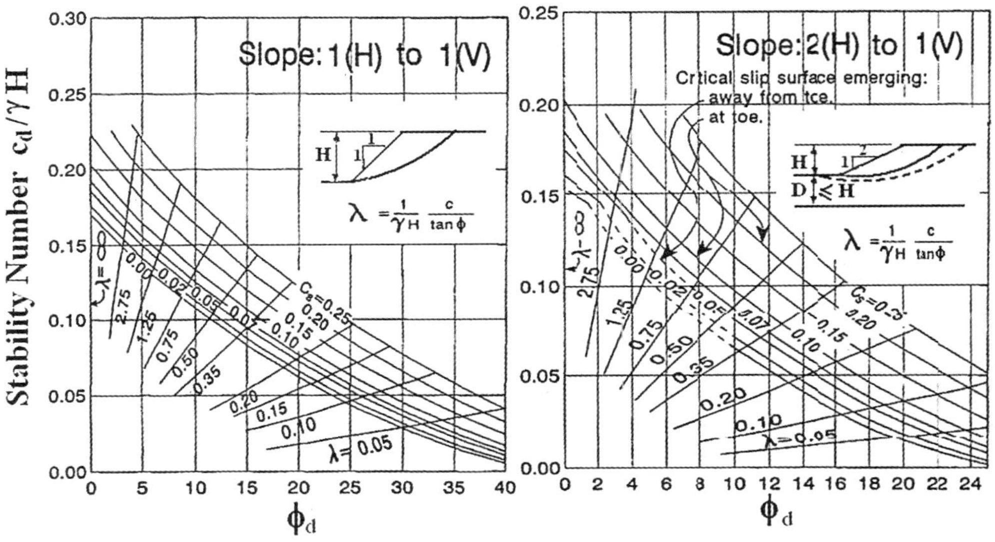

Fig. 7-5. Seismic stability charts for a simple slope (Leshchinski and San 1994. Reproduced with permission)

Without the use of λ, the chart is similar to Taylor's by assuming a value of ϕ_d and finding a corresponding value of c_d. Because the factors of safety, F_ϕ and F_c, are not equal, an iteration or graphical method must be used to determine the factor of safety with respect to shear strength. With the use of λ, the factor of safety with respect to shear strength can be determined directly without the need of iterations. Given the values of λ and C_s, a vertical line can be drawn from their point of intersection to determine ϕ_d, and a horizontal line to determine c_d. The two factors of safety, one based on F_ϕ and the other based on F_c, are automatically equal, so either one can be used to determine the factor of safety with respect to shear strength.

Example 7.3 Given a 2:1 slope with $H = 100$ ft (30.5 m), $c = 1000$ psf (48 kN/m^2), $\phi = 18°$, and $\gamma = 125$ pcf (19.7 kN/m^3), determine the factor of safety with respect to shear strength for $C_s = 0$ and $C_s = 0.2$.

Solution From Eq. (7-1), $\lambda = 1{,}000/(125 \times 100 \times \tan 18°) = 0.25$.

When $C_s = 0$: From Fig. 7-5, read vertically $\phi_d = 12°$, or $F = \tan 18°/\tan 12° = 1.53$. If read horizontally, $c_d/\gamma H = 0.051$, or $c_d = 0.051 \times 125 \times 100 = 637.5$ psf. $F = 1{,}000/637.5 = 1.56$, which checks with the 1.53 based on ϕ_d.

When $C_s = 0.2$: From Fig. 7-5, read vertically $\phi_d = 17.4°$, or $F = \tan 18°/\tan 17.4° = 1.04$. If read horizontally, $c_d/\gamma H = 0.078$, or $c_d = 0.078 \times 125 \times 100 = 975$ psf. $F = 1{,}000/975 = 1.03$, which checks with the 1.04 based on ϕ_d.

7.3 Triangular Fills on Rock or Stiff Slopes

Fig. 7-6 shows a triangular fill on a rock or stiff slope. The fill has a height, H, an angle of outslope, β, and a degree of natural slope, α. The natural slope is assumed to be much stiffer than the fill, so the failure surface will lie entirely within the fill. When $\alpha = 0$, the fill is placed on a level ground.

Based on the Fellenius or normal method, the factor of safety can be computed by Eq. (2-3). By dividing the equation into two parts, one due to cohesion and the other due to internal friction, Eq. (2-3) can be written as

$$F = \frac{\sum_{i}^{n} cb_i \sec\theta_i}{\sum_{i}^{n} W_i \sin\theta_i} + \frac{\sum_{1}^{n} W_i \cos\theta_i \tan\phi}{\sum_{1}^{n} W_i \sin\theta_i} \tag{7-2}$$

Because W_i is proportional to γH, Eq. (7-2) can be simplified as

$$F = \frac{cN_c}{\gamma H} + (\tan\phi)N \tag{7-3}$$

in which N_c and N are functions of geometry, independent of soil parameters. To apply Eq. (7-3), the location of the most critical circle must be known a priori. By assuming that the soil has cohesion, c, but no angle of internal friction, ϕ, the location of the most critical circle can be determined and used to evaluate N_c and N. This assumption is correct for total stress analysis with $\phi = 0$ but gives a slightly larger factor of safety for effective stress analysis with both c' and ϕ', because the most critical circle for soils with both cohesion and friction is different from that with cohesion only.

Values of N_c can be computed by LEAME or any other computer programs by assuming $H = 10$, $\gamma = 100$, $c = 1{,}000$, and $\phi = 0$. The factor of safety obtained by LEAME is actually the value of N_c, as can be seen by substituting these parameters into Eq. (7-3). Similarly, by assuming $c = 0$ and $\phi = 45°$, the factor of safety

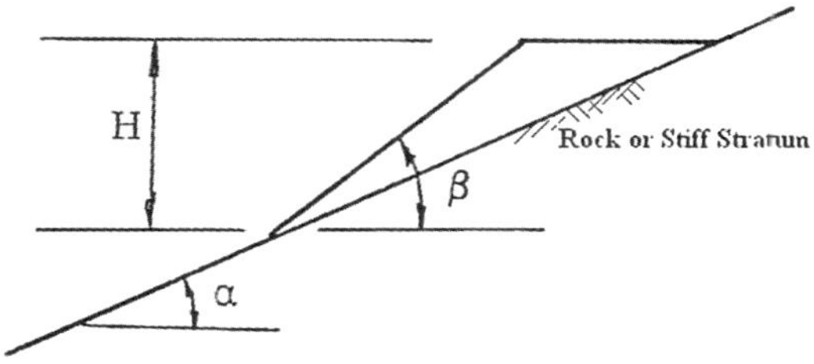

Fig. 7-6. *Triangular fill on rock slope*

is equal to N. Instead of using Eq. (7-3) directly, Huang (1977a, 1978b) used the following equation to determine F:

$$F = N_c\left[\frac{c}{\gamma H} + \frac{\tan\phi}{N_f}\right] \tag{7-4}$$

in which N_c = cohesion factor and N_f = friction factor defined as

$$N_f = \frac{N_c}{N} \tag{7-5}$$

Values of N_c and N_f for various combinations of α and β are presented in Figs. 7-7 and 7-8, respectively. Knowing N_c and N_f, the factor of safety can be computed by Eq. (7-4). For effective stress analysis with a pore pressure ratio, r_u, Eq. (7-4) becomes

$$F = N_c\left[\frac{c'}{\gamma H} + (1 - r_u)\frac{\tan\phi'}{N_f}\right] \tag{7-6}$$

Example 7.4 Given a triangular fill with $H = 30$ ft (9.1 m), $c = 800$ psf (38.3 kN/m^2), $\phi = 0$, $\gamma = 125$ pcf (19.7 kN/m^3), $\alpha = 15°$, and $\beta = 30°$, determine the factor of safety. If the degree of natural slope, α, is 0, determine the factor of safety and compare with that obtained by Fig. 7-1.

Solution With $\alpha = 15°$, and $\beta = 30°$, from Fig. 7-7, $N_c = 10.6$. With $H = 30$ ft, $c = 800$ psf, and $\gamma = 125$ pcf, from Eq. (7-4), $F = 800 \times 10.6/(125 \times 30) = 2.261$.

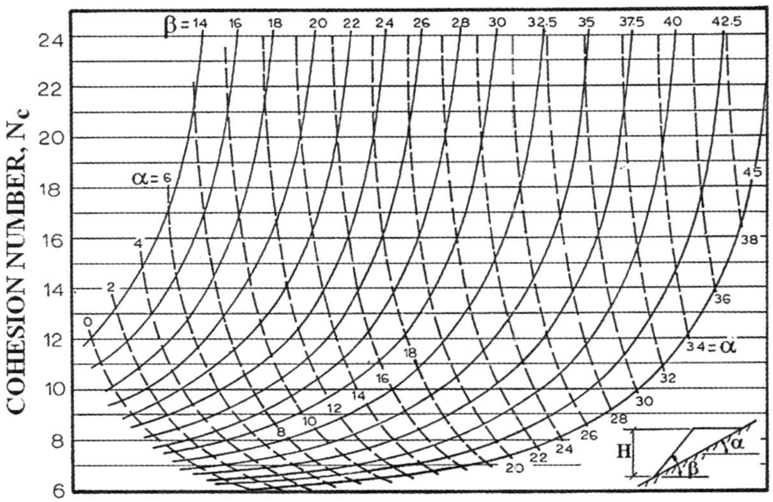

Fig. 7-7. Cohesion number for triangular fills (Huang 1977a. Reproduced with permission)

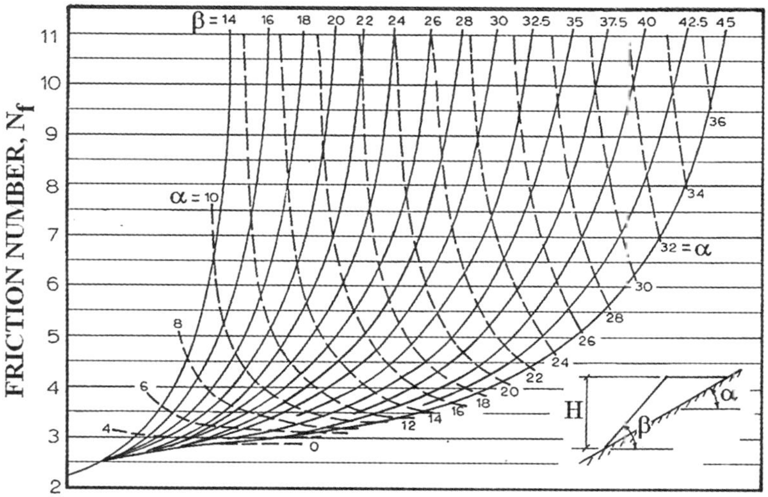

Fig. 7-8. Friction number for triangular fills (Huang 1977a. Reproduced with permission)

If $\alpha = 0$ and $\beta = 30°$, from Fig. 7-7, $N_c = 7.5$. From Eq. (7-4), $F = 800 \times 7.5/(125 \times 30) = 1.6$. With $D = 0$ and $\beta = 30°$, from Fig. 7-1, $c_d/(\gamma H) = 0.134$, so $c_d = 0.134 \times 125 \times 30 = 502.5$, and $F = 800/502.5 = 1.592$, which checks with the 1.6 obtained from Fig. 7-7.

When $\phi = 0$, Eq. (7-4) in conjunction with Fig. 7-7 should give a factor of safety very close to the minimum factor of safety obtained by LEAME, because the most critical circle actually is used to determine N_c. However, the use of the same circle to evaluate N_f is not theoretically correct and results in a factor of safety that is too high. Although a slightly higher factor of safety may be desirable and closer to the simplified Bishop method, to obtain more accurate and conservative results, a correction factor may be applied to the factor of safety obtained from Eq. (7-4).

Fig. 7-9 gives the correction factor, C_f, for the factor of safety computed by Eq. (7-4). The correction factor depends on the angle of outslope, β, the degree of natural slope, α, and the percent of cohesion resistance, P_c, which is defined as

$$P_c = \frac{\dfrac{c'}{\gamma H}}{\dfrac{c'}{\gamma H} + (1 - r_u)\dfrac{\tan \phi'}{N_f}} \tag{7-7}$$

The curves in Fig. 7-9 were obtained from a series of analyses by comparing the factor of safety from Eq. (7-6) with that from the LEAME computer program. The corrected factor of safety is the product of the correction factor and the factor

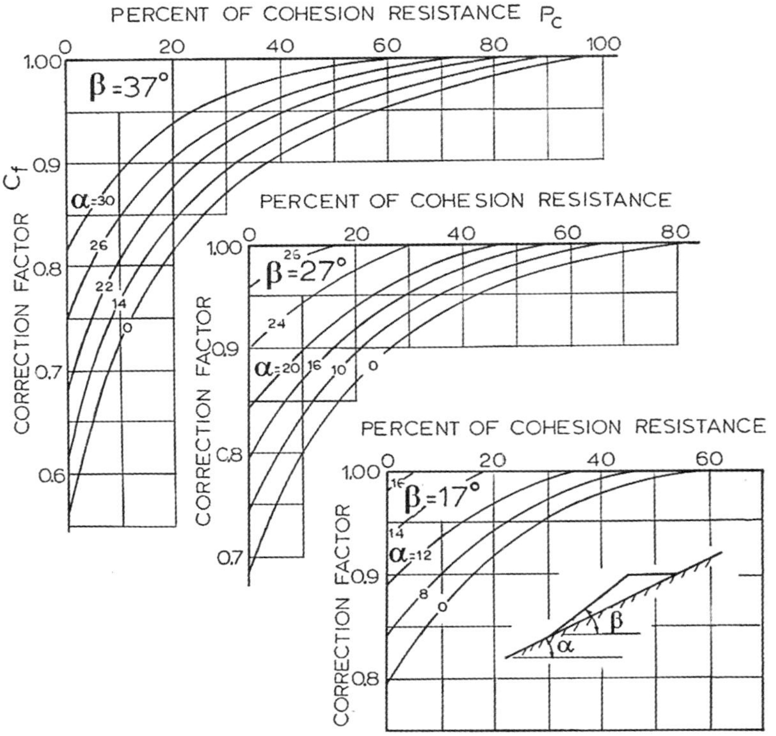

Fig. 7-9. Chart for correcting factor of safety (Huang 1977a. Reproduced with permission)

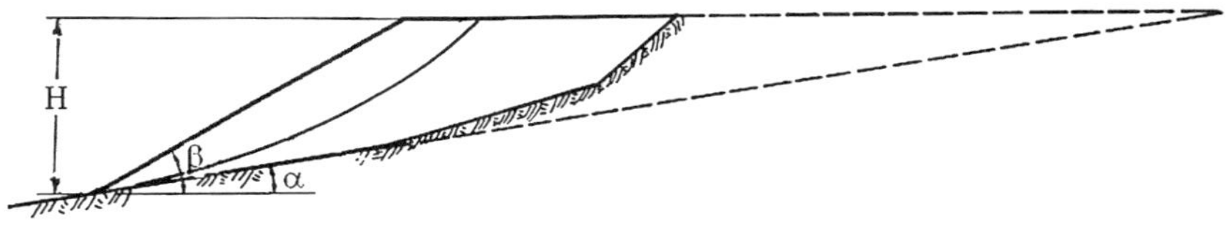

Fig. 7-10. Approximating of hollow fill by triangular fill

of safety from Eq. (7-6). Although Fig. 7-9 only gives the correction factor for three values of β, that is, 37°, 27°, and 17°, the correction factor for other values of β can be obtained by a straight-line interpolation.

Although the method presented here is based on the triangular cross section shown in Fig. 7-6, it also can be applied to the effective stress analysis of the slope shown in Fig. 7-10, where the rock slope is quite irregular. In such a case, α is the degree of natural slope at the toe. Because of the small cohesion used in effective stress analysis, the failure surface will be a shallow circle close to the surface of the slope near the toe and is independent of all slopes behind the toe.

Example 7.5 Fig. 7-11 shows the cross section of a triangular fill and the coordinates of boundary lines in parentheses. The soil has a cohesion of 200 psf (9.6 kN/m^2), a friction angle of 30°, and a total unit weight of 125 pcf (19.7 kN/m^3). If there is no seepage, determine the factor of safety by the normal method.

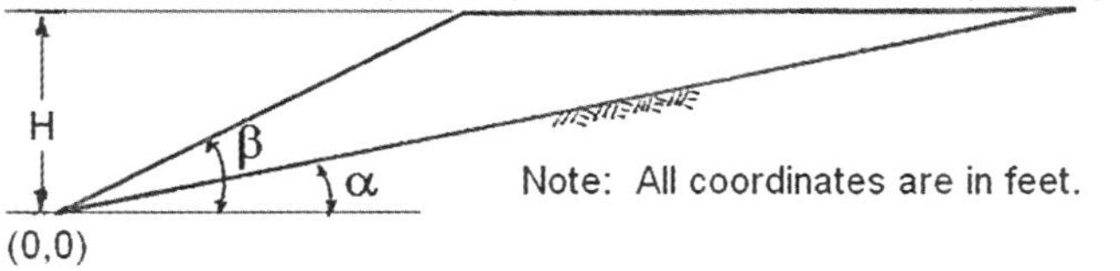

Fig. 7-11. *Example 7.5*

Solution From the coordinates, $H = 50$ ft (15.2 m), $\beta = \tan^{-1}(50/100) = 26.6°$, and $\alpha = \tan^{-1}(50/250) = 11.3°$. From Fig. 7-7, $N_c = 10.7$ and, from Fig. 7-8, $N_f = 4$. From Eq. (7-4), $F = 10.7[200/(125 \times 50) + (\tan 30°)/4] = 10.7(0.032 + 0.144) = 1.88$. From Eq. (7-7), $P_c = 0.032/(0.032 + 0.144) = 0.18$, or 18%. With $\beta = 26.6°$ and $\alpha = 11.3°$, from Fig. 7-9, $C_f = 0.89$. $F = 0.89 \times 1.88 = 1.67$. The factor of safety obtained by LEAME using the normal method is 1.663, which checks with the 1.67 obtained from the charts.

7.4 Trapezoidal Fills on Rock or Stiff Slopes

Fig. 7-12 shows the cross section of a trapezoidal fill with a height, H, an outslope S:1, an angle of natural slope, α, and a base width BH, where B is a ratio between base width and height. The triangular fill is a special case of trapezoidal fill when $B = 0$. By using the same procedure as in triangular fills, values of N_c and N_f for various combinations of S, B, and α are computed and presented in Figs. 7-13 and 7-14, respectively. Knowing N_c and N_f, the factor of safety can be computed by Eq. (7-6).

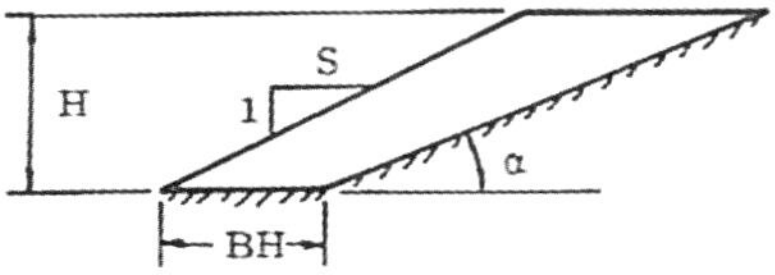

Fig. 7-12. *Trapezoidal fill on rock slope*

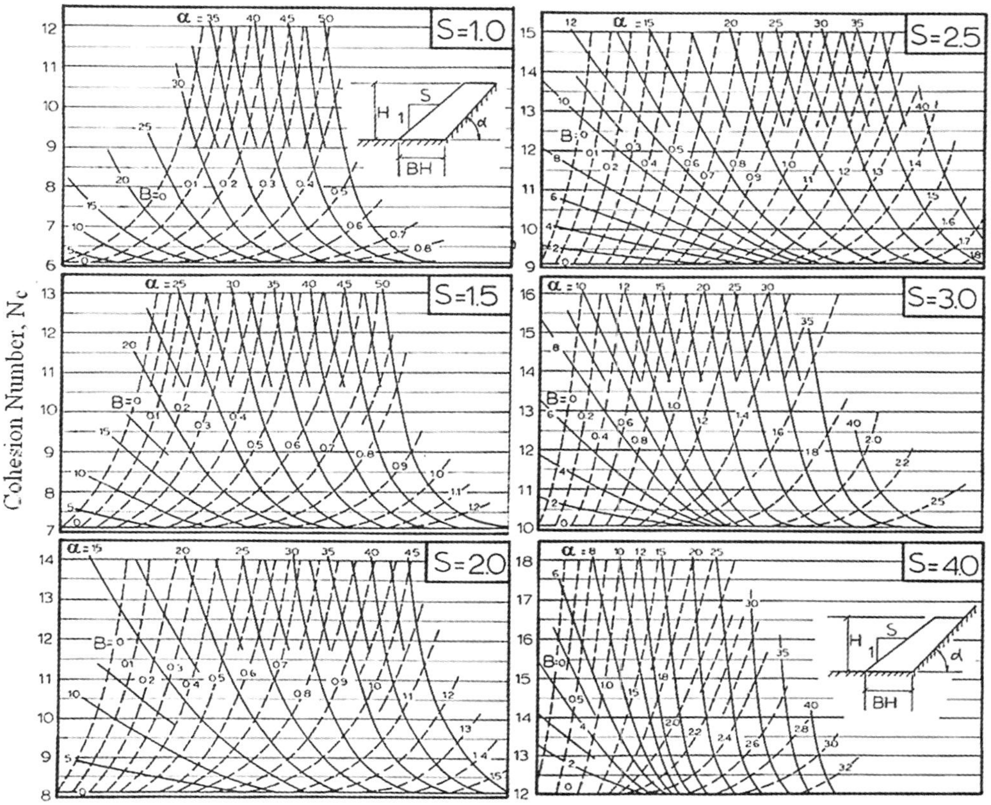

Fig. 7-13. Cohesion number for trapezoidal fills (Huang 1977a. Reproduced with permission)

Example 7.6 Given a trapezoidal fill with $S = 2$, $H = 30\,\text{ft}$ (9.1 m), $B = 0.4$, $\alpha = 15°$, $c = 800\,\text{psf}$ ($38.3\,\text{kN/m}^2$), and $\gamma = 125\,\text{pcf}$ ($19.7\,\text{kN/m}^3$), determine the factor of safety. Determine the factor of safety if $\alpha = 0$ and compare with that obtained by Fig. 7-1.

Solution With $S = 2$, $H = 30\,\text{ft}$, $B = 0.4$, $\alpha = 15°$, from Fig. 7-13, $N_c = 10.4$. From Eq. (7-4), $F = (800 \times 10.4)/(125 \times 30) = 2.22$.

If $\alpha = 0$, from Fig. 7-13, $N_c = 8.1$. From Eq. (7-4), $F = 800 \times 8.1/(125 \times 30) = 1.73$. From Fig. 7-1, with $D = 0$ and $\beta = \tan^{-1}0.5 = 26.6°$, $c_d/(\gamma H) = 0.124$, so $c_d = 0.124 \times 125 \times 30 = 465$ and $F = 800/465 = 1.72$, which checks with the 1.73 obtained from Fig. 7-13.

The friction number, N_f, is based on the circle with a center and radius the same as the most critical circle for $\phi' = 0$. In other words, the same circle is used for determining N_c and N_f. If α or c' is very small, the most critical circle for $\phi' = 0$ may be quite different from that for $\phi' \neq 0$, so a trial-and error procedure,

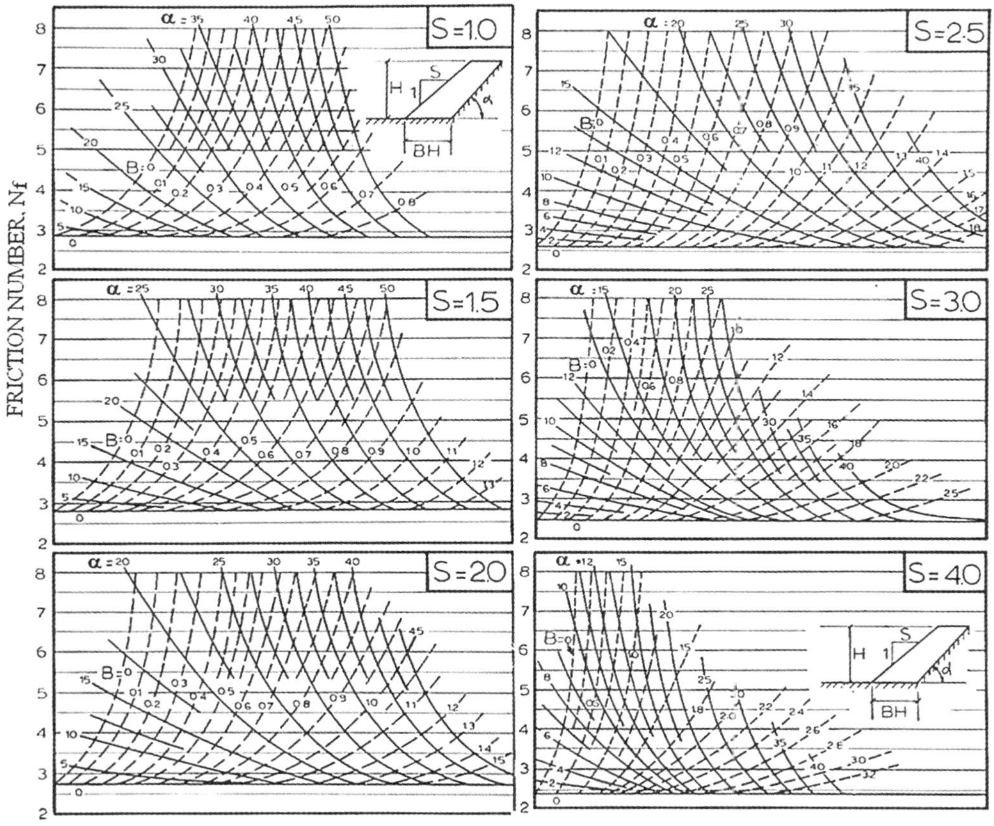

Fig. 7-14. Friction number for trapezoidal fills (Huang 1977a. Reproduced with permission)

as described in the following examples, should be used to determine the minimum factor of safety.

Example 7.7 Fig. 7-15 shows the cross section of a trapezoidal fill and the coordinates of boundary lines in parentheses. The soil has an effective cohesion of 200 psf (9.6 kN/m^2), an effective friction angle of 30°, and a total unit weight of 125 pcf (19.7 kN/m^3). If there is no seepage, determine the factor of safety.

Solution With $S = 2$, $B = 60/50 = 1.2$, and $\alpha = \tan^{-1}(50/90) = 29°$, from Fig. 7-13, $N_c = 8.5$ and, from Fig. 7-14, $N_f = 3.25$. From Eq. (7-6), $F = 8.5[200/(125 \times 50) + (\tan 30°)/3.25 = 1.78$. This factor of safety is based on the assumption that the most critical circle is tangent to the rock surface. For the effective stress analysis with a small cohesion and a large friction angle, the most critical circle may be a shallow circle, so several different locations of rock surfaces must be assumed to determine which gives the minimum factor of safety.

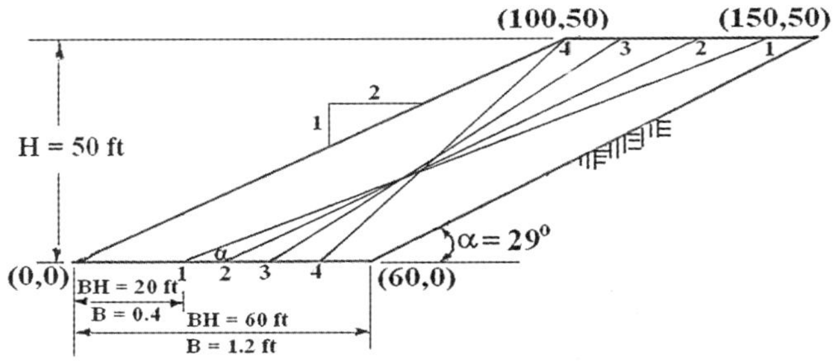

Fig. 7-15. Example 7.7

Four different values of *B*, that is, 0.4, 0.6, 0.8, and 1.0, are assumed. For each *B*, select the largest α that can be obtained from the charts, as shown in the following table:

Trial No.	*B*	α	N_c	N_f	*F*
1	0.4	22.5°	13.9	6.25	1.73
2	0.6	27.5°	13.3	6.4	1.63
3	0.8	35°	13.4	6.25	1.66
4	1.0	45°	13.7	5.9	1.77

It can be seen that the minimum factor of safety is 1.63 and occurs at trial No. 2. The factors of safety obtained by LEAME are 1.690 based on the simplified Bishop method and 1.625 based on the normal method.

Example 7.8 Fig. 7-16 shows an embankment on a horizontal ledge. The embankment is 50 ft (15.3 m) high with a side slope of 1.5:1. The soil has an effective cohesion of 300 psf (14.4 kN/m^2), an effective friction angle of 30°, and a total unit weight of 120 pcf (18.8 kN/m^3). The pore pressure ratio is assumed to be 0.1. Determine the factor of safety (a) using Figs. 7-7, 7-8, and 7-9, and (b) by a trial-and-error method using Figs. 7-13 and 7-14.

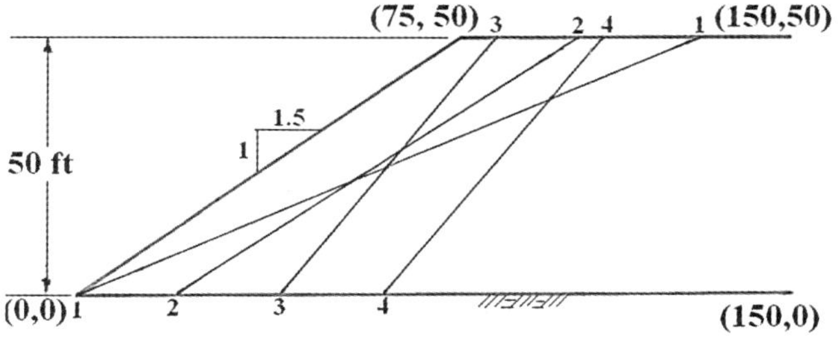

Fig. 7-16. Example 7.8

Solution (a) With $\alpha = 0$ and $\beta = \tan^{-1}(1/1.5) = 33.7°$, from Fig. 7-7, $N_c = 7.1$ and from Fig. 7-8, $N_f = 2.9$. From Eq. (7-6), $F = 7.1 \times [300/(120 \times 50) + (1 - 0.1)(\tan 30°)/2.9] = 7.1 \times (0.05 + 0.179) = 1.626$. From Eq. (7-7), $P_c = 0.05/(0.05 + 0.179) = 0.218$, or 21.8%. From Fig. 7-9, when $\beta = 27°$, $C_f = 0.87$ and, when $\beta = 37°$, $C_f = 0.82$. When $\beta = 33.7°$, $C_f = 0.87 - (33.7 - 27)(0.87 - 0.82)/10 = 0.837$. The factor of safety is $F = 0.837 \times 1.626 = 1.36$. The factor of safety by LEAME is 1.359 by the normal method.

(b) Starting from $B = 0$ and ending at $B = 1.2$, four values of B are tried, each with the largest possible α as can be obtained from the charts. Values of N_c are obtained from Fig. 7-13, N_f from Fig. 7-14, and F from Eq. (7-6). The results are presented in the following table:

Trial No.	B	α	N_c	N_f	F
1	0	22.5°	12.3	5.8	1.72
2	1.4	32.5°	12.0	6.7	1.53
3	0.8	50°	10.7	6.6	1.38
4	1.2	50°	7.4	3.6	1.44

It can be seen that the minimum factor of safety is 1.38, which checks with the 1.359 obtained from LEAME based on the normal method.

7.5 Effective Stress Analysis of Homogeneous Dams

Four sets of charts are presented in this section, each having their own advantages and limitations. A typical cross section of a dam is shown in Fig. 7-17. The dam has a height H and a ledge at a depth of DH below the toe, where D is the depth ratio.

Bishop's and Morgenstern's charts can be used to find the factors of safety at different D's and determine which is the minimum. Morgenstern's charts assume $D = 0$, whereas both Spencer's and Huang's charts assume $D = \infty$. Because the most critical failure surface in an effective stress analysis is usually a very shallow circle, the effect of D is not very significant. The use of $D = \infty$ always gives a slightly lower factor of safety and is therefore on the safe side. The slope of the dam is expressed as S:1, except in Spencer's charts where the slope angle β is used.

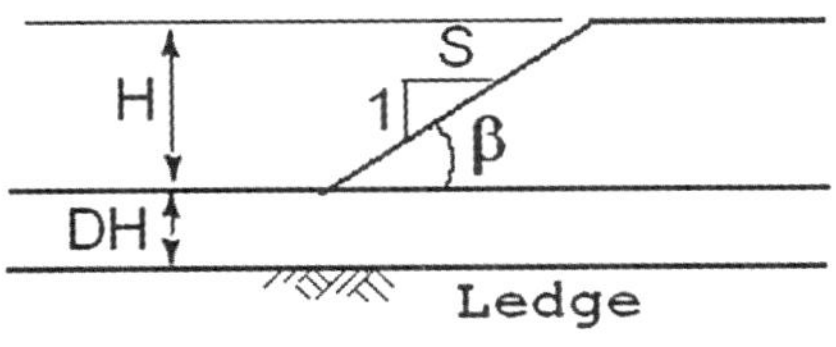

Fig. 7-17. Typical cross section of dams

7.5.1 Bishop's and Morgenstern's Charts

Fig. 7-18 shows the stability charts for effective stress analysis when $c'/\gamma H = 0$ and 0.025, and Fig. 7-19 shows those when $c'/\gamma H = 0.05$. The factor of safety is based on the simplified Bishop method (Bishop 1955) and can be expressed as

$$F = m - r_u n \qquad (7\text{-}8)$$

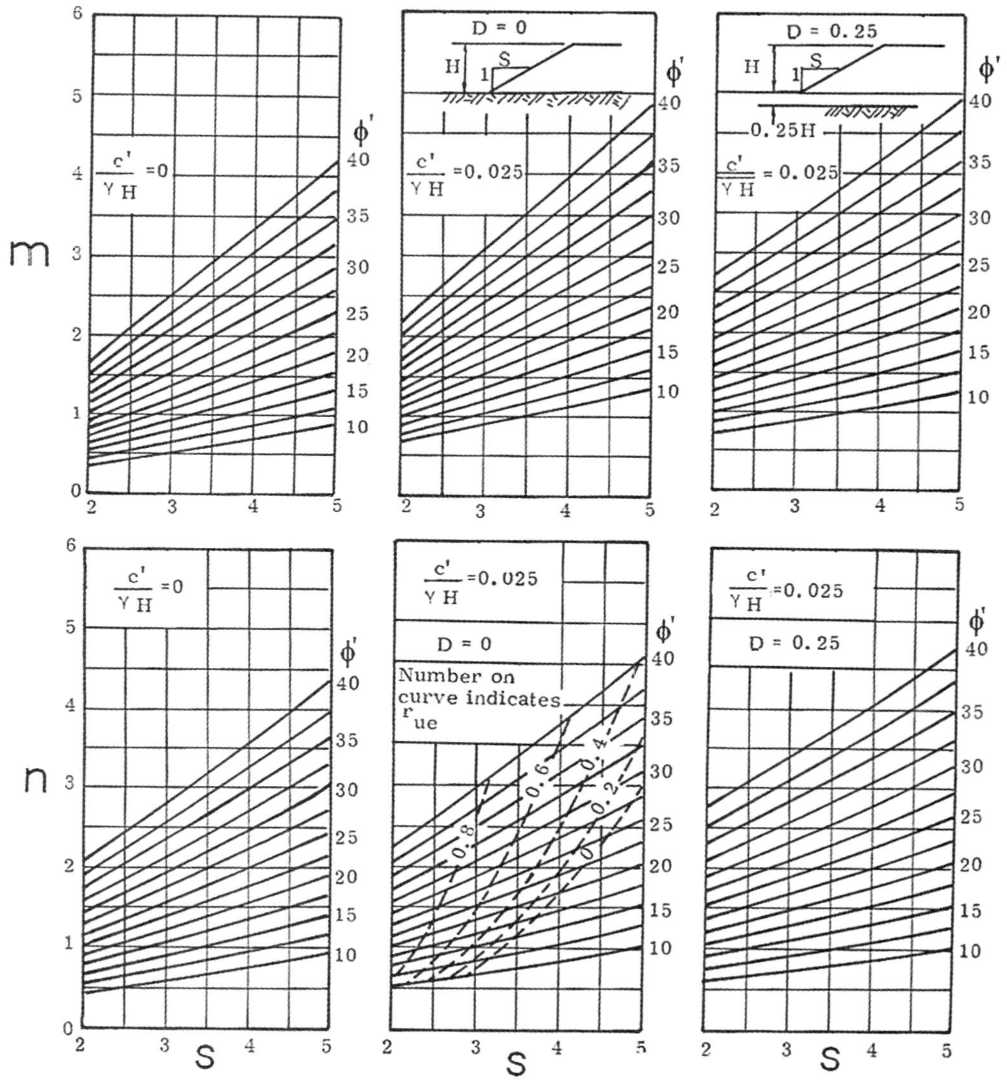

Fig. 7-18. Stability charts for c'/γH = 0 and 0.025 (Bishop and Morgenstern 1960. Reproduced with permission from ICE Publishing)

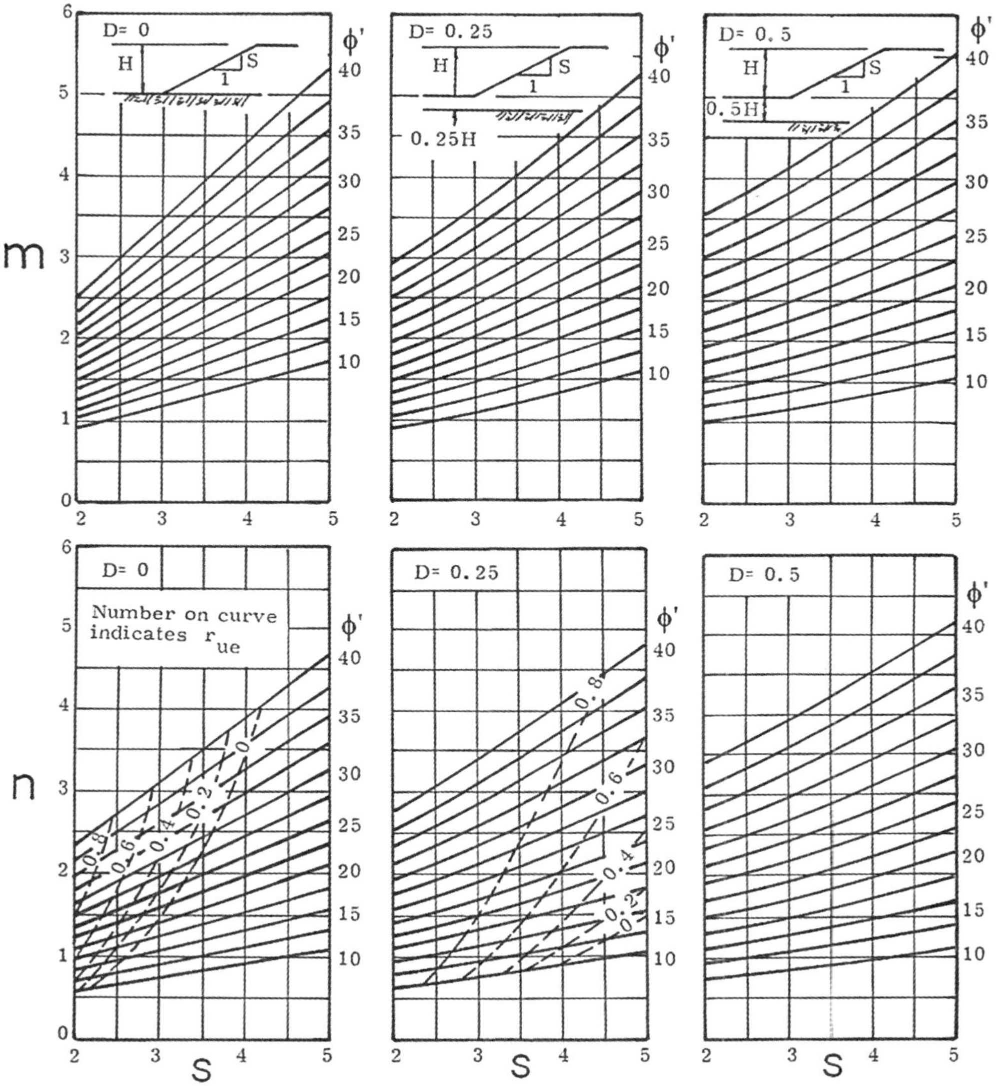

Fig. 7-19. Stability charts for c′/γH = 0.05 *(Bishop and Morgenstern 1960. Reproduced with permission from ICE Publishing)*

in which m and n are the stability coefficients determined from the charts. The values of m and n depend on the depth ratio, D. When $c'/\gamma H = 0.05$, the charts show three different depth ratios of 0, 0.25, and 0.5. When $c'/\gamma H = 0.025$, only two depth ratios of 0 and 0.25 are needed. The most critical circle is supposed to be a shallow circle and expected to fall within these depths. The charts for $c'/\gamma H = 0$ can be used to interpolate the factors of safety when $c'/\gamma H$ lies between 0 and 0.025. When $c'/\gamma H = 0$, it is not really necessary to use the charts because, similar to an infinite slope, the factor of safety can be computed by Eq. (6-5).

Example 7.9 Given the dam shown in Fig. 7-17 with $c'/\gamma H = 0$, $\phi' = 30°$, and $S = 3$ ($\beta = 18.4°$), determine the factors of safety for $r_u = 0$ and $r_u = 0.2$, using Fig. 7-18 and comparing with Eq. (6-5).

Solution When $r_u = 0$: With $c'/\gamma H = 0$, $\phi' = 30°$, and $S = 3$, from Fig. 7-18, $F = m = 1.74$. With $\phi' = 30°$ and $\beta = 18.4°$, from Eq. (6-5), $F = \tan 30°/\tan 18.4° = 1.73$, which checks closely with the 1.74 obtained from Fig. 7-18.

When $r_u = 0.2$: From Fig. 7-18, $m = 1.74$ and $n = 1.95$; and from Eq. (7-8), $F = 1.74 - 0.2 \times 1.95 = 1.35$. From Eq. (6-5), $F = (1 - 0.2) \times \tan 30°/\tan 18.4° = 1.38$. The difference between Fig. 7-18 and Eq. (6-5) is because of the difference in the definition of r_u. Eq. (6-5) is based on the normal method by reducing the normal force by $r_u W \cos\beta$, as shown in Fig. 6-1, whereas Fig. 7-18 is based on the simplified Bishop method by reducing the normal force by $r_u W \sec\beta$, as shown in Fig. 8-6. If r_u is defined by the simplified Bishop method, the r_u used in Eq. (6-5) should be multiplied by $\cos^2\beta$, or $F = 1.74 - 0.2 \times (\cos 18.4°)^2 \times 1.95 = 1.38$, which is exactly the same as that obtained from Fig. 7-18.

If the ledge is far from the surface, it is necessary to determine which depth ratio is most critical. This determination can be facilitated by using the line of equal pore pressure ratio, r_{ue}, on the chart defined as

$$r_{ue} = \frac{m_2 - m_1}{n_2 - n_1} \tag{7-9}$$

If the given value of r_u is greater than r_{ue} for the given section and strength parameters, the factor of safety determined with the greater depth ratio has a smaller value than the factor of safety determined with the smaller depth ratio because n_2 is greater than n_1. With the help of the lines of equal pore pressure ratio, the most critical depth ratio with the lowest factor of safety can be identified without having to compute the factor of safety for every depth ratio and determine which is most critical.

Example 7.10 Given $S = 4$, $H = 64$ ft (19.5 m), $DH = 30$ ft (9.2 m), $c' = 200$ psf (9.6 kPa), $\phi' = 30°$, $\gamma = 125$ pcf (19.6 kN/m^3), and $r_u = 0.5$, determine the factor of safety.

Solution Because $c'/\gamma H = 200/(125 \times 64) = 0.025$, the chart in Fig. 7-18 should be used. The depth ratio is $D = 30/64 = 0.47$, which is greater than 0.25, so the chart with $D = 0.25$ can be used to replace $D = 0.47$. Any circle with a D greater than 0.25 is expected to have a factor of safety equal to that with $D = 0.25$. The use of the lines of equal pore pressure ratio, as indicated by the dashed lines, gives $r_{ue} = 0.42$. Since $r_u > r_{ue}$, $D = 0.25$ is more critical than $D = 0$. When $D = 0.25$, $S = 4$, and $\phi' = 30°$, from Fig. 7-18, $m = 2.95$ and $n = 2.78$, or $F = 2.95 - 0.5 \times 2.78 = 1.56$.

It is interesting to compare the factors of safety between $D = 0$ and $D = 0.25$. When $D = 0$, from Fig. 7-18, $m = 2.89$ and $n = 2.63$, or $F = 2.89 - 0.5 \times 2.63 = 1.58$, which is only slightly greater than the 1.56 for $D = 0.25$. It can be seen that the effect of D on the factor of safety is quite small.

In this example, $c'/\gamma H$ is exactly equal to 0.025. If $c'/\gamma H$ is smaller than 0.025, the factor of safety for $c'/\gamma H = 0$ should also be determined The factor of safety can be computed by a straight-line interpolation between the two.

In Fig. 7-19, the case of $D = 0.5$ also applies to $D > 0.5$. If the given D is greater than 0.5, the lines of equal pore pressure ratio for $D = 0.25$ should be used to determine whether the case of $D = 0.5$ is more critical than $D = 0.25$. If so, the factor of safety for $D = 0.5$ gives the solution required. Otherwise, the line of equal pore pressure ratio for $D = 0$ should be used to determine whether $D = 0$ or $D = 0.5$ is the most critical.

Example 7.11 Same as Example 7.10 except that $c'/\gamma H$ is increased to 0.05; determine the factor of safety.

Solution With $S = 4$, $\phi' = 30°$, and $D = 0.25$, from the lines of equal pore pressure ratio in Fig. 7-19, $r_{ue} = 0.72$, which is greater than the given 0.5, so $D = 0.25$ is more critical than $D = 0.5$. Next, from the lines of equal pore pressure ratio for $D = 0$, it can be found that $r_{ue} = 0$, which is smaller than the given 0.5, so $D = 0.25$ is the most critical among the three. For $D = 0.25$, from Fig. 7-19, $m = 3.25$ and $n = 2.83$, or $F = 3.25 - 0.5 \times 2.83 = 1.84$.

If the lines of equal pore pressure ratio are not used, it will be necessary to compute the factors of safety for all three depth ratios to determine which is the most critical. To check that $D = 0.25$ does give the lowest factor of safety, the factors of safety at $D = 0$ and $D = 0.5$ are computed as follows: When $D = 0$, from Fig. 7-19, $m = 3.27$ and $n = 2.76$, or $F = 3.27 - 0.5 \times 2.76 = 1.89$. When $D = 0.5$, $m = 3.45$ and $n = 3.12$, or $F = 3.45 - 0.5 \times 3.12 = 1.89$. It can be seen that the minimum factor of safety does occur at $D = 0.25$ but the difference in factors of safety among the three depth ratios is not very significant.

7.5.2 Morgenstern's Charts

Morgenstern's charts (1963) can be used only to determine the factor of safety of a dam after a rapid drawdown. Fig. 7-20 shows the case considered. By assuming horizontal flow lines and vertical equipotential lines, the phreatic surface and the piezometric surface are identical.

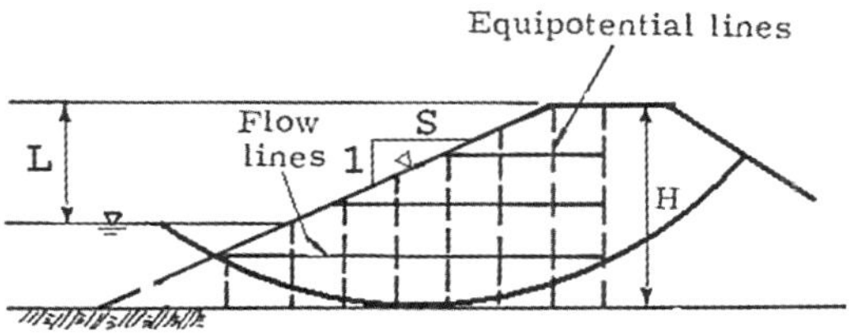

Fig. 7-20. Slope subject to rapid drawdown

It is assumed that an earth dam is placed on an impervious base. The original water level is at the same elevation as the top of the dam. Then the water level suddenly is lowered a distance L below the top of the dam to simulate rapid drawdown. The factor of safety is determined by the simplified Bishop method by assuming that the critical circle is tangent to the impervious base, that the flow lines are horizontal and the equipotential lines are vertical after rapid drawdown, and that the weight of the soil is twice the weight of water. The assumption of vertical equipotential lines indicates that the phreatic surface after the rapid drawdown is located along the surface of the slope.

Figs. 7-21, 7-22, and 7-23 show the factors of safety under rapid drawdown for $c'/\gamma H$ of 0.0125, 0.025, and 0.05, respectively. The factor of safety is plotted against the drawdown ratio, L/H, for various S and ϕ'. When the drawdown ratio is equal to 1, the circle tangent to the impervious base is the most critical. When the drawdown ratio is less than 1, several circles must be tried by assuming the impervious base at different elevations so that the one with the lowest factor of safety can be determined.

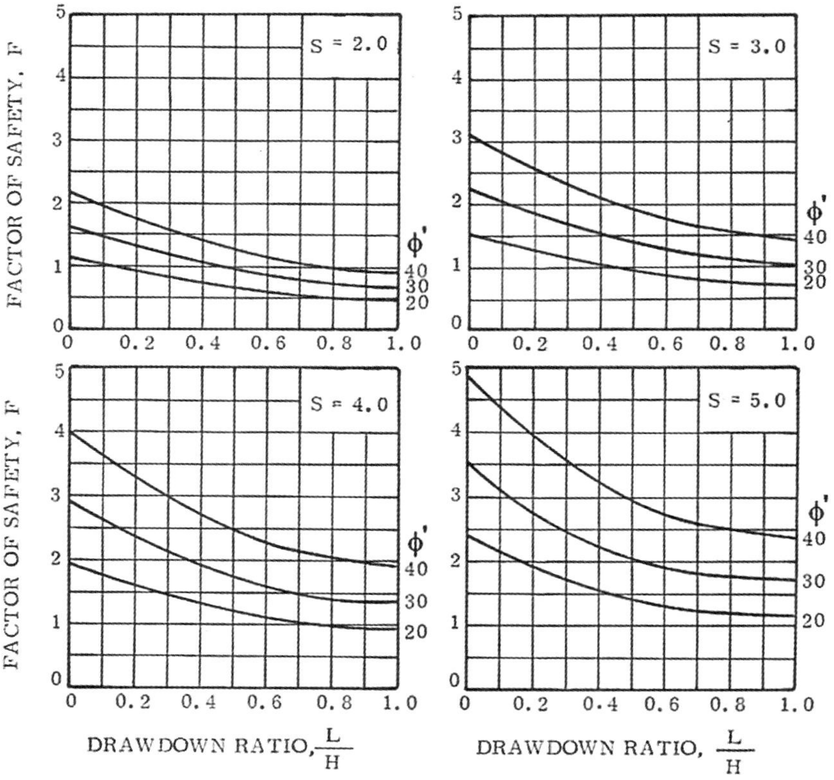

Fig. 7-21. Drawdown stability chart for c'/γH = 0.0125 (Morgenstern 1963. Reproduced with permission from ICE Publishing)

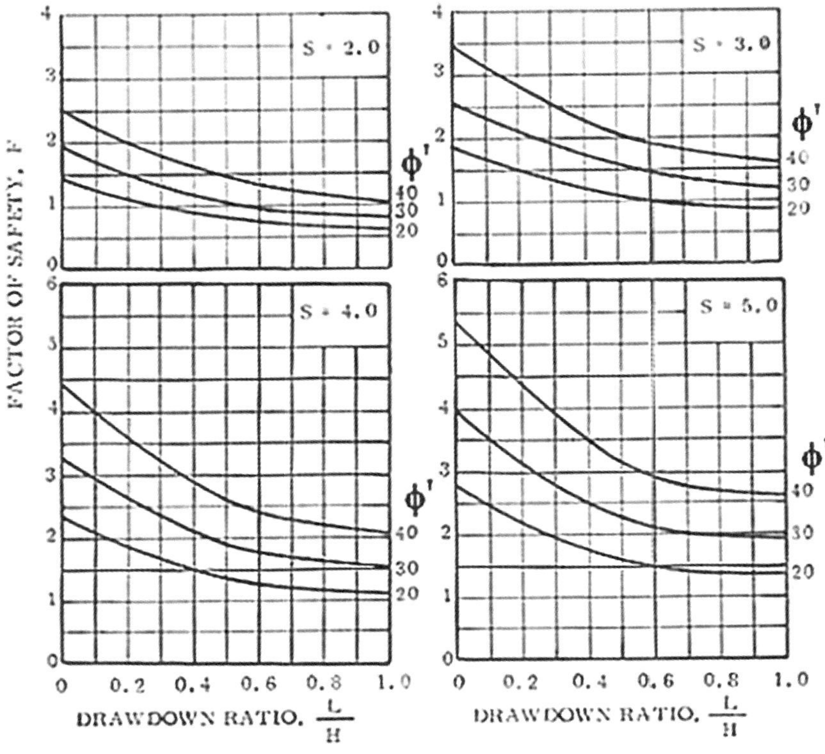

Fig. 7-22. Drawdown stability chart for c′/γH = 0.025 (Morgenstern 1963. Reproduced with permission from ICE Publishing)

Example 7.12 Given $S = 3$, $H = 65$ ft (19.8 m), $c' = 200$ psf (9.6 kPa), $\phi' = 30°$, and $\gamma = 124.8$ psf (19.6 kN/m^3), determine the factors of safety when $L/H = 1$ and $L/H = 0.5$.

Solutions When $L/H = 1$, $c'/\gamma H = 200/(124.8 \times 65) = 0.025$, $S = 3$, and $\phi' = 30°$, from Fig. 7-22, the factor of safety $F = 1.2$.

When $L/H = 0.5$, three different trials are needed to determine the minimum factor of safety, as shown in Fig. 7-24.

(a) Assume the circle tangent to the base with $H = 65$ ft and $L = 32.5$ ft. For $c'/\gamma H = 0.025$, $L/H = 0.5$, from Fig. 7-22, $F = 1.52$.

(b) Move the base up to the pool level and assume the circle tangent to the base with $H = L = 32.5$. For $c'/\gamma H = 0.05$, $L/H = 1$, from Fig. 7-23, $F = 1.48$.

(c) Place the base between (a) and (b) with $H = 48.75$ ft, $L = 32.5$ ft, $c'/\gamma H = 200/(124.8 \times 48.75) = 0.033$, and $L/H = 32.5/48.75 = 0.67$. The factor of safety can be determined by a straight-line interpolation of $c'/\gamma H$ between 0.025 and 0.05. From Fig. 7-22, with $c'/\gamma H = 0.025$, $F = 1.37$. From Fig. 7-23 with $c'/\gamma H = 0.05$, $F = 1.66$. By interpolation, $F = 1.37 + (1.66 - 1.37) \times 0.008/0.025 = 1.46$.

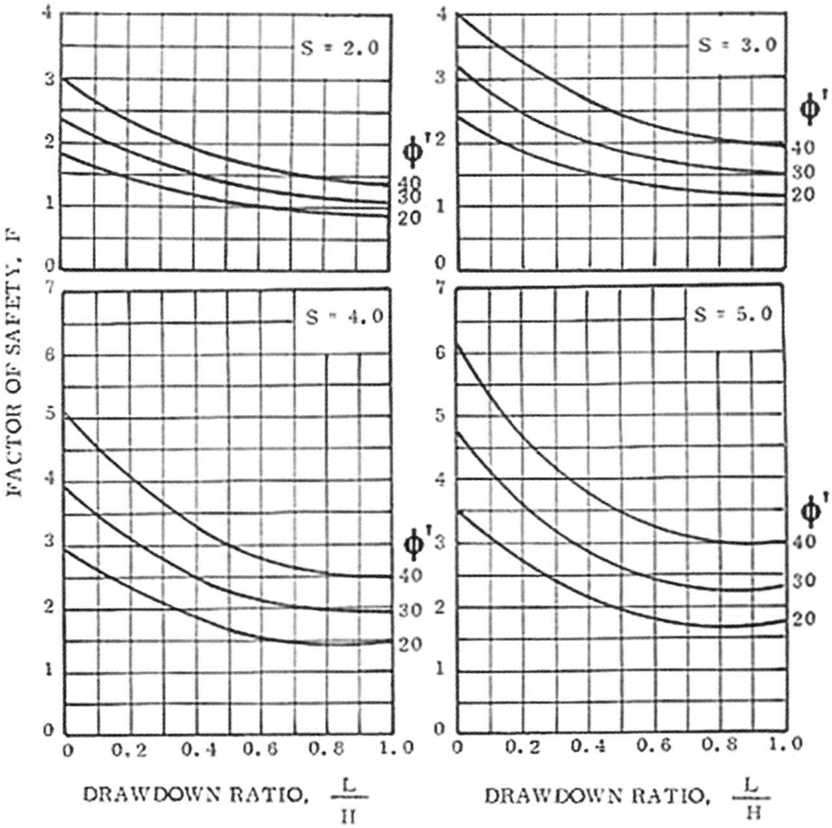

Fig. 7-23. Drawdown stability chart for c′/γH = 0.05 (Morgenstern 1963. Reproduced with permission from ICE Publishing)

The minimum factor of safety is 1.46. Although a slightly lower value could perhaps be found, further refinements are unwarranted. This example demonstrates that for partial drawdown, the critical circle often may lie above the base of the dam, and it is important to investigate several levels of tangency.

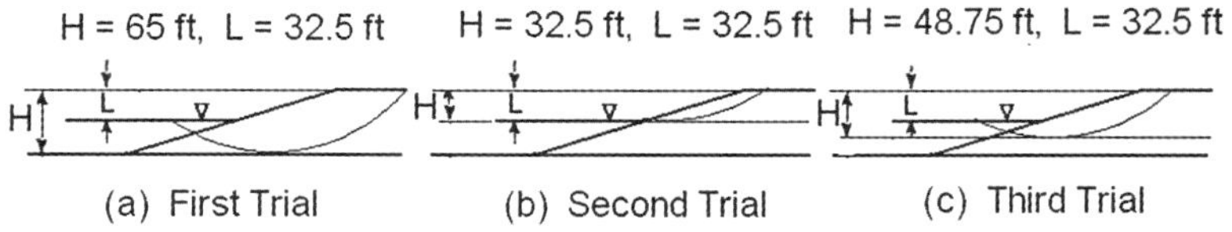

Fig. 7-24. Example 7.12

7.5.3 Spencer's Charts

Fig. 7-25 shows the Spencer's charts (1967) for determining the required slope angle when the factor of safety is given. This type of charts is very useful for preliminary design purposes, because the required factor of safety is always known a priori and the required slope angle is the answer to be sought. If the angle of slope is given, it is more convenient to use other charts, because the application of Spencer's charts requires the use of a trial-and-error procedure.

Spencer's charts are based on the original Spencer method, which assumes parallel interslice forces and satisfies both overall force and moment equilibrium. It checks well with the simplified Bishop method, which only satisfies the overall moment equilibrium, because the factor of safety based on moment equilibrium is insensitive to the direction of interslice forces. The charts use three different pore pressure ratios, that is, 0, 0.25, and 0.5, and assume that the ledge or firm stratum is at a great depth below the surface. In using the charts, it is necessary to find the developed friction angle defined as

$$\phi_d = \tan^{-1}(\tan\phi' / F) \tag{7-10}$$

Spencer (1967) also developed charts for locating the critical surface, which are not reproduced here. If the ledge is very close to the surface, the design based on Fig. 7-25 is somewhat conservative.

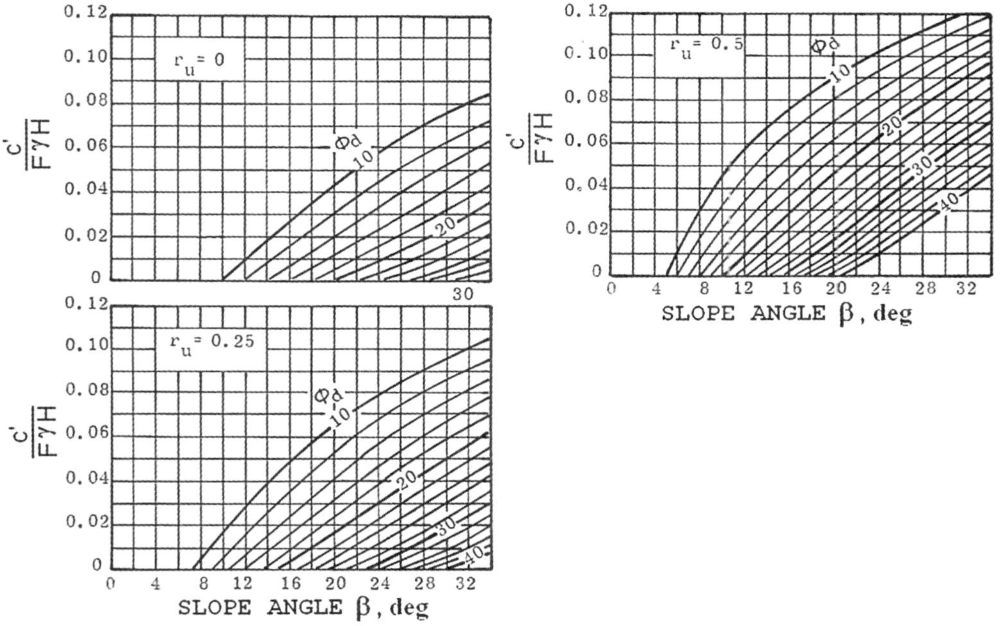

Fig. 7-25. Stability chart for different pore pressure ratios (Spencer 1967. Reproduced with permission from ICE Publishing)

Example 7.13 Same as Example 7.10 except that, instead of giving $S = 4$, a safety factor of 1.56 is given. Find the required S.

Solution With $H = 64$ ft, $c' = 200$ psf, $\gamma = 125$ pcf, and $F = 1.56$, $c'/F\gamma H = 200/(1.56 \times 125 \times 64) = 0.016$. With $\phi' = 30°$, $\phi_d = \tan^{-1}(\tan 30°/1.56) = 20.3°$, and $r_u = 0.5$, from Fig. 7-25, slope angle = 13.8°, or $S = 1/\tan 13.8° = 4.07$, which checks well with the given $S = 4$ in Example 7.10.

Note that the factor of safety obtained in Example 7.10 is based on the circle with a depth ratio, D, of 0.25, whereas the factor of safety used in Fig. 7-25 is the minimum among various depth ratios. If the minimum factor of safety in Example 7.10 is slightly smaller than 1.56, as it should be, the value of S will be slightly decreased and a better agreement between Examples 7.10 and 7.13 will be obtained.

7.5.4 Huang's Charts

Fig. 7-26 shows the stability charts for a homogeneous dam, as developed by Huang (1975). The factor of safety is based on the simplified Bishop method. Similar to Spencer's charts, it is assumed that the ledge is located at a great depth below the surface.

The left upper corner of Fig. 7-26 shows an earth embankment with a height, H, and an outslope of S:1. In the effective stress analysis, the soil has a small cohesion relative to the angle of internal friction, so the most critical failure surface may be a toe circle or a circle slightly below the toe. As long as the bedrock is at a considerable distance from the surface, the location of bedrock has no effect on the factor of safety.

In Fig. 7-26, the solid curves indicate zero pore pressure and the dashed curves indicate a pore pressure ratio of 0.5. The factor of safety for other pore pressure ratios can be obtained by a straight-line interpolation between the solid and dashed curves. The number on each curve is the cohesion factor (C.F.) in percent, which is equal to $100\ c'/\gamma H$.

For a given effective friction angle and a given cohesion factor, the factor of safety for a given slope can be determined directly from the charts. These charts cannot be applied to total stress analysis because, when $\phi = 0$, the most critical circle will be a deep circle tangent to the bedrock. Because the depth to bedrock is not given, the factor of safety cannot be determined. This is why all curves stop at $\phi' = 5°$ and should not be extended to $\phi' = 0$.

Fig. 7-27 presents a practical example for the application of stability charts. This dam, which provides the water supply to Springfield, Kentucky, failed by sliding away some of the material on the downstream face. The location of the failure surface is very close to the most critical circle obtained by LEAME, as indicated in the figure. This provides a good opportunity to back-calculate the shear strength of the soil in the field. The failure can be considered as a full-scale model test. When the dam failed, the factor of safety should have decreased to 1.

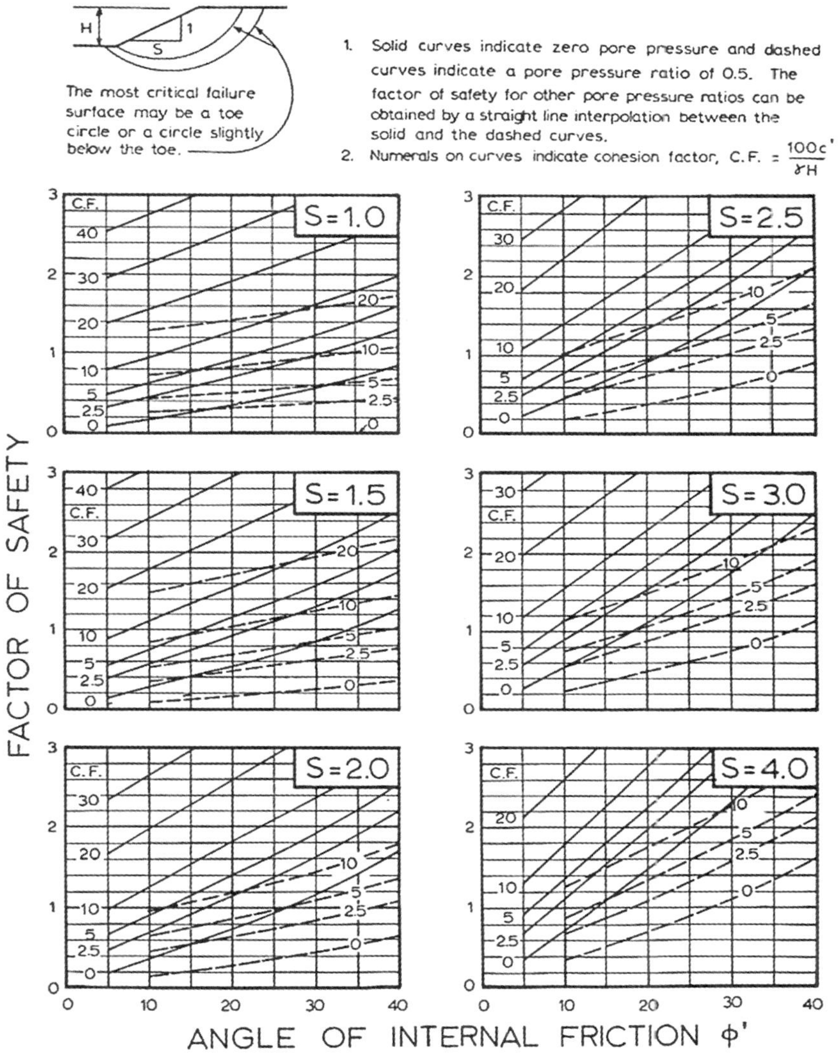

Fig. 7-26. Stability charts for effective stress analysis of earth dam (Huang 1975, © National Academy of Sciences, Washington, DC. Reproduced with permission of the Transportation Research Board, Washington, DC)

As shown in the figure, the original downstream slope is not uniform, being flatter at the toe than at the top. However, it can be changed to a uniform slope by approximating the slope at the toe with a horizontal line and an inclined line, so that the cut is equal to the fill. The downstream slope is 1.75:1, and the height is 37 ft (11.3 m). The phreatic surface is determined theoretically by drawing a flownet, as described in Section 4.1.1. By using LEAME and assuming an effective cohesion of 200 psf (9.6 kPa) and an effective friction angle of 25°, a safety

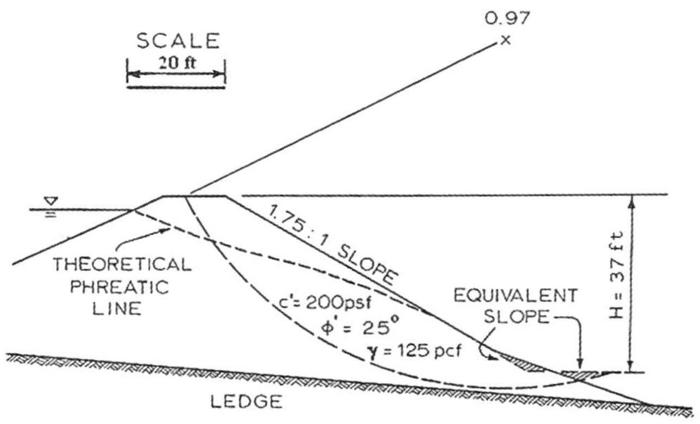

Fig. 7-27. Stability of Springfield Dam in Kentucky
Note: 1 ft = 0.395 m; 1 psf = 47.9 Pa; 1 pcf = 157.1 N/m³

factor of 0.97 was obtained. This indicates that the assumed shear strength is reasonable, because it yields a factor of safety close to 1. Therefore, this shear strength can be used for the redesign of the dam. Unfortunately, there was an office building not far from the dam, so the downstream slope could not be flattened. Finally, a rock berm at a slope of 1:1 was constructed to increase the factor of safety.

Example 7.14 For the dam shown in Fig. 7-27 with H = 37 ft (11.3 m), S = 1.75, c' = 200 psf (9.6 kPa), ϕ' = 25°, γ = 125 pcf (19.7 kN/m^3), and the given phreatic surface, determine the factor of safety.

Solution Cohesion factor = 100 × 200/(125 × 37) = 4.32. To determine the pore pressure ratio, it is necessary to know the percentage of fill under water. It is estimated that 75% of the area is below and 25% is above the water table. The detailed calculations are presented in Fig. 7-28.

For an outslope of 1.5:1, a cohesion factor of 4.32, and a friction angle of 25°, as indicated by the chart on the left of Fig. 7-28, the factor of safety is 1.3 from the solid curves, where no fill is under water, and 0.7 from the dashed curves, where the entire fill is under water, so the factor of safety is $F = 0.75 \times 0.7 + 0.25 \times 1.3 = 0.85$.

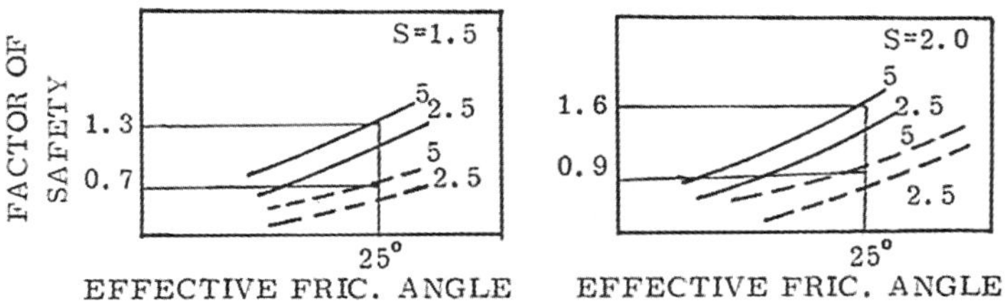

Fig. 7-28. Example 7.14

For an outslope of 2:1, as indicated by the chart on the right of Fig. 7-28, the factor of safety is 1.6 from the solid curves and 0.9 from the dashed curves, so the factor of safety is $F = 0.75 \times 0.9 + 0.25 \times 1.6 = 1.07$.

The actual outslope is 1.75:1, which is the average of 1.5:1 and 2:1, so the average factor of safety is $F = (0.85 + 1.07)/2 = 0.96$.

7.5.5 Comparison of Charts

The advantages and limitations of the various charts are as follows:

1. Bishop's and Morgenstern's charts are applicable no matter whether the ledge is on, near to, or far below the toe. The factors of safety for different depth ratios, D, can be determined from the charts. With the help of the lines of equal pore pressure ratio, the most critical depth ratio with the lowest factor of safety can be identified easily without needing to compute the safety factor of every depth ratio and determining which is most critical. The charts can be applied to the case of full rapid drawdown by assuming $r_u = 0.5$. Due to the limitation of $c'/\gamma H$ to 0.05, the charts cannot be used for total stress analysis with relatively high cohesion. If $c'/\gamma H$ is not exactly equal to 0.025 or 0.05, some kind of interpolation is needed.
2. Morgenstern's charts can be used only to analyze full or partial drawdown. Although the charts assume that the dam is placed directly on a ledge with $D = 0$, it can give an approximate but slightly higher factor of safety even when the ledge is at great depth below the toe. To determine the lowest factor of safety for partial drawdown, at least three circles, each with the ledge at a different height, must be tried. The charts can be used only when $c'/\gamma H$ is not greater than 0.05, and need interpolations if $c'/\gamma H$ is not exactly equal to 0.0125, 0.025, or 0.05.
3. Spencer's charts assume a ledge far below the toe but can give conservative results if the ledge is at or close to the toe. Instead of checking the factor of safety for a given slope, the required slope angle for a given factor of safety can be read directly from the charts. This feature makes the charts particularly useful for preliminary design, because the required factor of safety is usually known before the slope angle is determined. If the slope angle is given, the factor of safety can be determined by a trial-and-error procedure, which is quite cumbersome. The charts can be used for the case of full rapid drawdown by assuming $r_u = 0.5$.
4. Similar to Spencer's charts, Huang's charts also assume a ledge far below the toe and will give a lower and more conservative factor of safety if the ledge is at or close to the toe. Given the slope S, the friction angle, ϕ', and the cohesion factor, $100\ c'/\gamma H$, the factor of safety can be read directly from the charts with no further calculations needed. These charts are good supplements to Bishop's and Morgenstern's charts by adding a case of $D = \infty$, so if D is great than 0.25 or 0.5, the minimum factor of safety can

be determined right away and no comparison of different depth ratios is needed. The curves for $r_u = 0$ and $r_u = 0.5$ and those for different cohesion factors are all placed on the same chart, so the factor of safety for any given r_u and cohesion factor can be interpolated visually. The charts can be used for full rapid drawdown by assuming $r_u = 0.5$. The extension of the charts to very large cohesion factors allows the use of the charts for total stress analysis with a large cohesion and a relatively small friction angle.

It is interesting to compare the results obtained by different methods, as illustrated by the following example.

Example 7.15 Given $H = 48$ ft (14.6 m), $D = 0$, $S = 3$, $c' = 300$ psf (14.4 kPa), $\phi' = 30°$, and $\gamma = 125$ pcf (19.7 kN/m^3), compare the factors of safety after full rapid drawdown, as obtained by all four methods.

Solution

$$c'/\gamma H = 300/(125 \times 48) = 0.05$$

Bishop's and Morgenstern's charts: With $D = 0$, $S = 3$, and $\phi' = 30°$, from Fig. 7-19, $m = 2.57$ and $n = 2.17$. For $r_u = 0.5$, $F = 2.57 - 0.5 \times 2.17 = 1.49$.

Morgenstern's charts: With $S = 3$, $L/H = 1$, and $\phi' = 30°$, from Fig. 7-23, $F = 1.49$, which checks with Bishop's and Morgenstern's charts, because both assume $D = 0$.

Huang's charts: With $S = 3$, cohesion factor C.F. $= 100c'/(\gamma H) = 100 \times 300/(125 \times 48) = 5$, and $\phi' = 30°$, from the dashed curves in Fig. 7-26, $F = 1.46$, which is slightly smaller than 1.49, because Huang's charts assume $D = \infty$ instead of $D = 0$.

Spencer's charts: $\tan\phi' = \tan 30° = 0.577$ and $\beta = \tan^{-1}(0.333) = 18.4°$. The factor of safety can be determined by trial and error. Several values of F are tried to make β as close to 18.4° as possible.

Assume $F = 1.5$, with $c'/F\gamma H = 0.05/1.5 = 0.0333$, $\phi_d = \tan^{-1}(0.577/1.5) = 21°$, and $r_u = 0.5$, from Fig. 7-25, $\beta = 18.0°$.

Assume $F = 1.45$, with $c'/F\gamma H = 0.05/1.45 = 0.0345$, $\phi_d = \tan^{-1}(0.577/1.45) = 21.7°$, and $r_u = 0.5$, from Fig. 7-25, $\beta = 18.8°$.

Assume $F = 1.47$, with $c'/F\gamma H = 0.05/1.47 = 0.034$, $\phi_d = \tan^{-1}(0.577/1.47) = 21.4°$, and $r_u = 0.5$, from Fig. 7-25, $\beta = 18.4°$.

The factor of safety by Spencer's charts is 1.47, which is slightly smaller than 1.49, because Spencer's charts are also based on $D = \infty$.

Due to the small scale of the charts, it is difficult to read them to two decimal points. However, this example does indicate that all four charts yield about the same result, and the location of the ledge has very little effect on the factor of safety obtained.

7.6 Effective Stress Analysis of Nonhomogeneous Dams

All the charts presented in the previous section are based on the assumption that the dam and foundation have the same soil parameters. This is usually not true in reality, so the case shown in Fig. 7-29, where the dam and foundation have different effective cohesions, c', effective friction angles, ϕ', and pore pressure ratios, r_u, is of practical interest.

For a given circle, similar to Eq. (2-13), the static factor of safety based on the normal method can be expressed as

$$F = \frac{\sum_{i=1}^{n} [c'b_i \sec\theta_i + (1 - r_u)W_i \cos\theta_i \tan\phi']}{\sum_{i=1}^{n} W_i \sin\theta_i} \tag{7-11}$$

The three soil parameters, c', ϕ', and r_u, are shown clearly in the equation, whereas all the remaining terms, other than the unit weight, are related to geometry, independent of the soil properties. Because the soil unit weight does not change significantly, an average unit weight can be assumed. Note that W_i is proportional to the soil unit weight, γ, and height, H. By dividing both the numerator and denominator by γH, Eq. (7-11) can be written as

$$F = \frac{c_1'}{\gamma H} N_{c1} + \frac{c_2'}{\gamma H} N_{c2} + (1 - r_{u1})(\tan\phi_1')N_{f1} + (1 - r_{u2})(\tan\phi_2')N_{f2} \tag{7-12}$$

in which subscripts 1 and 2 refer to soils 1 and soil 2, respectively, N_c = cohesion number, and N_f = friction number. These numbers are the collection of geometric terms, such as b_i and θ_i, and can be determined by LEAME or any other computer programs using the Fellenius or normal method.

To determine N_{c1}, input $c_2' = 0$, $\phi_1' = 0$, $\phi_2' = 0$, $c_1' = 12{,}500$, $\gamma = 125$, and $H = 100$. The factor of safety thus obtained is equal to N_{c1}, as can be seen by substituting the stated values into Eq. (7-12). To determine N_{f1}, input $c_1' = 0$, $c_2' = 0$, $\phi_1' = 45°$, r_{u1} = 0, and $\phi_2' = 0$, so $F = N_{f1}$. The same procedure can be applied to soil 2 by exchanging the subscripts.

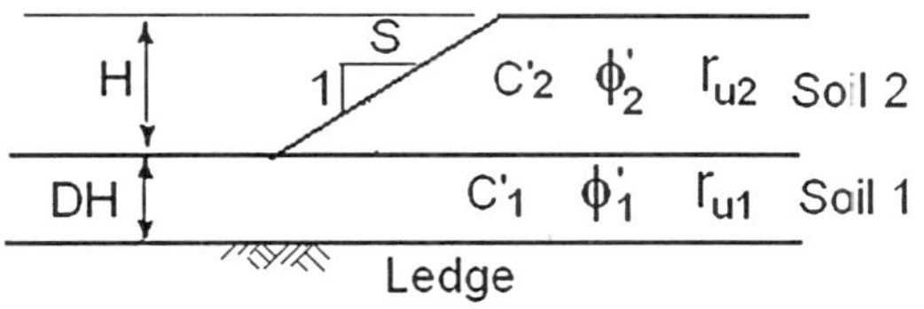

Fig. 7-29. Soil parameters for a nonhomogeneous dam

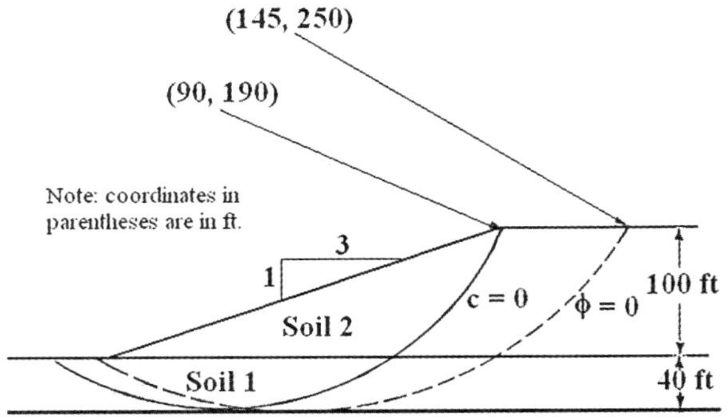

Fig. 7-30. Location of critical circles
Note: 1 ft = 0.305 m

For a given slope, S, and depth ratio, D, it is necessary to know the location of the most critical circle. Unfortunately, the location of the most critical circle depends on the shear strength of the two soils. Fig. 7-30 shows two extreme cases, one with $\phi = 0$ and the other with $c = 0$. By using LEAME and assuming that soils 1 and 2 are identical, the most critical center for $c = 0$, as indicated by the solid curve, is always located at (90, 190), regardless of the magnitude of ϕ; that for $\phi = 0$, as indicated by the dashed curve, is always located at (145, 250), regardless of the magnitude of c. The use of these two circles for a given set of soil parameters may result in two widely different factors of safety. For example, the critical center for $c = 0$ and $\phi = 30°$ should be at (90, 190), and the factor of safety should be 2.111; however, if the center at (145, 250) were used, the factor of safety would be 2.412. Fortunately, in the effective stress analysis, the shear strength contributed by the effective cohesion is very small and, therefore, has very little effect on the location of the most critical circle. A comparison between $c = 0$ and $\phi = 30°$ and the most general case of $c = 200$ psf (9.6 kPa) and $\phi = 30°$ shows that the most critical center is the same and the factor of safety is increased only slightly, from 2.111 to 2.248. To be more realistic, the case of $c_1' = c_2' = 200$ psf (9.6 kPa) and $\phi_1' = \phi_2' = 30°$ was used to locate the most critical center.

Fig. 7-31 shows the charts for N_c and N_f, as determined by LEAME. When $D > 0$, the factor of safety is determined by Eq. (7-12). When $D = 0$, there is only one soil and the factor of safety is determined by

$$F = \frac{c'}{\gamma H} N_c + (1 - r_u)(\tan\phi')N_f \tag{7-13}$$

Eq. (7-13) also can be applied to a homogeneous dams, where $N_c = N_{c1} + N_{c2}$ and $N_f = N_{f1} + N_{f2}$.

It should be mentioned that theoretically there is a slight variation in N_{c1} among the various depth ratios, D. Because the variation is small and has very little effect on the factor of safety, the average of the six depth ratios was used

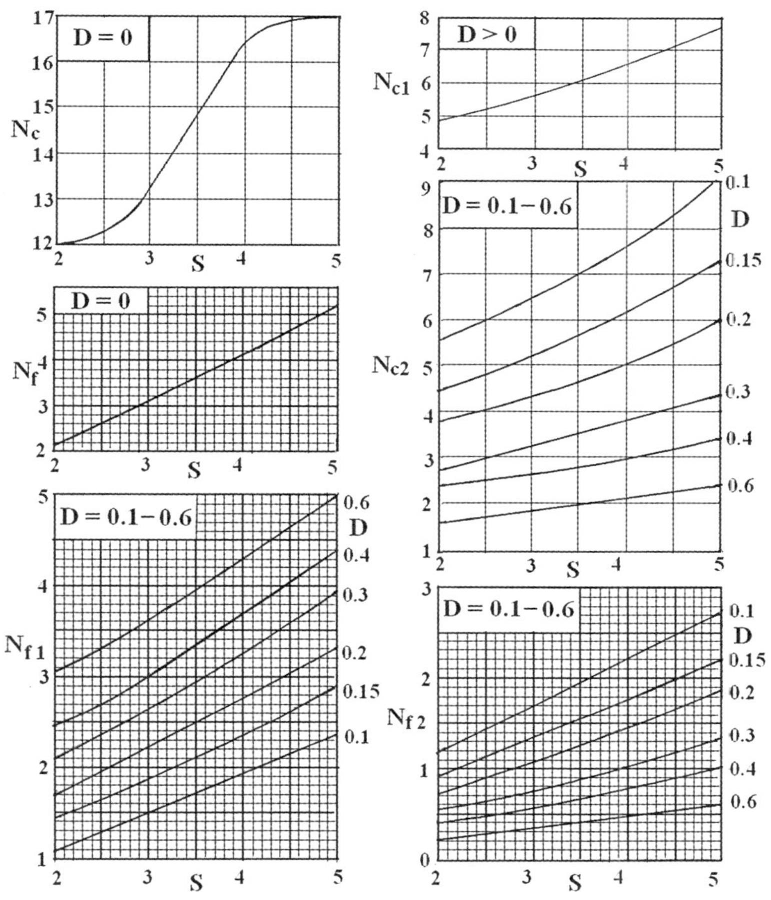

Fig. 7-31. *Cohesion and friction numbers for various depth ratios*

to plot N_{c1}. The coordinates for N_c, N_{c1}, and N_{c2} are not finely divided, because their contribution to the factor of safety is quite small, so an estimate of their values to one decimal point is sufficient.

Example 7.16 In Fig. 7-29, given $S = 3.5$, $D = 0.4$, $H = 100\,\text{ft}$ (30.5 m), $\gamma = 125\,\text{pcf}$ (19.7 kN.m^3), $c'_1 = 100$ psf (4.8 kPa), $\phi'_1 = 25°$, $r_{u1} = 0.4$, $c'_2 = 200$ psf (9.6 kPa), $\phi'_2 = 35°$, $r_{u2} = 0.2$, determine the factor of safety.

Solution $c'_1/\gamma H = 100/(125 \times 100) = 0.008$, $c'_2/\gamma H = 0.016$, $(1 - r_{u1})\tan\phi'_1 = (1 - 0.4)$ $\tan 25° = 0.28$, $(1 - r_{u2})\tan\phi_2 = (1 - 0.2)\tan 35° = 0.56$.

When $D = 0.4$, from Fig. 7-31, $N_{c1} = 6.1$, $N_{c2} = 2.8$, $N_{f1} = 3.34$, $N_{f2} = 0.66$, from Eq. (7-12), $F = 0.008 \times 6.1 + 0.016 \times 2.8 + 0.28 \times 3.34 + 0.56 \times 0.66 = 1.398$.

To be sure that $D = 0.4$ is the most critical circle, the factor of safety for $D = 0.3$ also must be determined. When $D = 0.3$, from Fig. 7-31, $N_{c1} = 6.1$, $N_{c2} = 3.5$,

$N_{f1} = 2.95$, $N_{f2} = 0.9$, from Eq. (7-12), $F = 0.008 \times 6.1 + 0.016 \times 3.5 + 0.28 \times 2.95 + 0.56 \times 0.9 = 1.43$, which is greater than the 1.398 for $D = 0.4$.

The minimum factor of safety is 1.398, which checks with the 1.420 obtained by LEAME, using the normal method. If the simplified Bishop method is used, the factory of safety is 1.574, so the use of normal method is more conservative.

A major difference between this section and Sections 7.3 and 7.4 lies in the soil used to evaluate these cohesion and friction numbers. In the previous sections, a soil with $\phi = 0$ is used to determine the location of the most critical circle, so the charts are quite accurate for total stress analysis with $\phi = 0$. For effective stress analysis with $\phi \neq 0$, a correction factor or a trial-and-error procedure must be applied to determine the minimum factor of safety. In this section, a soil with $c' = 200\,\text{psf}$ ($9.6\,\text{kN/m}^2$) and $\phi = 30°$ is used to determine the location of the most critical circle, so Fig. 7-31 can be used for effective stress analysis but not for total stress analysis. For total stress analysis, Fig. 7-35 based on $\phi = 0$, as presented in Section 7.8, should be used.

Huang (1979, 1980) also presented a method and a series of charts for determining both the static and the seismic factors of safety of nonhomogeneous dams consisting of a large number of soil layers. This information was included in the previous book (Huang 1983) but was purposely deleted from this volume because the method is too cumbersome to use. With the LEAME computer program readily available, it is no longer necessary to use any stability charts for preliminary design and estimating purposes unless they are very simple and easy to use.

7.7 Total Stress Analysis of Dams with $\phi = 0$

For a homogeneous dam with $\phi = 0$, Eq. (7-6) can be simplified to

$$F = \frac{c}{\gamma H} N_c \tag{7-14}$$

in which N_c is a cohesion number. By definition, the factor of safety can be expressed as

$$F = \frac{c}{c_d} \tag{7-15}$$

in which c = cohesion of the soil and c_d = developed cohesion, or the cohesion actually developed. Equating Eqs. (7-14) and (7-15),

$$N_c = \frac{\gamma H}{c_d} \tag{7-16}$$

It can be seen that the cohesion number, N_c, is the reciprocal of the stability number shown in Figs. 7-1 and 7-3.

Fig. 7-32 shows the charts for total stress analysis of dams with $\phi = 0$ (Huang 1975). These charts are based on a homogeneous simple slope, as shown in the upper left corner of the figure. The dam has a height, H, and an outslope S:1. A ledge is located at a depth of DH below the toe, where D is the depth ratio. The center of the circle is at a horizontal distance XH and a vertical distance YH from the top edge of the dam. When the critical circle passes below the toe, it can be proved easily that the center of the critical circle lies on a vertical line intersecting the slope at midheight, or $X = 0.5S$. This type of failure surface is called a midpoint circle, the results of which are shown by the solid curves in Fig. 7-32. If the

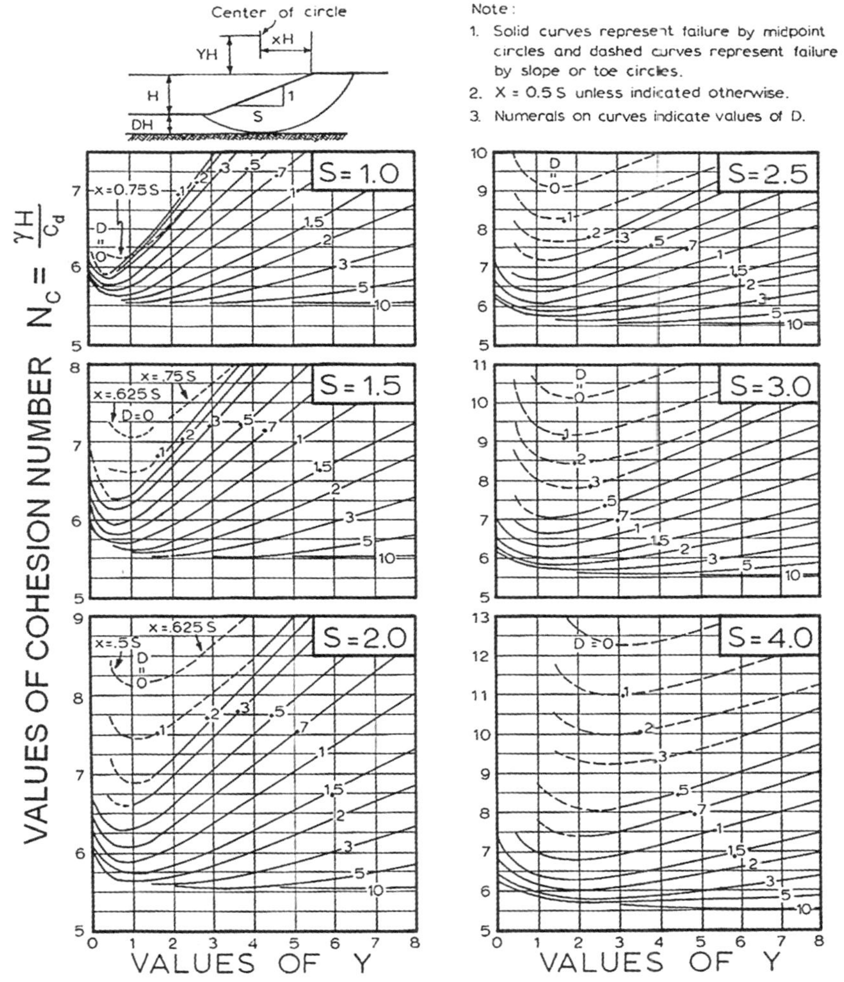

Fig. 7-32. *Stability charts for total stress analysis of earth dams (Huang 1975, © National Academy of Sciences, Washington, DC. Reproduced with permission of the Transportation Research Board, Washington, DC)*

depth ratio, D, is small, the failure surface may intersect the slope at or above the toe. This type of failure surface is called a toe or slope circle, the results of which are shown by the dashed curves. It can be seen from the figure that in most cases the most critical circle occurs when $X = 0.5S$ except for $D = 0$ and $S \leq 2$, where the value of X specially is noted.

It can be seen from the figure that the deeper the circle, the smaller the cohesion number and the smaller the factor of safety. Therefore, the critical circle is always tangent to the ledge. The charts are different from Taylor's in that the cohesion numbers for various circles with centers at different coordinates (XH, YH) are shown, whereas Taylor's only shows the stability number for the most critical circle. Given the location of the circle, the chart can be applied to both homogeneous and nonhomogeneous dams, as illustrated by the following examples.

Example 7.17 The homogeneous slope is the same as in Example 7.1, with H = 40 ft (12.2 m), $\beta = 22.5°$ ($S = 2.5$), $D = 1.5$, c = 1,200 psf (57.5 kPa), and γ = 120 pcf (18.9 kN/m^3). Determine the factor of safety.

Solution With $S = 2.5$ and $D = 1.5$, from Fig. 7-32, the minimum factor of safety occurs at $Y = 1.2$ with $N_c = 5.8$. From Eq. (7-14), $F = 1{,}200 \times 5.8/(120 \times 40) = 1.45$, which checks with the 1.46 obtained from Taylor's chart.

Example 7.18 Fig. 7-33 shows the cross section of a nonhomogeneous dam. The dam is 20 ft (6.1 m) high and has an outslope of 3:1 and a cohesion of 1,500 psf (71.9 kPa). The foundation consists of one 40-ft (12.2-m) soil layer having a cohesion of 800 psf (38.3 kPa) and one 20-ft (6.1 m) soil layer having a cohesion of 100 psf (4.8 kPa), which is underlain by a ledge. Although the unit weights for different soils are generally not the same, an average unit weight of 130 pcf (20.4 kN/m^3) is assumed. Determine the factor of safety.

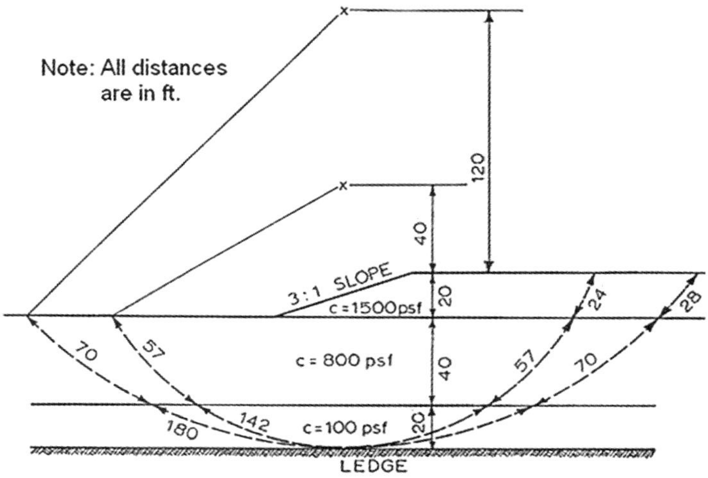

Fig. 7-33. Example 7.18

Solution With $S = 3$ and $D = 60/20 = 3$, from Fig. 7-32 the most critical circle for a homogeneous dam is a midpoint circle with a center located at $YH = 2 \times 20 =$ 40 ft above the top of the dam, and a minimum cohesion number of 5.7 is obtained. The average cohesion can be determined by measuring the length of the arc through each soil layer, or

$$c = \frac{24 \times 1,500 + 2 \times 57 \times 800 + 142 \times 100}{24 + 2 \times 57 + 142} = 505 \text{ psf}$$

From Eq. (7-14), $F = 505 \times 5.7/(130 \times 20) = 1.11$.

If a circle with a larger radius, say $Y = 6$ or $YH = 6 \times 20 = 120$ ft, is used, from Fig. 7-32, $N_c = 6.05$. The average cohesion is

$$c = \frac{28 \times 1,500 + 2 \times 70 \times 800 + 180 \times 100}{28 + 2 \times 70 + 180} = 494 \text{ psf}$$

The factor of safety is $F = 494 \times 6.05/(130 \times 20) = 1.15$, which is slightly greater than the 1.11 obtained previously, so the minimum factor of safety is 1.11. It can be seen that the use of the critical center based on a homogeneous dam still yields a smaller factor of safety. This is usually the case, so only one or two circles need to be tried. The factor of safety obtained by LEAME is 1.077.

In the previous example, the weaker layer lies at the bottom directly above the ledge, so it is apparent that the most critical circle should be tangent to the ledge. If the bottom layer is stronger than the top layer, the factors of safety at two different depth ratios, one tangent to the bottom of each layer, must be tried to determine which is more critical.

7.8 Total Stress Analysis of Triangular Fills on Soil Slopes

Fig. 7-34 shows a triangular fill having a height, H, an angle of outslope, β, and a degree of natural slope, α. The fill is built of soil 2 with an undrained shear

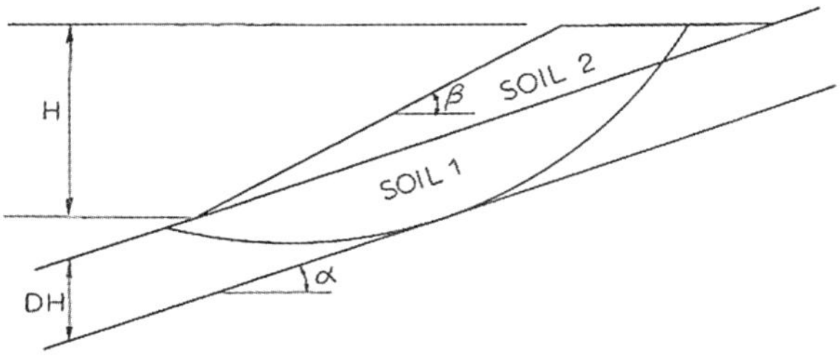

Fig. 7-34. Triangular fill on soil slope

strength, or cohesion, of c_2, whereas the natural slope is formed by soil 1 with an undrained shear strength of c_1. Below soil 1 lies the bedrock or stiff material. It is assumed that soils 1 and 2 have the same unit weight, γ. If the two unit weights are different, an average unit weight should be used.

In Fig. 7-34, the thickness of soil 1 is DH, where D is the depth ratio and H is the height of the fill. When $D = 0$, the circle is tangent to the natural slope and the case of triangular fills on rock or stiff slopes applies. For $D > 0$, unless soil 2 is much weaker than soil 1, the most critical circle with the lowest factor of safety is always tangent to the rock.

The factor of safety for frictionless material also can be expressed in the following form (Huang 1977b):

$$F = N_c\left[\frac{c_1}{\gamma H}(1 - L_f) + \frac{c_2}{\gamma H}L_f\right] \tag{7-17}$$

in which L_f = length factor defined as

$$L_f = \frac{L_2}{L_1 + L_2} \tag{7-18}$$

in which L_1 = length of the failure arc in soil 1 and L_2 = length of the failure arc in soil 2.

Values of N_c and L_f for various combinations of α and β are presented in Fig. 7-35 with D values of 0.2, 0.4, and 0.6. When $D = 0$, Eq. (7-6) in conjunction with Fig. 7-7 can be used instead.

Example 7.19 For the slope shown in Fig. 7-34 with $\alpha = 15°$, $\beta = 35°$, $\gamma = 125$ pcf (19.6 kN/m^3), c_1 = 1,200 psf (57.5 kN/m^2), and c_2 = 1,500 psf (71.8 kN/m^2), determine the factor of safety.

Solution $c_1/\gamma H = 1{,}200/(125 \times 50) = 0.192$; $c_2/\gamma H = 1{,}500/(125 \times 50) = 0.24$; when $D = 0.6$, from Fig. 7-35, $N_c = 5.4$ and $L_f = 0.11$; from Eq. (7-17), $F = 5.4 \times [0.192 \times (1 - 0.11) + 0.24 \times 0.11] = 1.065$, which checks with the 1.067 obtained by LEAME.

In the previous book (Huang 1983), charts for N_{f1} and and N_{f2}, similar to Eq. (7-12), were presented for the effective stress analysis of triangular fills on soil slopes. Because the location of most critical circle is based on soils with $\phi = 0$ and the actual soils have both c and ϕ, the factor of safety thus obtained is inaccurate, so those charts for effective stress analysis are not presented here.

Huang (1977b) also presented stability coefficients for sidehill benches in a series of tables with D ranging from 0 to 1, α from 0° to 30°, and β from 5° to 40°. Each combination consists of three cases: case 1 for soils with $\phi = 0$, case 2 for soils with $\phi = 30°$ and $c/\gamma H = 0.025$, and case 3 for soils with $c = 0$. Depending on the shear strength of the actual soils, one of these cases can be selected to compute the factor of safety.

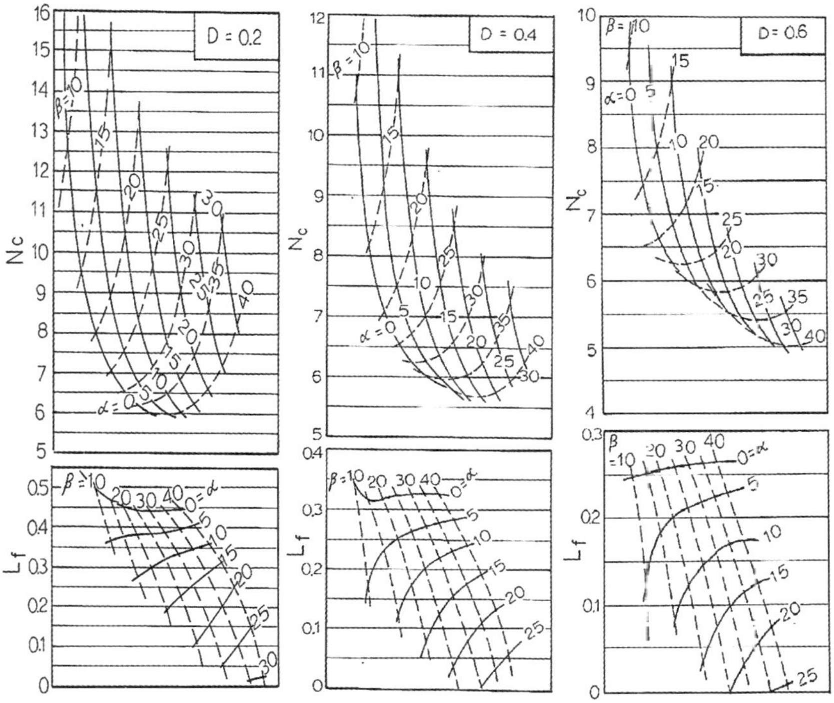

Fig. 7-35. Charts for total stress analysis of triangular fills on soil slopes (Huang 1978b)

7.9 Friction Circle Method

It was shown in Fig. 1-3 that a circular failure surface with $\phi = 0$ is statically determinate, because the three forces due to weight, cohesion, and normal reaction can be determined by considering both force and moment equilibrium. If the soil in the slope has both cohesion and internal friction, the problem becomes statically indeterminate and the friction circle method, originally proposed by Taylor (1937), can be used. Although this method can be applied only to a homogeneous slope and is of limited utility, an understanding of the method will give insight into the problems of slope stability.

Fig. 7-36 shows the forces in a stability analysis by the friction circle method. A circular failure surface of radius R and a concentric circle of radius $R\sin\phi_d$ are shown, where ϕ_d is the developed friction angle. Any line tangent to the inner circle must intercept the failure circle at an obliquity ϕ_d. This inner circle is called the friction circle. The forces considered in this analysis include the driving force, D, which consists of weight, seismic force, and neutral force; the resultant force owing to cohesion, T; and the resultant of normal and frictional forces along the failure arc, P. The magnitude and the line of application of D are known. The magnitude of T is cL_c/F, where L_c is the chord length, and F is the unknown factor

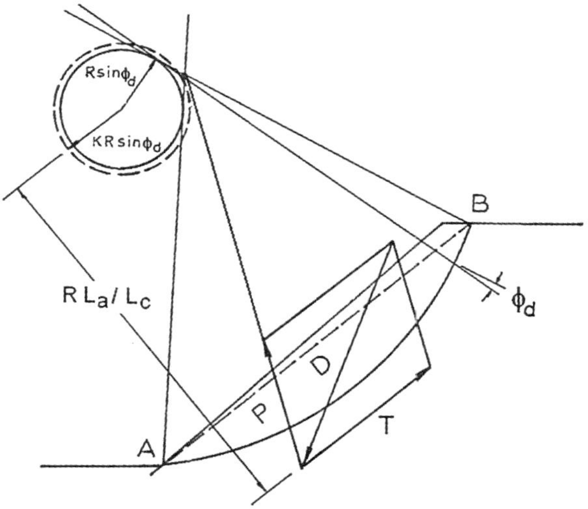

Fig. 7-36. Forces in friction circle method

of safety. As explained in Section 1.2.1, the line of application of T is parallel to chord AB at a distance of RL_a/L_c from the center of circle, where L_a is the arc length. To satisfy moment equilibrium, the three forces, D, T, and P, must meet at the same point. The problem now at hand is how to determine the direction of P. Once the direction of P is known, a parallelogram can be drawn, and the magnitude of T, as well as P, can be determined. The direction of P cannot be determined from statics unless a distribution of the normal stress along the failure arc is assumed.

One possible, although somewhat trivial, assumption is that all of the normal stress is concentrated at a single point along the failure arc. In such a case, P is tangent to the friction circle, and a lower bound of F is obtained. Another assumption is that the normal stress is concentrated entirely at the two end points of the failure arc. In this case, the resultant of these two end forces is tangent to a circle slightly larger than the friction circle with a radius of $KR\sin\phi_d$, where K is a coefficient greater than unity, and an upper bound of F is obtained. Taylor (1937) computed the factor of safety by assuming the normal stress is distributed uniformly or as a half sine curve. He found that the coefficient K depends on the central angle, as shown in Fig. 7-37. Intuitively, the use of a half sine curve with a maximum normal stress at center and zero stress at both ends should provide a quite realistic value for the factor of safety. The direction of P shown in Fig. 7-36 is based on the assumption that the forces are concentrated at the two end points, so their resultant should pass through the intersection of the two tangents to the friction circle.

Whitman and Moore (1963) applied different normal stress assumptions to determine the factor of safety of the slope shown in Fig. 7-38 by the friction circle method. By assuming that the soil has an effective cohesion of 90 psf (4.3 kPa), an effective friction angle of 32°, and a total unit weight of 125 pcf (19.7 kN/m^3),

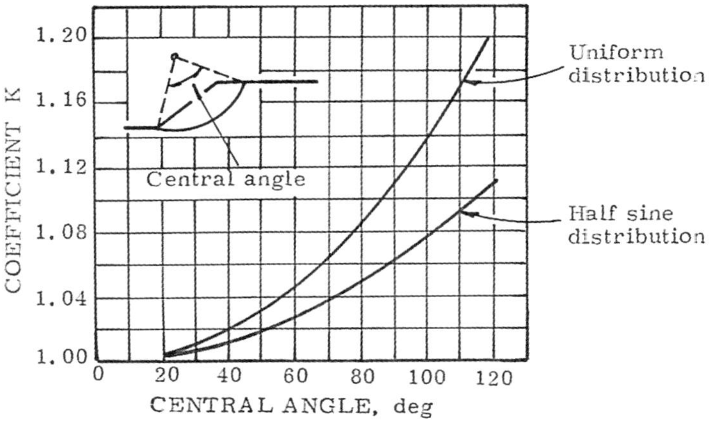

Fig. 7-37. Coefficient K of friction circle (Taylor 1937)

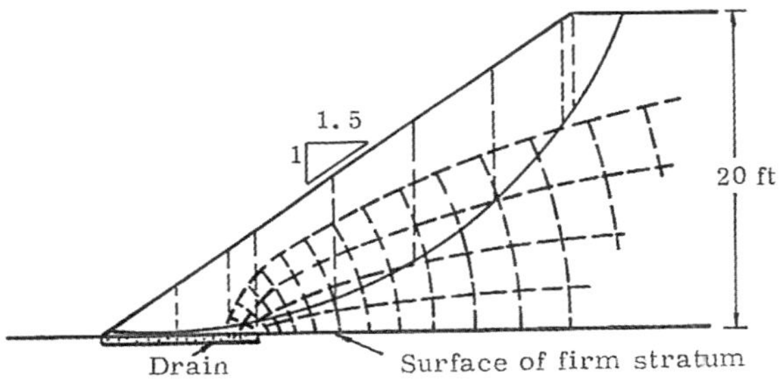

Fig. 7-38. Slope analyzed by friction circle method (Whitman and Moore 1963, © Associacao Brasileira de Mecanica dos Solos)
Note: 1 ft = 0.305 m

they found that the upper and lower bounds for the factor of safety are 1.60 and 1.27, respectively. Assuming that the normal effective stresses are distributed as a half sine curve, the factor of safety is 1.34.

To use the method, a safety factor with respect to the friction angle, F_ϕ, is assumed and the developed friction angle determined by

$$\phi_d = \tan^{-1}\left(\frac{\tan\phi}{F_\phi}\right) \tag{7-19}$$

Based on the central angle, a friction circle with a radius of $KR\sin\phi_d$ can be constructed and the magnitude of T determined. The factor of safety with respect to cohesion is

$$F_c = \frac{cL_c}{T} \tag{7-20}$$

As explained in Section 1.2.1, cL_c is the resisting force due to cohesion along the failure arc and T is the driving force along the failure arc to keep equilibrium.

To determine the factor of safety with respect to shear strength, three friction circles must be drawn to obtain three pairs of F_ϕ and F_c. A graphical method then can be used so that $F_\phi = F_c$, as illustrated by the following example.

Example 7.20 Fig. 7-39(a) shows a slope to be analyzed by the friction circle method. The soil has a cohesion of 400 psf (19.2 kPa), a friction angle of 20°, and a total unit weight of 125 pcf (19.7 kN/m^3). Assuming the normal stress as a half sine distribution, determine the factor of safety.

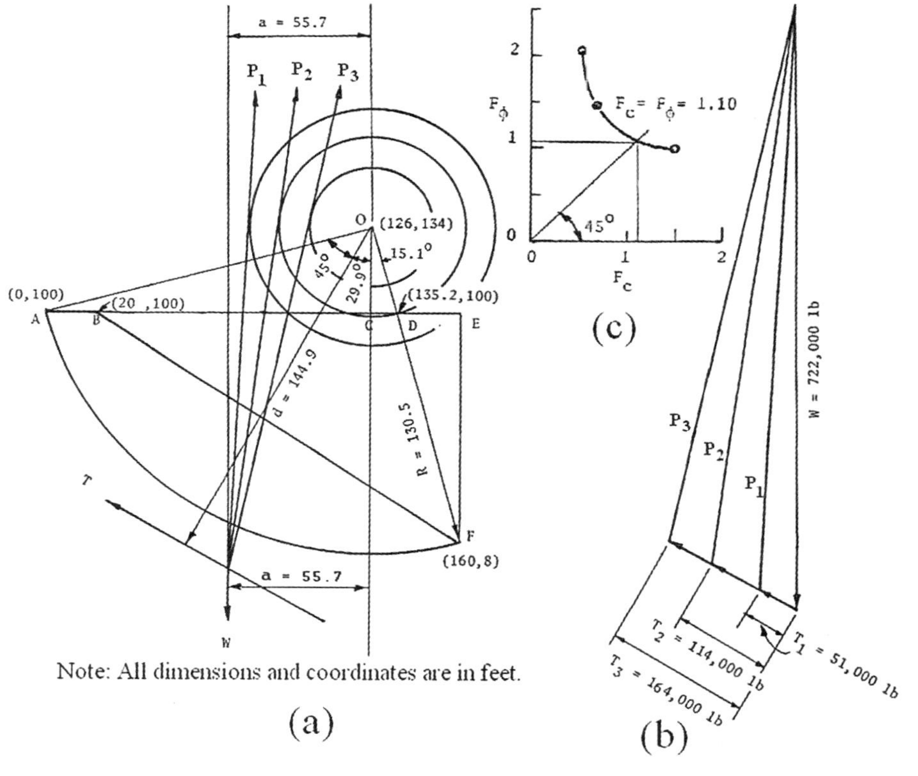

Fig. 7-39. Example 7.20

Solution First, determine the magnitude and line of application of weight, W. From Fig. 7-39(a), the area of failure mass = OAF − OAC − OCD − BEF + DEF = $\pi(130.5)^2/4 - 0.5 \times 34 \times 126 - 0.5 \times 34 \times 9.2 - 0.5 \times 92 \times 140 + 0.5 \times 92 \times 24.8 = 13{,}376 - 2142 - 156 - 6440 + 1141 = 5779\ \text{ft}^2$. $W = 5{,}779 \times 125 = 722{,}375\ \text{lb}$.

The center of gravity of area OAF is on the bisector of the central angle at a distance of b from the center. From Eq. (1-10),

$$b = \frac{4}{3} \times 130.5 \left[\sin\left(\frac{90^\circ}{2}\right) \Big/ \left(\frac{\pi}{2}\right) \right] = 78.33 \text{ ft}$$

Take the moment at center $M = 125 \times [13{,}376 \times 78.33 \times \sin 29.9 - 2142 \times 126/3 - 156 \times (-9.3)/3 - 6440 \times (126 - 113.3) + 1141 \times (-9.2 - 2 \times 24.8/3)] = 125 \times (522{,}286 - 89{,}964 + 484 - 81{,}788 - 29{,}362) = 40{,}207{,}000$ ft-lb. The distance from W to the center of the circle is $a = 40{,}207{,}000/722{,}375 = 55.7$ ft.

The resultant force due to cohesion, T, is parallel to chord BF at a distance of RL_a/L_c from the center. With $L_a = 2\pi \times 130.5/4 = 204.99$ ft (62.5 m) and $L_c = [(160)^2 + (100 - 8)^2]^{0.5} = 184.56$ ft (56.3 m), $d = 130.5 \times 204.99/184.56 = 144.9$ ft.

The location of W and T is plotted in Fig. 7-39(a). The resultant of normal and frictional forces, P, must pass through the intersection point of W and T and be tangent to the friction circle.

By assuming F_ϕ of 1, 1.4, and 2, respectively, three friction circles can be drawn. With a central angle of 90°, from Fig. 7-37, $K = 1.062$. The radii, R_f, of the friction circles are computed as follows:

When $F_\phi = 1$, from Eq. (7-19), $\phi_d = \phi = 20°$, $R_f = 1.062 \times 130.5 \times (\sin 20°) = 47.4$ ft.
When $F_\phi = 1.4$, $\phi_d = \tan^{-1}(\tan 20°/1.4) = 14.6°$, $R_f = 1.062 \times 130.5 \times (\sin 14.6°) = 34.9$ ft.
When $F_\phi = 2$, $\phi_d = \tan^{-1}(\tan 20°/2) = 10.3°$, $R_f = 1.062 \times 130.5 \times (\sin 10.3°) = 24.8$ ft.

Next, draw the force diagram shown in Fig. 7-39(b). With the magnitude of W known, draw P and T parallel to those in Fig. 7-39(a). The magnitudes of T for the three values of F_ϕ can be scaled from the diagram and are noted in the figure. The factors of safety with respect to cohesion, F_c, are computed as follows:

When $F_\phi = 1$, $T1 = 51{,}000$ lb, from Eq. (7-20), $F_c = 400 \times 184.56/51{,}000 = 1.45$.
When $F_\phi = 1.4$, $T2 = 114{,}000$ lb, $F_c = 400 \times 184.56/114{,}000 = 0.65$.
When $F_\phi = 2$, $T2 = 164{,}000$ lb, $F_c = 400 \times 184.56/164{,}000 = 0.45$.

Finally, plot F_ϕ versus F_c, as shown in Fig. 7-39(c), and draw a smooth curve through the three points. The intersection of the 45° line with the curve gives $F = F_\phi = F_c = 1.10$.

7.10 Logarithmic-Spiral Method

In using the friction circle method or the method of slices, the distribution of forces along the failure arc or on both sides of a slice must be assumed arbitrarily. This difficulty can be overcome if a logarithmic spiral is used as a failure surface. No matter what the magnitude of normal forces on the failure surface may be, the property of the logarithmic spiral is such that the resultant of the normal and

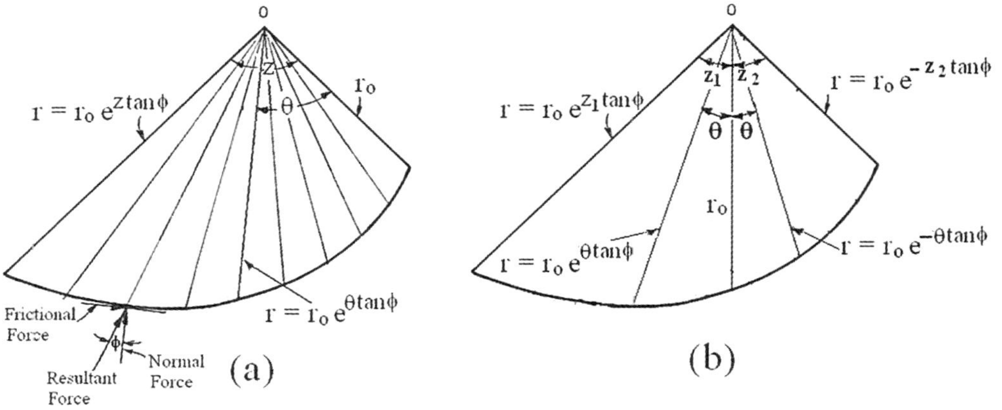

Fig. 7-40. *Characteristics of a logarithmic spiral*

frictional forces always will pass through the origin of the spiral. Consequently, when a moment is taken about the origin, the combined effect of normal and frictional forces is nil, and only the weight and cohesion moments need to be considered. This logarithmic-spiral method was first suggested by Taylor (1937) for stability analysis.

7.10.1 Factor of Safety with Respect to Cohesion

Fig. 7-40(a) illustrates the construction of a logarithmic spiral. The equation of a logarithmic spiral in polar coordinates can be expressed as

$$r = r_o e^{\theta \tan \phi} \tag{7-21}$$

in which r = radius from origin to logarithmic spiral, r_o = initial radius, θ = angle between the initial radius and radius r, in radians, and ϕ = angle of internal friction of the soil. Starting from the center O, a number of radial lines can be drawn. The first line has a length r_o and all the other lines, each making a different angle θ with the first line, have a length computed by Eq. (7-21). The shape of the logarithmic spiral is controlled by the friction angle, ϕ. The larger the ϕ, the more weight is placed near the toe, and the smaller the overturning moment. The figure is based on a ϕ of 20°. Note that a line normal to the logarithmic spiral always makes an angle of 20° with the radial line. Any radial line can be assigned as r_o, as shown in Fig. 7-40(b). Eq. (7-21) applies if θ is measured clockwise, or the length of the radial lines increases with the increase in θ. If θ is measured counterclockwise, or the length of the radial lines decreases with the increase in θ, θ is considered negative, so a negative sign should be placed before θ in Eq. (7-21).

Fig. 7-41 shows a logarithmic spiral passing through the toe of a simple slope with an angle, i, and a height, H. The origin, O, of the logarithmic spiral is located by two arbitrary angles t and z, in which z is the angle between the initial radius,

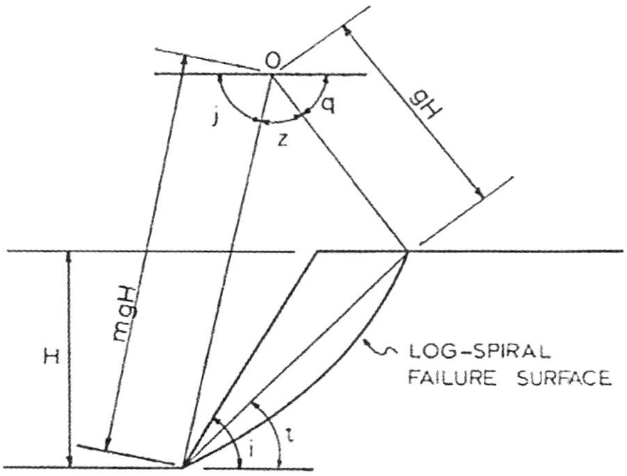

Fig. 7-41. Logarithmic spiral passing through toe

gH, and the final radius, mgH. The following equations were presented by Taylor (1937):

$$m = e^{z \tan \phi} \quad (7\text{-}22)$$

in which z = central angle in radians.

$$g = \frac{1}{\sin t \sqrt{1 + m^2 - 2m \cos z}} \quad (7\text{-}23)$$

$$j = t + \sin^{-1} \frac{\sin z}{\sqrt{1 + m^2 - 2m \cos z}} \quad (7\text{-}24)$$

$$q = \pi - j - z \quad (7\text{-}25)$$

in which t, z, j, and q are angles in radians or in degrees, whichever is more convenient. If in degrees, the π in Eq. (7-25) should be changed to 180°. The moment, M_w, about the origin due to the weight of soil mass is

$$M_w = \frac{\gamma H^3 g^3}{3}\left[\frac{m^3 \sin j - \sin q - 3\tan\phi\left(m^3 \cos j + \cos q\right)}{9\tan^2\phi + 1}\right] - \frac{\gamma H^3}{6}\left[g^3 \sin^3 q\left(\cot^2 q - \cot^2 j\right) + \cot^2 i - 3mg \cos j\left(\cot i - \cot j\right) - \cot^2 j\right] \quad (7\text{-}26)$$

in which γ = unit weight of soil. The moment, M_c, due to the cohesion along the logarithmic spiral is

$$M_c = \frac{c_d g^2 H^2}{2 \tan \phi}\left(m^2 - 1\right) \quad (7\text{-}27)$$

in which c_d is the developed cohesion, or the cohesion actually developed over the failure surface. Let $M_w = M_c$, the developed cohesion, c_d, can be determined. The factor of safety with respect to cohesion is

$$F = \frac{c}{c_d} \tag{7-28}$$

Example 7.21 Eqs. (7-23), (7-24), (7-26), and (7-27) were presented by Taylor (1937). Prove that these equations are theoretically correct.

Solution Based on the law of cosine, or the well-known trigonometry formula $a^2 = b^2 + c^2 - 2bc \cos A$, from ΔOAB in Fig. 7-42(a),

$$\left(\frac{H}{\sin t}\right)^2 = (gH)^2 + (mgH)^2 - 2m(gH)^2 \cos z \quad \text{or} \quad \frac{1}{\sin^2 t} = g^2\left(1 + m^2 - 2m\cos z\right)$$
$$\text{so } g = \frac{1}{\sin t\sqrt{1 + m^2 - 2m\cos z}} \tag{7-29}$$

Based on the law of sine, $a/\sin A = b/\sin B$, from ΔOAB in Fig. 7-42(a),

$$\frac{gH}{\sin(j-t)} = \frac{\dfrac{H}{\sin t}}{\sin z} \quad \text{or} \quad \sin(j-t) = g\sin z \sin t \quad \text{or} \quad \sin(j-t) = \frac{\sin z}{\sqrt{1 + m^2 - 2m\cos z}}$$
$$\text{so } j = t + \sin^{-1}\frac{\sin z}{\sqrt{1 + m^2 - 2m\cos z}} \tag{7-30}$$

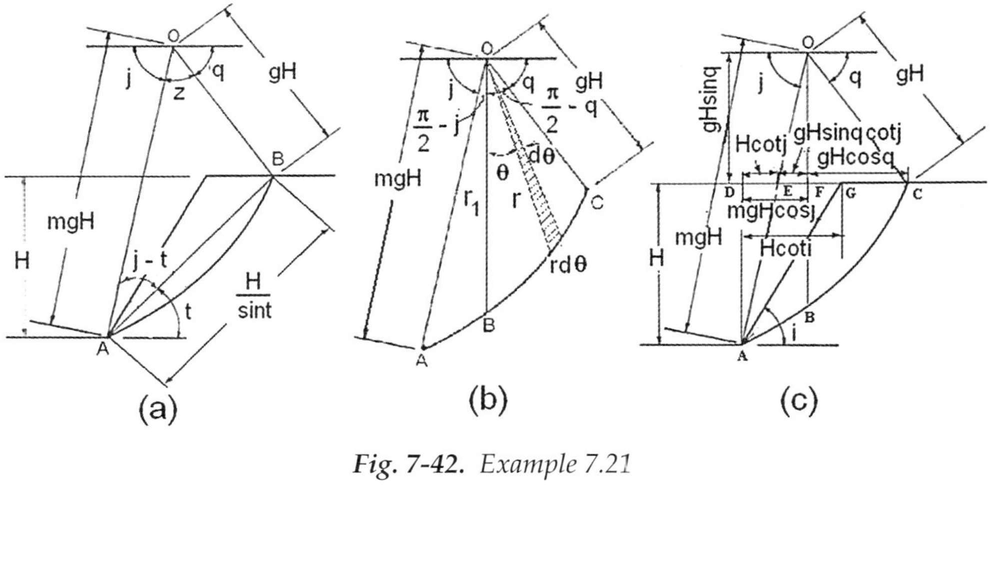

Fig. 7-42. Example 7.21

To determine M_w, Eq. (7-26) is divided into two parts. The first part is the moment due to area OABC, as shown in Fig. 7-42(b), and the second part is the moments due to several triangular areas, as shown in Fig. 7-42(c). The moment of the failure mass is the difference between the two parts.

In the first part, area OABC is divided into two areas, OAB and OBC, by a vertical line with a length r_1. From Eq. (7-21), $r_1 = gHe^{(\pi/2-q)\tan\phi}$. The moment due to OBC causes the mass to overturn and is positive, whereas that due to OAB is negative.

First, consider the moment due to the weight of area OBC. The hatched triangle shown in Fig. 7-42(b) has an area of $r^2 d\theta/2$ and a moment arm of $2r\sin\theta/3$, so the moment due to weight of area OBC can be obtained by integration:

$$M_{OBC} = \frac{\gamma}{3}\int_0^{\frac{\pi}{2}-q} r^3 \sin\theta d\theta = \frac{\gamma r_1^3}{3}\int_0^{\frac{\pi}{2}-q} e^{-3\theta\tan\phi}\sin\theta d\theta$$

$$= \frac{\gamma g^3 H^3}{3}\left[\frac{e^{3\left(\frac{\pi}{2}-q\right)\tan\phi} e^{-3\theta\tan\phi}(-3\tan\phi\sin\theta - \cos\theta)}{9\tan^2\phi+1}\right]_0^{\frac{\pi}{2}-q}$$

$$= \frac{\gamma g^3 H^3}{3}\left(\frac{-3\tan\phi\cos q - \sin q + e^{3\left(\frac{\pi}{2}-q\right)\tan\phi}}{9\tan^2\phi+1}\right)$$

Next, consider the moment due to weight of area OAB:

$$M_{OAB} = \frac{\gamma}{3}\int_0^{\frac{\pi}{2}-j} r^3 \sin\theta d\theta = \frac{\gamma r_1^3}{3}\int_0^{\frac{\pi}{2}-j} e^{3\theta\tan\phi}\sin\theta d\theta$$

$$= \frac{\gamma g^3 H^3}{3}\left[\frac{e^{3\left(\frac{\pi}{2}-q\right)\tan\phi} e^{3\theta\tan\phi}(3\tan\phi\sin\theta - \cos\theta)}{9\tan^2\phi+1}\right]_0^{\frac{\pi}{2}-j}$$

$$= \frac{\gamma g^3 H^3}{3}\left(\frac{m^3(3\tan\phi\cos j - \sin j) + e^{3\left(\frac{\pi}{2}-q\right)\tan\phi}}{9\tan^2\phi+1}\right)$$

$$M_{OABC} = M_{OBC} - M_{OAB} = \frac{\gamma H^3 g^3}{3}\left[\frac{m^3\sin j - \sin q - 3\tan\phi(m^3\cos j + \cos q)}{9\tan^2\phi+1}\right]$$

which checks with the first part of Eq. (7-26).

In the second part of Eq. (7-26), the moments due to the weights of the triangular areas are

$$\begin{aligned} M_{OFC} + M_{OEF} + M_{ADG} - M_{ADE} &= \frac{\gamma}{2}\Bigg[(gH\sin q)(gH\cos q)\left(\frac{gH\cos q}{3}\right) \\ &+ (gH\sin q)(gH\sin q\cot j) \\ &\left(-\frac{gH\sin q\cot j}{3}\right) + H(H\cot i)\left(\frac{H\cot i}{3} - mgH\cos j\right) \\ &- H(H\cot j)\left(\frac{H\cot j}{3} - mgH\cos j\right) \\ &= \frac{\gamma H^3}{6}[g^3\sin q\cos^2 q - g^3\sin^3 q\cot^2 j + \cot^2 i \\ &- 3mg\cot i\cos j - \cot^2 j + 3mg\cot j\cos j] \\ &= \frac{\gamma H^3}{6}[g^3\sin^3 q(\cot^2 q - \cot^2 j) + \cot^2 i \\ &- 3mg\cos j(\cot i - \cot j) - \cot^2 j] \end{aligned}$$

which checks with the second part of Eq. (7-26).

Eq. (7-27) is much easier to derive. The shear resistance along the failure surface for a small incremental length $r d\theta$ is $c_d r d\theta$. The moment about the center is $c_d r^2 d\theta$, where $r = gHe^{\theta\tan\phi}$. The moment due to cohesion can be determined by integration:

$$\begin{aligned} M_c &= \int_0^z c_d (gHe^{\theta\tan\phi})^2 d\theta = c_d g^2 H^2 \int_0^z e^{2\theta\tan\phi} d\theta = \frac{c_d g^2 H^2}{2\tan\phi}\left[e^{2\theta\tan\phi}\right]_0^z \\ &= \frac{c_d g^2 H^2}{2\tan\phi}(e^{2z\tan\phi} - 1) = \frac{c_d g^2 H^2}{2\tan\phi}(m^2 - 1) \end{aligned}$$

which checks with Eq. (7-27).

Taylor's method assumes that the angle of internal friction, ϕ, although fully mobilized, is still not sufficient to resist the overturning moment, so part of the shear resistance is carried by the cohesion. If the angle of internal friction of the soil exceeds the angle of the slope, the developed cohesion, c_d, becomes negative, and a usable factor of safety cannot be obtained. Thus, the method is applicable only to relatively steep slopes or to moderate slopes with weak soils.

Example 7.22 Fig. 7-43 shows a 1.5:1 slope. The soil in the slope has a cohesion of 600 psf (28.7 kPa), a friction angle of 20°, and a total unit weight of 125 pcf (19.7 kN/m^3). A logarithmic spiral passing through the toe is defined by an angle, t, of 29.9° and a central angle, z, of 90°, or $\pi/2$. If the angle of internal friction is developed fully, determine the factor of safety with respect to cohesion.

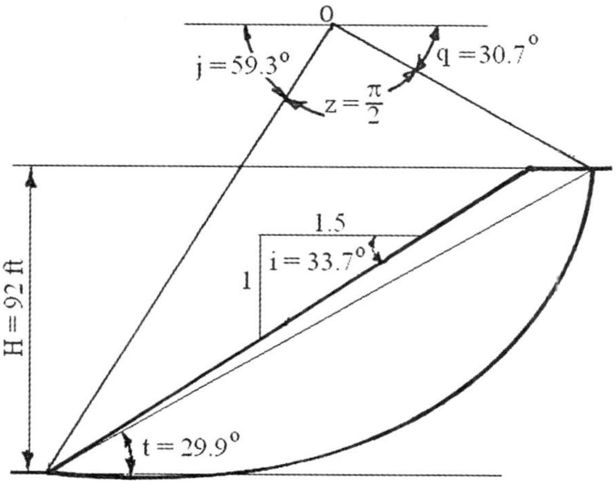

Fig. 7-43. Example 7.22

Solution From Eq. (7-22), $m = e^{0.5\pi \tan 20} = 1.771$ and $m^3 = 5.555$. From Eq. (7-23),

$$g = \frac{1}{\sin 29.9° \sqrt{1 + (1.771)^2 - 2 \times 1.771 \times \cos 90°}} = \frac{1}{0.498 \times 2.034} = 0.987$$

From Eq. (7-24), $j = 29.9° + \sin^{-1}(\sin 90°/2.034) = 59.3°$. From Eq. (7-25), $q = 180° - 59.3° - 90° = 30.7°$. From Eq. (7-26),

$$M_w = \frac{125(92)^3(0.987)^3}{3}\left[\frac{5.555 \sin 59.3° - \sin 30.7° - 3\tan 20°(5.555\cos 59.3° + \cos 30.7°)}{9\tan^2 20° + 1}\right]$$

$$- \frac{125(92)^3}{6}[(0.987)^3 \sin^3 30.7°(\cot^2 30.7° - \cot^2 59.3°) + \cot^2 33.7°$$

$$- 3 \times 1.771 \times 0.987 \cos 59.3°(\cot 33.7° - \cot 59.3°) - \cot^2 59.3°)]$$

$$= 31,196,344 \times (0.230)/2.192 - 16,222,667[0.128(2.836 - 0.353)$$

$$+ 2.248 - 2.677(1.499 - 0.594) - 0.353]$$

$$= 3,273,339 + 3,404,505 = 6,677,844 \text{ ft-lb}$$

From Eq. (7-27),

$$M_c = \frac{c_d(0.987)^2(92)^2}{2\tan 20°}\left[(1.771)^2 - 1\right] = 24,199 c_d$$

Let $M_w = M_c$, $c_d = 6,677,844/24,199 = 275$ psf, so the factor of safety = 600/275 = 2.182.

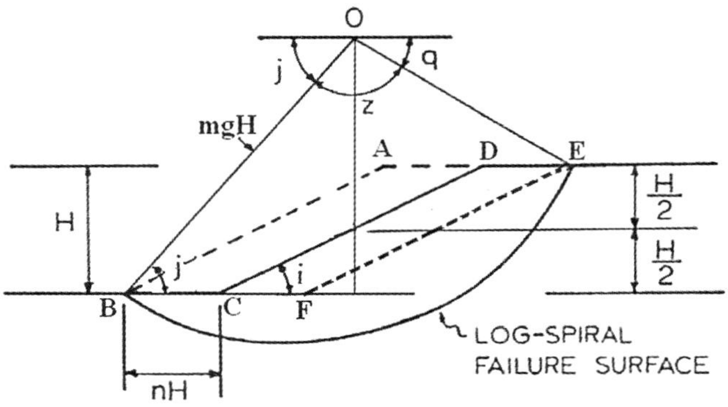

Fig. 7-44. Logarithmic spiral passing below toe

The logarithmic spiral passing through the toe may not have the lowest factor of safety. It is therefore necessary to examine a failure surface passing below the toe, as shown in Fig. 7-44. It can be seen easily that the most dangerous situation occurs when the origin of the logarithmic spiral lies vertically above the mid-height of the slope, because the removal of area ABCD reduces the counterweight and decreases the factor of safety. However, further removal of area CDEF reduces the overturning moment and increases the factor of safety. From geometry, the distance nH, from the failure surface to the toe, is

$$nH = mgH\cos j - \frac{H}{2}\cot i \tag{7-31}$$

The increase in overturning moment, M'_w, due to the removal of area ABCD is

$$M'_w = \frac{\gamma}{2}H(nH)^2 = \frac{\gamma}{2}H^3\left(mg\cos j - \frac{1}{2}\cot i\right)^2 \tag{7-32}$$

Let $M_w + M'_w = M_c$; the developed cohesion, c_d, can be determined and the factor of safety obtained.

Example 7.23 The logarithmic spiral in Example 7.22 passes through the toe and is not the most critical, because its origin does not lie vertically above the mid-height of the slope. Determine the most critical location of the logarithmic spiral and the minimum factor of safety with respect to cohesion.

Solution The dashed line in Fig. 7-45 is the location of the original slope surface. To obtain a lower factor of safety, the origin of the logarithmic spiral must move outward toward the toe, or the slope surface moves inward a distance of nH. From Eq. (7-31), $nH = 1.771 \times 0.987 \times 92 \times \cos 59.3° - 0.5 \times 92 \times \cot 33.7° = 13.1$ ft, so the logarithmic spiral must move horizontally toward the toe a distance of 13.1 ft.

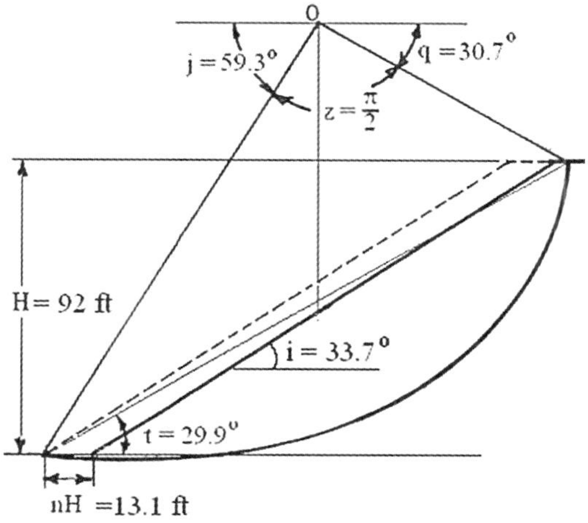

Fig. 7-45. Example 7.23

From Eq. (7-32), $M'_w = 0.5 \times 125 \times 92 \times (13.1)^2 = 986{,}758$ ft-lb. From Example 7.22, M_w = 6,677,844 ft-lb and M_c = 24,199c_d. Let $M_w + M'_w = M_c$, or 6,677,844 + 986,758 = 24,199c_d, so c_d = 7,664,602/24,199 = 316 psf. The factor of safety = 600/316 = 1.899.

7.10.2 Factor of Safety with Respect to Shear Strength

The factor of safety determined by Taylor, as described, is that with respect to cohesion, instead of that with respect to shear strength (i.e., both cohesion and angle of internal friction). In view of the fact that the design and evaluation of slopes by most engineering organizations are based on the factor of safety with respect to shear strength, a modification of Taylor's method is needed. Furthermore, modern construction usually requires greater safety with the angle of slope generally smaller than the angle of internal friction of the soil, thus making Taylor's method inapplicable. This problem can be overcome if a factor of safety greater than 1 is applied to both cohesion and internal friction.

To determine the factor of safety with respect to strength, all the equations derived by Taylor are still valid, except that any terms involving $\tan\phi$ must be divided by F and the term c_d in Eq. (7-27) must be replaced by c/F. Thus, Eqs. (7-22), (7-26), and (7-27) become

$$m = e^{\left(\frac{z \tan\phi}{F}\right)} \tag{7-33}$$

$$M_w = \frac{\gamma H^3 g^3}{3}\left[\frac{m^3 \sin j - \sin q - \dfrac{3\tan\phi}{F}\left(m^3 \cos j + \cos q\right)}{\dfrac{9\tan^2\phi}{F^2} + 1}\right] \tag{7-34}$$

$$M_c = \frac{cg^2H^2}{2\tan\phi}(m^2 - 1) \tag{7-35}$$

Similar to the previous procedure, by letting $M_w + M'_w = M_c$, the factor of safety, F, can be determined by trial and error using a spreadsheet.

Example 7.24 Same as Example 7.23, but determine the factor of safety with respect to shear strength.

Solution Table 7-1 is the spreadsheet for computing the factor of safety by trial and error. The first two rows are the input parameters. The variables to be evaluated are listed in the first column, and the equations to be used are listed in the last column. Several factors of safety were assumed. The purpose herein is to find a factor of safety that gives the value of $M_w + M'_w - M_c$ as close to 0 as possible. It was found that when $F = 1.303$, Value = –12,766; when $F = 1.304$, Value = 16,382. Because –12,766 is closer to 0 than is 16,382, the factor of safety is 1.303.

Table 7-1. *Spreadsheet for Analyzing Logarithmic-Spiral Failure Surface*

c in psf = 750	**tan ϕ = 0.364**		**γ in pcf = 125**		**H in ft = 92**	
t in rad = 0.522	**z in rad = 1.571**		**i in rad = 0.588**			
Assumed F	1.200	1.300	1.302	1.303	1.304	
m	1.610	1.553	1.551	1.551	1.550	Eq. 7-22
g	1.058	1.086	1.086	1.087	1.087	Eq. 7-23
j	1.078	1.094	1.094	1.095	1.095	Eq. 7-24
q	0.493	0.476	0.476	0.476	0.476	Eq. 7-25
M_w	12,210,214	14,374,447	14,414,633	14,434,684	14,454,706	Eq. 7-26
M_c	15,550,916	14,501,651	14,482,003	14,472,197	14,462,403	Eq. 7-27
M'_w	155,734	26,808	25,424	24,747	24,080	Eq. 7-32
Value = $M_w + M'_w - M_c$	–3,184,968	–100,396	–41,946	–12,766	16,382	
				F = 1.303		

Huang and Avery (1976) developed a computer program using a logarithmic spiral to determine the factor of safety of homogeneous slopes. Let $M_w + M'_w = M_c$, and a quadratic equation in the following general form is obtained:

$$AF^2 + BF + C = 0 \tag{7-36}$$

in which the coefficients, A, B, and C, are functions of not only γ, H, c, ϕ, t, and z but also of F itself. Eq. (7-36) can be solved by an iterative method. First, a value of F was assumed and a new F was computed by

$$F = \sqrt{\frac{-(BF + C)}{A}} \tag{7-37}$$

By using the new F as the assumed F, another new F was obtained. The process was repeated until the difference between the new F and the assumed F became negligible.

The program consists of three do loops: one for angle t, one for angle z, and another for F. Angle t starts from large to small at a specified interval, with the first angle slightly smaller than the angle of the slope. After a starting angle, z, and a specified interval are assigned, the factor of safety for the starting z and the next decreasing z will be computed. If the latter is smaller than the former, the factor of safety for each successive decreasing z will be determined until a lowest value is obtained. If greater, the movement will be in the opposite direction until a lowest factor of safety is found. Using the t and z with the lowest factor of safety as a new starting angle and an interval one-fourth of the original, the process is repeated until a new lowest factor of safety is obtained.

The result of the study by Huang and Avery shows that the factor of safety for a simple slope can be determined effectively by the logarithmic-spiral method. The disadvantage of this method lies in the requirement that the angle of internal friction of the soils be constant throughout the slope. However, the simple fact that the method satisfies moment equilibrium with no further assumptions other than the logarithmic-spiral failure surfaces makes possible the use of the method as a yardstick to check the accuracy of the other methods.

Summary

1. Given the slope angle, β, and the depth ratio, D, Taylor's chart for homogeneous slope with $\phi = 0$ can be used to find the stability number, $c_d/\gamma H$. By dividing the allowable cohesion, c, with the developed cohesion, c_d, the factor of safety can be found. The chart also identifies three types of failure circles: midpoint circle, toe circle, and slope circle. For a midpoint circle, the chart can give the distance, nH, from the toe to the point where the failure circle appears on the ground surface. If loads are placed adjacent to the toe to prevent failure by the midpoint circle, the failure surface will be changed to a toe circle and a reduced stability number can be read from the chart.

2. Similar to the case of $\phi = 0$, Taylor's chart for homogeneous slopes with both c and ϕ also identifies three types of failure circles. Given a developed friction angle, ϕ_d, a corresponding stability number, $c_d/\gamma H$, can be read from the chart. By assuming that the friction angle be fully developed, or $\phi_d = \phi$, the developed cohesion, c_d, can be found, and the factor of safety with respect to cohesion, F_c, can be computed by c/c_d. By assuming that the cohesion is developed fully, or $c_d = c$, the developed friction angle, ϕ_d can be found, and the factor of safety with respect to friction angle, F_ϕ, can be determined by $\tan\phi/\tan\phi_d$. To determine the factor of safety with respect to shear strength, a trial-and-error or graphical procedure can be used.

3. Charts based on the Fellenius or normal method are presented to determine the factor of safety of triangular fills on rock or stiff slopes. The rock surface or natural ground is assumed to be much stiffer than the fill, so all circular failure surfaces will lie entirely within the fill. Based on the dimensions of the fill, a cohesion number, N_c, and a friction number, N_f, can be found from the charts, and the factor of safety can be computed. Because the most critical circle used for developing these charts is based on $\phi = 0$, the charts give an accurate factor of safety for total stress analysis with $\phi = 0$ but may result in a factor of safety that is too high for effective stress analysis with both c' and ϕ'. To obtain more accurate results, a correction factor must be applied.

4. Similar charts are presented for trapezoidal fills on rock or stiff ground surfaces. The charts give an accurate factor of safety for total stress analysis with $\phi = 0$ but a slightly larger factor of safety for effective stress analysis with both c' and ϕ'. A factor of safety slightly higher than that by the normal method is desirable, because it checks more closely with the well-recognized simplified Bishop method. To obtain more accurate results, a trial-and-error process must be used to locate the minimum factor of safety.

5. Four sets of charts are presented for the effective stress analysis of homogeneous dams. The charts by Bishop and Morgenstern (1960) can be used for dams with $D = 0$, 0.25, and 0.5, where D is the depth ratio or the depth to bedrock divided by the height of the dam. By comparing the factors of safety with different depth ratios, the minimum factor of safety can be determined. The charts by Huang (1975) supplement Bishop's and Morgenstern's by adding a case of $D = \infty$, so if D is greater than 0.25 or 0.5, the minimum factor of safety can be determined right away and no comparison of different depth ratios is needed. The charts by Spencer (1967) also assume $D = \infty$ and can be used to determine the slope angle, β, if the factor of safety is given. This feature is different from all the other charts where the slope angle is given and the factor of safety is to be found. The charts by Morgenstern (1963) are different from the others, because they can be used only to determine the factor of safety due to rapid drawdown. Although all these charts can be used for full drawdown by assuming a pore pressure ratio of 0.5, Morgenstern's charts are the only ones suitable for analyzing the case of partial drawdown. In spite of the difference in the assumed location of the ledge (some at or near to the bottom of the dam and some at a great depth), Example 7.15 clearly shows that they all yield about the same factor of safety. In view of the preliminary nature of the charts, the difficulty in reading them accurately, and the insensitivity of the safety factor to the location of the ledge, it appears reasonable to assume that the ledge is located at a great depth, such as in the charts by Spencer and Huang, so a lowest and most conservative factor of safety can be obtained.

6. A new method is presented for determining the factor of safety of earth dams with two sets of soil parameters, one for the foundation and one for the embankment. Based on the normal method, the factor of safety can be expressed by Eq. (7-12), which contains two cohesion numbers, N_{c1} and N_{c2}, and two friction numbers, N_{f1} and N_{f2}. Note that subscripts 1 and 2 refer to the foundation and the embankment, respectively. Charts for determining these numbers in terms of slope, S, and depth ratio, D, are presented in Fig. 7-31. Similar to the use of Bishop's and Morgenstern's charts, the factors of safety at several depth ratios must be computed to determine which is minimum. This new method is different from the previous method in determining the location of the most critical circle. The new method assumes that the soils have an effective cohesion $c_1' = c_2' = 200$ psf (9.6 kPa) and an effective friction angle $o_1' = \phi_2' = 30°$, whereas the previous method assumes $\phi_1' = \phi_2' = 0$. As a result, these new charts are more accurate for effective stress analysis with a small cohesion and large friction angle, and the previous charts are more accurate for total stress analysis with a large cohesion and a small friction angle.

7. For the total stress analysis of earth dams with $\phi = 0$, the charts developed by Huang (1975) can be used. Given the slope, S, and the depth ratio, D, the cohesion number, N_c, which is the reciprocal of the stability number by Taylor's chart, can be found from the charts. The charts are different from Taylor's in that the cohesion numbers for various circles with centers at different coordinates (XH, YH) are shown, whereas Taylor's only shows the stability number for the most critical circle. Knowing the location of the circle, it is possible to apply the charts to nonhomogeneous dams consisting of several different soil layers. For a given circle, N_c can be found from Fig. 7-32, and the average cohesion, c, can be determined by measuring the length of the arc through each layer and finding the weighted average, so the factor of safety can be computed by Eq. (7-14). Theoretically, several circles, each with a center at a different YH, should be tried to determine which is most critical. Fortunately, the most critical circle for a nonhomogeneous dam is usually the same as that for a homogeneous dam, so only one or two circles need to be tried to determine which is most critical.

8. Charts are presented for the total stress analysis of sidehill benches with the soil in the bench being different from the soil in the natural ground. By assuming the natural ground as horizontal, the charts also can be used for the total stress analysis of earth dams and embankments.

9. The friction circle method is basically a graphical method for determining the safety factor of a homogeneous slope with both a cohesion and a friction angle. To make the problem statically determinate, an assumption of a normal stress distribution along the failure arc must be made. The most reasonable assumption is that the normal stress is distributed as a half sine curve with its maximum at the center and zero at both ends. Given

the developed friction angle, ϕ_d, the developed cohesion, c_d, can be determined. To determine the factor of safety with respect to shear strength, three different friction circles, each with a different ϕ_d, must be constructed, and the corresponding c_d is determined. The method is cumbersome to use and is of little utility. However, this method is of historical significance, because it was the earliest method suggested by Taylor (1937) in developing the stability charts presented in Figs. 7-1 and 7-3.

10. Similar to the friction circle method, the logarithmic-spiral method can be used to determine the safety factor of a homogeneous slope with both a cohesion and an angle of internal friction. However, the method is algebraic and can be programmed for a computer. The equations used in the method are derived, and the procedures to obtain the factor of safety are illustrated. Although the method is of limited applications and only can be applied to a simple embankment composed of one soil and not subjected to any pore pressure, the simple fact that it satisfies moment equilibrium with no further assumptions other than the logarithmic-spiral failure surfaces makes it possible to use this method as a yardstick to check the accuracy of the other methods.

Problems

7.1 An embankment with a height of 50 ft and a slope of 2:1 is placed on a 40-ft soil foundation underlain by rock, as shown in Fig. P7-1. If the embankment and the foundation consist of the same soil with a total unit weight of 125 pcf and the embankment fails immediately after construction, what should be the undrained shear strength of the soil? Where should the failure surface appear at the ground, as indicated by the distance, x, from the toe?
[Answer: 1,038 psf, 45 ft]

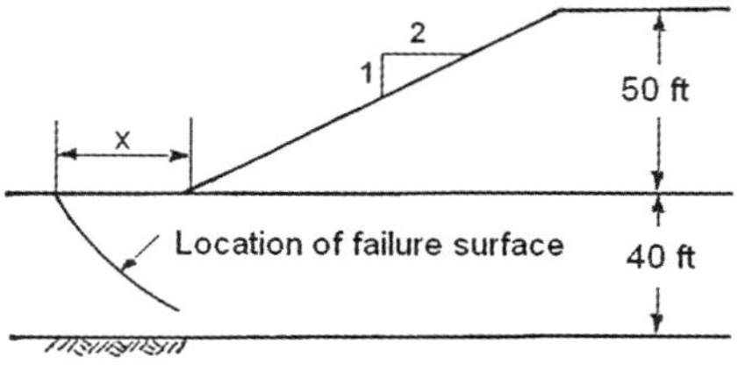

Fig. P7-1.

7.2 The cross section is the same as that shown in Fig. P7-1. If the soil has a cohesion of 500 psf, a friction angle of 20°, and a total unit weight of 125 pcf, determine the factor of safety with respect to shear strength using Fig. 7-3.
[Answer: 1.62]

7.3 Same as Problem 7.2, but determine the factor of safety using Fig. 7-5.
[Answer: 1.62]

7.4 Fig. P7-4 shows the dimensions of an embankment on a horizontal ledge. If the soil in the embankment has an undrained shear strength of 1,800 psf and a total unit weight of 120 pcf, determine the factor of safety using Fig. 7-7 and compare it with that by Fig. 7-1.
[Answer: 1.42, 1.40]

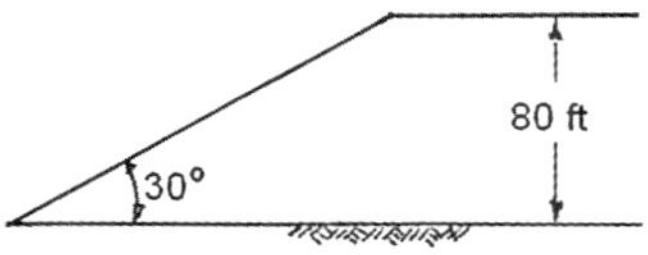

Fig. P7-4.

7.5 Fig. P7-5 shows the dimensions of a trapezoidal fill on rock. If the fill has an undrained shear strength of 1,800 psf and a total unit weight of 120 pcf, determine the factor of safety.
[Answer: 2.10]

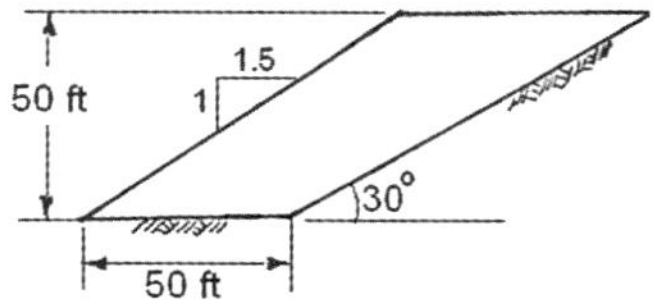

Fig. P7-5.

7.6 The cross section of the fill is shown in Fig. P7-4. If the soil has a cohesion of 800 psf, a friction angle of 25°, and a total unit weight of 120 pcf, determine the factor of safety by Figs. 7-7, 7-8, and 7-9.
[Answer: 1.73]

7.7 The cross section of a trapezoidal fill is shown in Fig. P7-5. If the soil has a cohesion of 800 psf, a friction angle of 25°, and a total unit weight of 120 pcf, determine the factor of safety by Figs. 7-13 and 7-14 using several combinations of B and α.
[Answer: 1.90]

7.8 Fig. P7-8 shows the dimensions of an embankment placed directly on a ledge and the soil parameters for stability analysis. Determine the factor of safety using Fig. 7-31.
[Answer: 1.51]

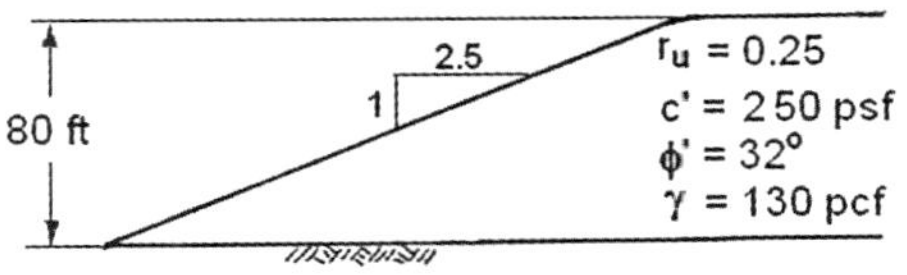

Fig. P7-8.

7.9 Fig. P7-9 shows the cross section of an earth dam placed directly on a soil foundation underlain by rock. The dam and foundation are formed by the same material with the soil parameters shown in the figure. Determine the factor of safety using (a) Fig. 7-18, (b) Fig. 7-25, and (c) Fig. 7-26.
[Answer: 1.22, 1.23, 1.21]

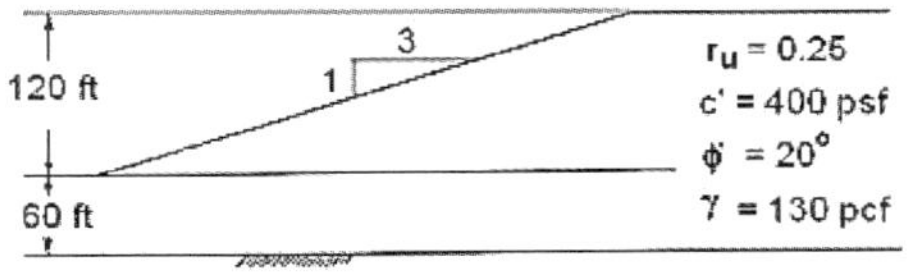

Fig. P7-9.

7.10 Fig. P7-10 shows the cross section of an earth dam and the soil parameters. The original pool elevation was at the top of the dam and was rapidly lowered 30 ft. Determine the factor of safety immediately after the drawdown.
[Answer: 1.30]

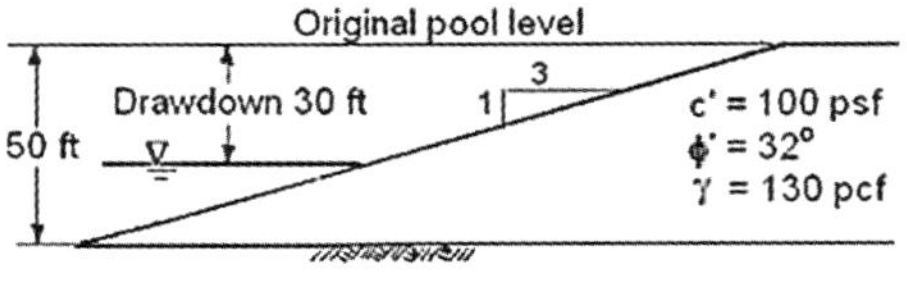

Fig. P7-10.

7.11 Fig. P7-11 shows the cross section of an embankment on a soil foundation. The shear strength and pore pressure ratio of the embankment are different from those of the foundation soil, as indicated in the figure. Determine the factor of safety using Fig. 7-31.
[Answer: 1.11]

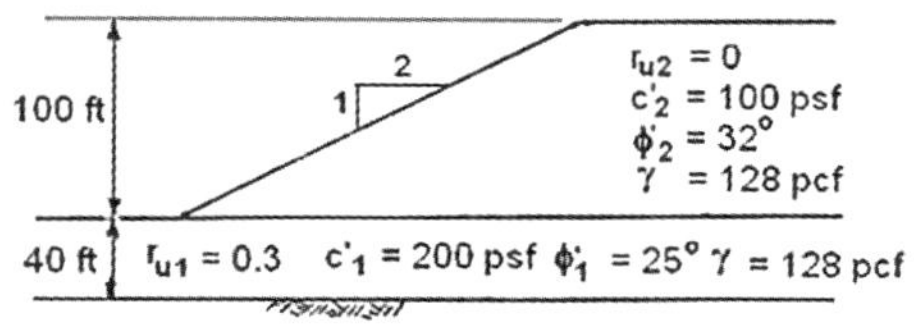

Fig. P7-11.

7.12 Fig. P7-12 shows the cross section of an earth dam and the soil parameters for total stress analysis. Determine the factor of safety using (a) Fig. 7-31, (b) Fig. 7-32, and (c) Fig. 7-35. Comment on the accuracy of each chart by pointing out which is most accurate and which is least accurate. Why?
[Answer: 1.80, 1.76, 1.72]

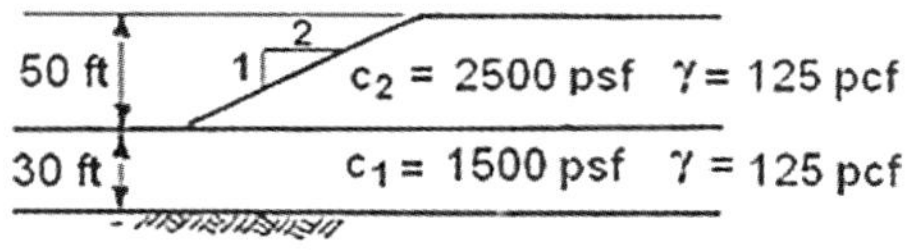

Fig. P7-12.

7.13 Fig. P7-13 shows the dimensions of a slope and the location of the circular failure surface. If the soil has a cohesion of 1,000 psf, a friction angle of 15°, and a total unit weight of 125 pcf, determine the factor of safety with respect to shear strength by the friction circle method.
[Answer: 1.37]

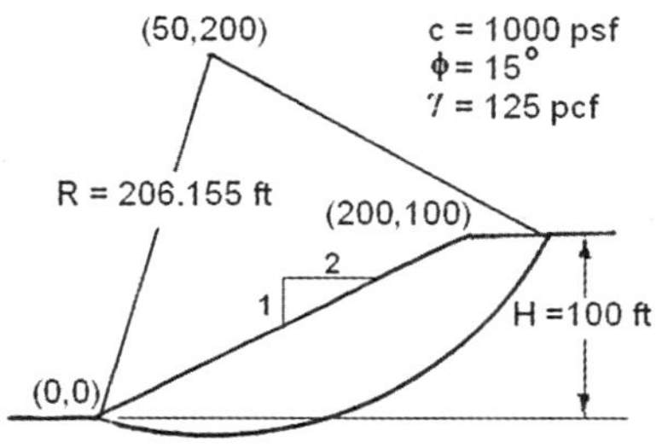

Fig. P7-13.

7.14 Fig. P7-14 shows the dimensions of a slope, the location of the logarithmic spiral, and the shear strength parameters of the soil. Determine the factor of safety with respect to cohesion.
[Answer: 29.237]

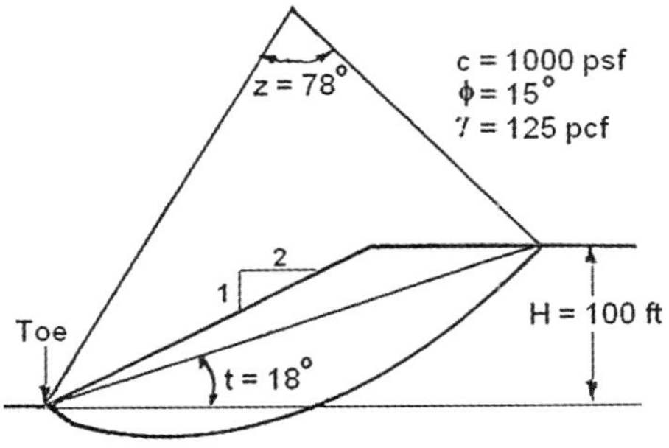

Fig. P7-14.

7.15 The logarithmic spiral is the same as Problem 7.14, but the origin is moved forward above the midpoint of the slope, as shown in Fig. P7-15. Determine the factor of safety with respect to cohesion.
[Answer: 3.003]

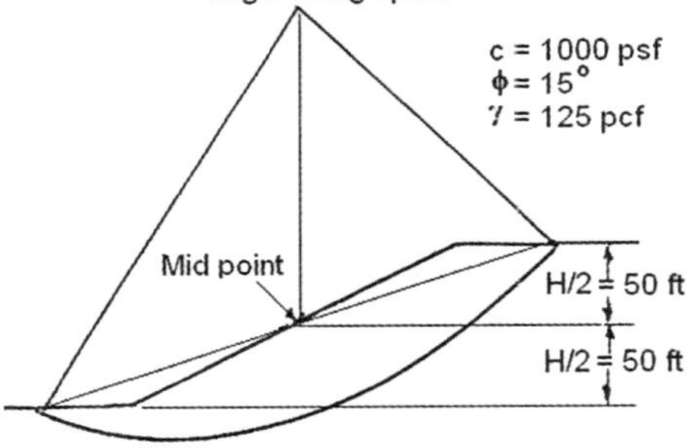

Fig. P7-15.

7.16 Same as Problem 7.15, but determine the factor of safety with respect to shear strength.

[Answer: 1.434]

Chapter 8

Method of Slices

The method of slices is a very powerful tool that can be used to analyze slopes of any configuration consisting of different soils and groundwater conditions. The problem is statically indeterminate and, to solve it statically, some simplifying assumptions must be made. In this chapter, four different limit equilibrium methods will be discussed: normal, simplified Bishop, original Spencer, and Spencer. All of these methods are incorporated in the LEAME computer software described in the companion volume to this book. The simplified Bishop method can be used for circular failure surfaces, and the Spencer method is recommended for noncircular and composite failure surfaces. The original Spencer method can be used as a check by simply changing an input parameter. The important equations used in LEAME will be derived, and some special techniques to solve these equations also will be discussed. The use of spreadsheets to compute the factors of safety will be demonstrated.

8.1 Overall Moment Equilibrium

Because the equations of overall moment equilibrium are used in the first three methods (normal, simplified Bishop, and original Spencer), they will be derived in this section before each method is discussed. Furthermore, an analysis of overall moment equilibrium will display all the external forces involved.

Fig. 8-1 shows the cross section of a circular failure surface. The failure mass is divided into a number of slices, one of which is designated as slice i. This slice has a width, b_i, a total weight, W_i, and an angle, θ_i, at the bottom. The slice is subjected to a horizontal seismic force, C_sW_i, at the midheight of the slice, a normal force due to water pressure, U_i, on the failure surface, and a line load, L_i, on the surface of the slope. Note that C_s is the seismic coefficient and that slice i is only one of the many slices in a given slope. By changing the subscript i, these forces and weights can be applied to other slices as well. Because the failure surface cuts through bodies of water on both sides of the slope, a horizontal force, P_1, due to water pressure is applied on the left side and P_2 on the right side. The moment arms from the center to each of the forces are shown in the figure.

According to Mohr-Coulomb theory, the shear strength, s, of a soil can be expressed as

$$s = c' + (\sigma_n - u)\tan\phi' \tag{8-1}$$

in which c' = effective cohesion, σ_n = total normal stress on the failure surface, u = pore water pressure on the failure surface, and ϕ' = effective angle of internal friction. After reducing the shear strength by a factor of safety and multiplying by the area of the failure surface, the shear force, T_i, at the bottom of slice i can be written as

$$T_i = \frac{c_i' b_i \sec\theta_i + N_i' \tan\phi_i'}{F} \tag{8-2}$$

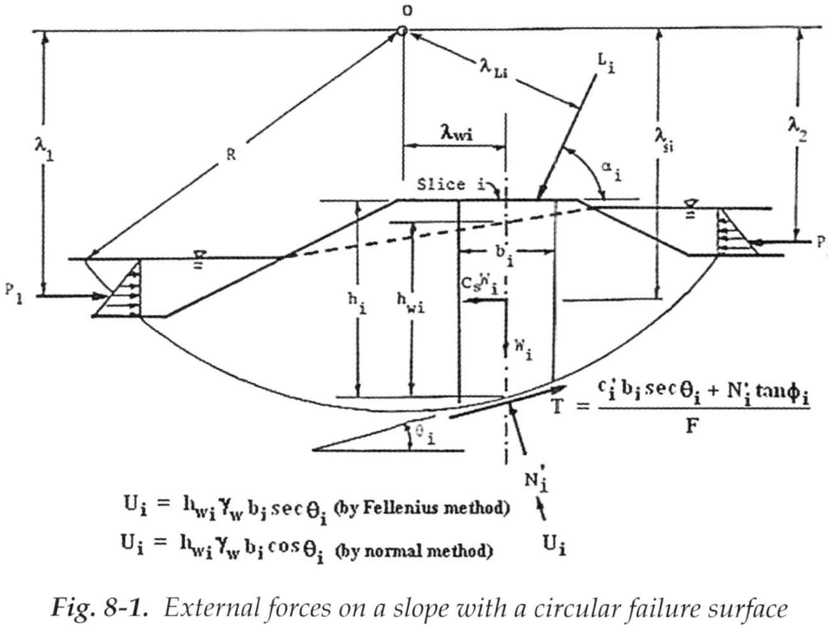

Fig. 8-1. *External forces on a slope with a circular failure surface*

in which N_i' is the effective force normal to the failure surface. The force normal to the failure surface due to pore water pressure is U_i and can be computed by

$$U_i = u_i b_i \sec\theta_i \tag{8-3}$$

The pore water pressure, u_i, at the bottom of slice i can be determined from the location of the phreatic surface. By drawing a vertical line through the middle of the slice, the height of the phreatic surface above the bottom of slice, h_{wi}, can be determined and the pore water pressure computed by

$$u_i = \gamma_w h_{wi} \tag{8-4}$$

in which γ_w = unit weight of water. In the normal method, $\sec\theta_i$ in Eq. (8-3) must be changed to $\cos\theta_i$ to avoid the possibility of negative N_i', as discussed in Section 2.4.1.

If a pore pressure ratio, r_u, is specified,

$$u_i = r_u \gamma h_i \tag{8-5}$$

in which γ = average unit weight of slice and h_i = height of the slice. If, in addition to the phreatic surface, an excess pore pressure is generated during construction by the weight of new fill only, then h_i in Eq. (8-5) should be the height of the slice in the new fill rather than the total height of the slice. The sum of N_i' and U_i is the total force normal to the failure surface.

Referring to Fig. 8-1 for a circular failure surface and summing the moments about point O for all slices results in the following equilibrium equation:

$$\sum RW_i \sin\theta_i - R\sum \frac{c_i' b_i \sec\theta_i + N_i' \tan\phi_i'}{F} + \sum C_s W_i \lambda_{si} + \sum L_i \lambda_{Li} - P_1\lambda_1 + P_2\lambda_2 = 0 \tag{8-6}$$

in which R = radius of the circle. Because the normal forces, N_i' and U_i, pass through the moment center, they do not contribute to the overall moment. Also, the forces on both sides of each slice are internal forces and do not enter into the equation, because whenever there is a force on one slice, there is an equal but opposite force on the adjoining slice, thus neutralizing their effect. A rearrangement of terms in Eq. (8-6) yields

$$F = \frac{R\sum (c_i' b_i \sec\theta_i + N_i' \tan\phi_i')}{\sum RW_i \sin\theta_i + C_s \sum W_i \lambda_{si} + \sum L_i \lambda_{Li} - P_1\lambda_1 + P_2\lambda_2} \tag{8-7}$$

With the exception of the effective normal force, N_i', all terms in Eq. (8-7) are either given or can be calculated from geometry. Because N_i' depends on the forces between two slices and is statically indeterminate, some simplifying assumptions must be made to solve N_i'. For example, the Fellenius or normal method assumes that there is no force between two slices, so N_i' can be determined simply by considering the equilibrium of all forces in the normal or N_i' direction. The simplified Bishop method assumes that the force between two slices is horizontal, so N_i' can be determined directly by considering the equilibrium of all forces in the vertical direction. The original Spencer method assumes that all the interslice forces make an angle δ with the horizontal, so N_i' can be determined by summing all forces in a direction perpendicular to δ. Eq. (8-7) can be used in the normal, simplified Bishop, and original Spencer methods to determine the factor of safety for circular failure surfaces.

For noncircular failure surfaces, an arbitrary point, O, must be selected as a moment center, as shown in Fig. 8-2. Since there is no fixed radius, R, and the normal forces, N_i' and U_i, may not pass through the moment center, R must be replaced by λ_{Ti}, and a term $(N_i'+U_i)\lambda_{Ni}$ must be added to the overturning moment. Thus, Eq. (8-7) should be modified as

$$F = \frac{\sum \lambda_{Ti}(c_i' b_i \sec\theta_i + N_i' \tan\phi_i')}{\sum W_i\lambda_{wi} + C_s \sum W_i\lambda_{si} + \sum L_i\lambda_{Li} - P_1\lambda_1 + P_2\lambda_2 - \sum (N_i'+U_i)\lambda_{Ni}} \tag{8-8}$$

For composite failure surfaces, Eq. (8-7) applies to the circular part and Eq. (8-8) to the noncircular part.

To ensure that the arbitrary selection of a moment center has no effect on the factor of safety, one basic requirement is that the overall force equilibrium must be satisfied in two perpendicular directions. The normal method considers the force equilibrium of each slice in a direction normal to the failure surface, but the direction changes from slice to slice, so there is not a single direction in which the force equilibrium is satisfied. The simplified Bishop method considers the force equilibrium of each slice in the vertical direction. If the force equilibrium in the vertical direction is satisfied for each slice, the overall force equilibrium in

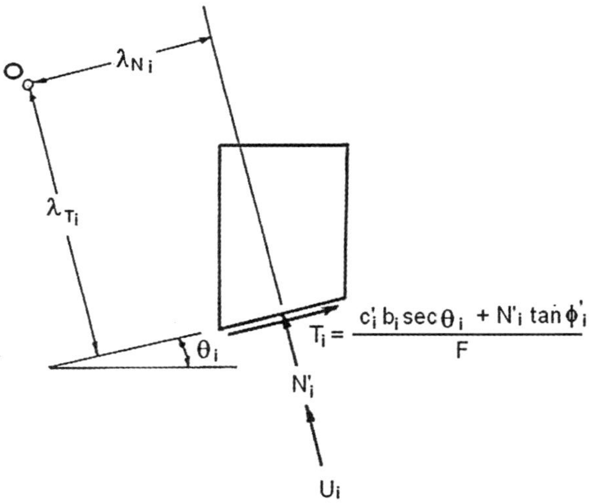

Fig. 8-2. Moments due to normal and shear forces on a noncircular failure surface

the vertical direction is satisfied automatically. Because the horizontal force equilibrium is not satisfied for each slice, two moment centers with different y coordinates may yield two different factors of safety, although the factor of safety is not affected by the difference in x coordinates. Therefore, neither the normal nor the simplified Bishop methods is suitable for analyzing noncircular and composite failure surfaces, and the use of Spencer's method, which satisfied all equations of equilibrium, is recommended.

8.2 Normal Method

The normal method (Bailey and Christian 1969), designated as method 1, is similar to the well-known Fellenius method in which the forces between two slices are assumed to be zero. Fig. 8-3 shows one of the slices for a noncircular failure surface. Theoretically, many of the variables should have a subscript i. For simplicity, the subscript purposely is omitted hereafter for all the equations that follow. The case of a noncircular failure surface is illustrated, because it is more complex than a circular failure surface. By assuming $\lambda_N = 0$ and $\lambda_T = R$, the equations derived from the noncircular failure surface can be applied to a circular surface as well.

Summing the forces in a direction normal to the failure surface,

$$N' = W'\cos\theta - C_s W\sin\theta + L\sin\alpha\cos\theta - L\cos\alpha\sin\theta \tag{8-9}$$

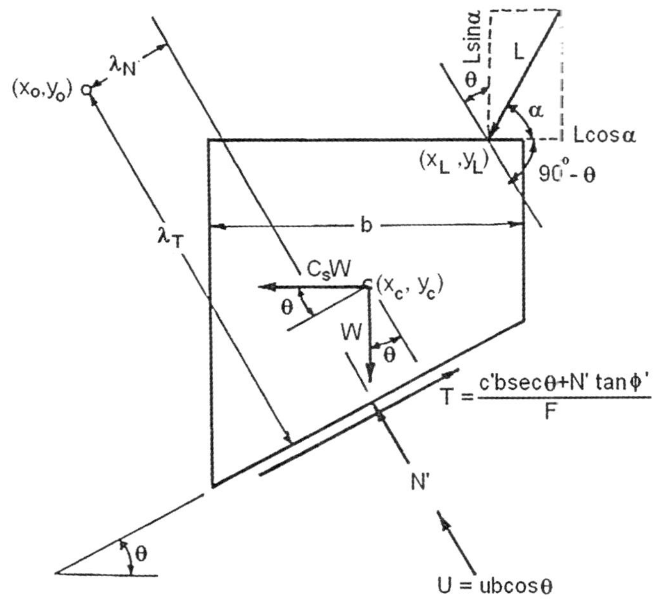

Fig. 8-3. Forces on a slice based on the normal method

in which α = angle of inclination of the line load, L, and W' = submerged weight, or

$$W' = W - ub \tag{8-10}$$

If N_i' obtained from Eq. (8-9) is negative, the frictional resistance no longer exists, and $\tan\phi'$ in Eq. (8-7) or (8-8) should be assigned 0. Knowing N_i', Eq. (8-7) can be used to determine the factor of safety for circular failure surfaces and Eq. (8-8) for noncircular failure surfaces. In Eq. (8-8), $W_i\lambda_{wi} = \lambda_T W_i \sin\theta_i + \lambda_N W_i \cos\theta_i$ and $N_i' + U_i = W_i \cos\theta_i$, so Eq. (8-8) is exactly the same as Eq. (8-7). Therefore, for the normal method, Eq. (8-7) also can be applied to noncircular failure surfaces by simply replacing R by λ_{Ti}.

To obtain a unique factor of safety, the total number of equations must be equal to the total number of unknowns. With a total of n slices, the number of equations and unknowns are tabulated as follows:

Eqs.		**Unknowns**	
Description	**No.**	**Description**	**No.**
Σ forces normal to failure surface = 0	n	Effective normal forces, N'	n
Overall moment equilibrium	1	Factor of safety, F	1
Total	$n+1$	Total	$n+1$

The application of the normal method is quite straightforward. First, determine N' by Eq. (8-9) and then compute F by Eq. (8-7) or (8-8). Every term in the equations is given or can be computed from geometry, and no iterations are needed. No matter what method is specified, LEAME always uses the normal method to determine the initial factor of safety for the first iteration.

Example 8.1 Fig. 8-4 shows the dimensions of a slope with a circular failure surface. The failure mass is divided into five slices of equal width. The centerlines of slices are shown in dashed lines with the coordinates (x_m, y_m) in parentheses at the lower end points. The soil has a cohesion, c, of 500 psf (23.9 kPa), a friction angle, ϕ, of 18°, and a total unit weight, γ, of 125 pcf (19.7 kN/m^3). Determine the factor of safety by the normal method, or method 1.

Solution The solution is presented in Table 8-1. Each column is explained here:

1. To simplify the calculations, especially later by the more complex methods, only five slices are used. The same problem was solved by LEAME using both five and 10 slices. The factor of safety with five slices is 1.617, which is not too much different from the 1.630 by 10 slices.
2. Each slice has a width, b, of 40 ft. This does not occur in real cases because, to obtain more accurate results, a slice must be subdivided at each break point on the slope surface. This example is so designed that the break

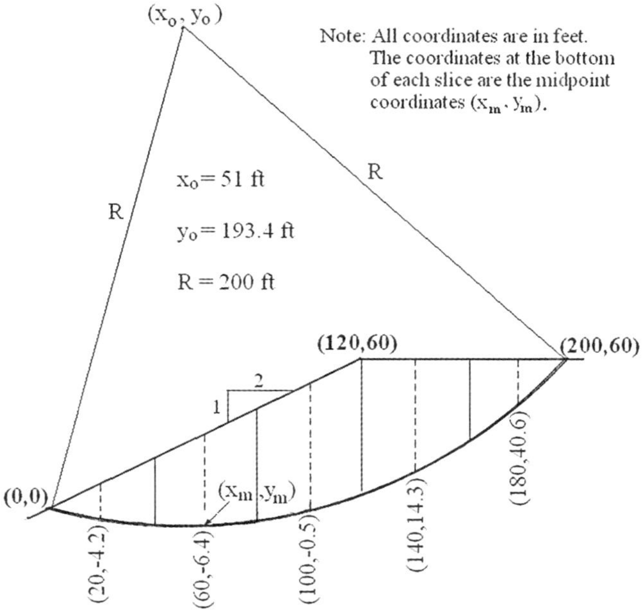

Fig. 8-4. Example 8.1

Table 8-1. *Analysis of Circular Failure Surface by Normal Method*

Slice No. ft (1)	Width b ft (2)	Height h lb (3)	Weight W deg (4)	Angle θ ft-lb (5)	Driving M_o ft-lb (6)	Resisting M_r ft-lb (7)
1	40	14.2	71,000	–8.9	–2,196,887	8,607,000
2	40	36.4	182,000	2.6	1,651,213	15,819,000
3	40	50.5	252,500	14.2	12,388,023	20,033,200
4	40	45.7	228,500	26.4	20,319,828	17,766,000
5	40	19.4	97,000	40.2	12,521,879	10,051,500
				Sum	44,684,056	72,276,700

point (120, 60) is located exactly on the boundary between two slices, so no subdivision of the slice is needed.

3. The height, h, of each slice is the length of the dashed line, or the difference in the y coordinates between the two end points. With an outslope of 2:1, the y coordinates at the upper end point is $y = 0.5x$ and those at the lower end point are shown in the figure.
4. The weight, W, of each slice is equal to $bh\gamma$.
5. The angle, θ, at the bottom of each slice can be determined from geometry. It can be proved easily that $\theta = \tan^{-1} [(x_m - x_o)/(y_o - y_m)$.

6. The driving moment, $M_o = RW\sin\theta$. The values of W, θ, and M_o listed in columns 4 to 6 can be used later in other examples.
7. The resisting moment, M_r, is computed by $M_r = R(cb\sec\theta + W\cos\theta\tan\phi)$

The factor of safety by the normal method is $F = 72{,}276{,}700/44{,}684{,}056 = 1.618$, which checks with the 1.617 obtained by LEAME.

The next example involves the use of the normal method to determine the factor of safety for a noncircular failure surface. Although the normal method is not applicable to noncircular failure surfaces, some of the information provided by the example can be used later in other examples where more refined methods are involved.

Example 8.2 Fig. 8-5 shows the dimensions of a slope with a noncircular failure surface. Similar to Example 8.1, the failure mass is divided into five slices, as shown by the solid lines. The centerline of each slice is shown in dashed lines with the coordinates (x_m, y_m) at the lower end points. The soil parameters are the same as Example 8.1. Determine the factor of safety by the normal method, or method 1.

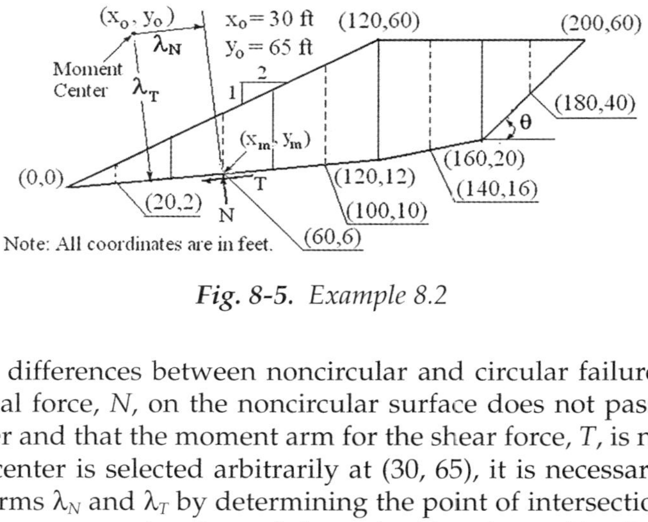

Fig. 8-5. Example 8.2

Solution The differences between noncircular and circular failure surfaces are that the normal force, N, on the noncircular surface does not pass through the moment center and that the moment arm for the shear force, T, is not a constant. If a moment center is selected arbitrarily at (30, 65), it is necessary to compute the moment arms λ_N and λ_T by determining the point of intersection of two perpendicular lines, one passing through (x_m, y_m) with a slope of $\tan\theta$, and the other passing through (x_o, y_o) with a slope of $-1/\tan\theta$. The general equation for a line with a slope of $\tan\theta$ is

$$y - x\tan\theta = c \tag{8-11}$$

If the line passes through point (x_m, y_m), the constant, c, can be obtained by substituting $x = x_m$ and $y = y_m$ into Eq. (8-11), or

$$y - x\tan\theta = y_m - x_m\tan\theta \tag{8-12}$$

Similarly, the equation for a line passing through (x_o, y_o) with a slope of $-1/\tan\theta$ is

$$y + \frac{x}{\tan\theta} = y_o + \frac{x_o}{\tan\theta} \tag{8-13}$$

$$\text{or} \quad y = y_o + \frac{x_o - x}{\tan\theta} \tag{8-14}$$

Substituting Eq. (8-14) into Eq. (8-12),

$$x = \frac{y_o - y_m + x_m \tan\theta + \dfrac{x_o}{\tan\theta}}{\tan\theta + \dfrac{1}{\tan\theta}} \tag{8-15}$$

Given the value of x, y can be obtained by substituting x into Eq. (8-14). Knowing the point of intersection (x, y), λ_N and λ_T can be computed by

$$\lambda_N = \pm\sqrt{(x - x_m)^2 + (y - y_m)^2} \tag{8-16}$$

where λ_N is positive if $x < x_m$ and negative if $x > x_m$.

$$\lambda_T = \sqrt{(x - x_o)^2 + (y - y_o)^2} \tag{8-17}$$

The solution is presented in Table 8-2.

The factor of safety is $F = 23{,}556{,}000/13{,}919{,}692 = 1.692$, which checks with the 1.686 obtained by LEAME. Note that column 6 is computed by Eq. (8-15), column 7 by Eq. (8-14), column 8 by Eq. (8-16), and column 9 by Eq. (8-17); and that the column 10 overturning moment $M_o = W(x_m - x_o) - \lambda_N W\cos\theta$, and the column 11 resisting moment $M_r = \lambda_T(cb\sec\theta + W\cos\theta\tan\phi)$.

Table 8-2. *Analysis of Noncircular Failure Surface by Normal Method*

Slice No. (1)	b ft (2)	h ft (3)	W lb (4)	θ deg (5)	x ft (6)	y ft (7)	λ_N ft (8)	λ_T ft (9)	M_o ft-lb (10)	M_r ft-lb (11)
1	40	8.0	40,000	5.7	36.14	3.60	−16.22	61.71	245,592	2,037,100
2	40	24.0	120,000	5.7	36.14	3.60	23.98	61.71	728,100	3,638,000
3	40	40.0	200,000	5.7	36.14	3.60	64.18	61.71	1,227,500	5,238,900
4	40	44.0	220,000	11.3	43.64	−3.26	98.27	69.61	2,999,700	6,299,100
5	40	20.0	100,000	45.0	117.50	−22.50	88.39	123.74	8,718,800	6,342,900
							Sum		13,919,692	23,556,000

In the example, it is really not necessary to compute λ_N because W can be resolved at (x_m, y_m) into $W\sin\theta$ and $W\cos\theta$. Because $W(x_m - x_o) = \lambda_T W\sin\theta + \lambda_N W\cos\theta$, a simpler equation $M_o = \lambda_T W\sin\theta$ may be used. However, this is not true if there are seismic or other forces or if the simplified Bishop or the original Spencer method is employed. The values of λ_N and λ_T listed in the table can be used later in Example 8.4.

8.3 Simplified Bishop Method

The simplified Bishop method (Bishop 1955), designated as method 2, is the most widely used method recognized by the engineering profession. It is recommended for use with circular failure surfaces and should yield a factor of safety very close to the more refined methods. By assuming the forces between two slices as horizontal and considering the vertical equilibrium of each slice, the effective normal force, N', can be determined even without knowing the magnitude of the horizontal forces on both sides of the slice.

Fig. 8-6 shows one of the slices to be analyzed by the simplified Bishop method. Based on the equilibrium of all forces in the vertical direction,

$$W - \frac{c'b\sec\theta + N'\tan\phi'}{F}\sin\theta + L\sin\alpha - N'\cos\theta - ub = 0 \tag{8-18}$$

$$\text{or} \quad N' = \frac{F(W' + L\sin\alpha) - c'b\tan\theta}{F\cos\theta + \sin\theta\tan\phi'} \tag{8-19}$$

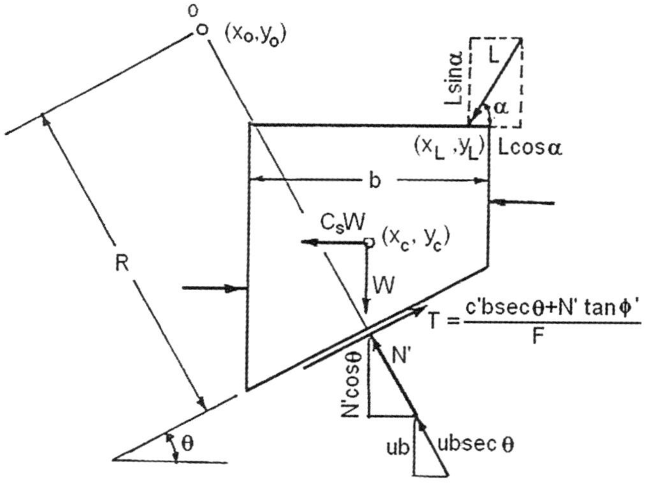

Fig. 8-6. Forces on a slice based on the simplified Bishop method

in which $W' = W - ub$. Knowing N', the factor of safety can be determined from Eq. (8-7) for circular failure surfaces and from Eq. (8-8) for noncircular failure surfaces. If N' is negative, $\tan\phi'$ in Eqs. (8-7), (8-8), and (8-19) should be changed to 0. The number of equations and unknowns are tabulated as follows:

EQS.		UNKNOWNS	
Description	**No.**	**Description**	**No.**
Σ forces in vertical direction = 0	n	Effective normal forces, N'	n
Overall moment equilibrium	1	Factor of safety, F	1
Total	$n+1$	Total	$n+1$

It can be seen from Eq. (8-7) or (8-8) and (8-19) that F depends on N', whereas N' depends on F. Thus, an iteration method must be used to solve F. First, N' is computed by Eq. (8-19), using the factor of safety determined by the normal method as the first trial. Based on the N' thus determined, a new F is computed by Eq. (8-7) or (8-8). The process is repeated until the difference between the assumed F and the new F is reduced to a specified tolerance. By using Newton's method of tangent, as described in Section 8.6.1, the factor of safety converges very rapidly, usually within two or three iterations.

Example 8.3 Based on the information provided by Example 8.1, determine the factor of safety by the simplified Bishop method, or method 2.

Solution The simplified Bishop method requires iterations and is much easier to solve by a spreadsheet, such as that in Table 8-3.

Items W, θ, and the driving moment, which are needed to compute the factor of safety, F, by the simplified Bishop method, are provided in Example 8.1. The normal force, N, is obtained from Eq. (8-19) by assuming an initial F of 1.618, as determined by the normal method. The resisting moment is the numerator of Eq. (8-7) and can be expressed as $R(c \times b \times \sec\theta + N \times \tan\phi')$. Dividing the sum of the resisting moment by the sum of the driving moment gives a new F of 1.688. Using the new F as the assumed F, the process is repeated until the factor of safety converges to 1.693 at the fourth iteration, which checks with the 1.692 obtained by LEAME. Special techniques are available to speed up the convergence, as illustrated by Example 8.6.

The iterative procedures illustrated by Table 8-3 are similar to those programmed in LEAME. When spreadsheets are used, it is really not necessary to use four sets of normal force, N, and resisting moment to determine the four factors of safety. By use of trial and error, only one set of data will be sufficient. Any factor of safety can be assumed, and a new factor of safety is computed. Using the computed factor of safety as the assumed factor of safety, the process is repeated until the factor of safety converges to 1.693, as shown by the last column in the table. If the factor of safety does not converge but oscillates back

Table 8-3. *Spreadsheet for Circular Failure Surface by Simplified Bishop Method*

ASSUMED FACTOR OF SAFETY				1.618		1.688		1.692		1.693	
No.	**W**	**θ**	**Driving M.**	**N′**	**Resisting M.**	**N′**	**Resisting M.**	**N′**	**Resisting M.**	**N′**	**Resisting M.**
1	71,000	−8.9	−2,201,000	76,221	9,001,918	76,034	8,989,740	76,024	8,989,086	76,023	8,989,050
2	182,000	2.6	1,638,000	179,984	15,700,217	180,076	15,706,138	180,080	15,706,457	180,081	15,706,474
3	252,500	14.2	12,372,500	244,793	20,033,677	245,416	20,074,178	245,450	20,076,362	245,452	20,076,480
4	228,500	26.4	20,336,500	225,750	19,135,840	226,868	19,208,488	226,928	19,212,414	226,932	19,212,626
5	97,003	40.2	12,513,000	96,880	11,532,662	97,961	11,602,896	98,020	11,606,701	98,023	11,606,906
	Sum		44,659,000		75,404,313		75,581,441		75,591,020		75,591,537
Computed factor of safety					1.688		1.692		1.693		1.693

and forth, instead of using the computed factor of safety as the assumed factor of safety, a slightly smaller value midway between the computed and assumed factors of safety should be used to ensure convergence.

8.4 Original Spencer Method

The original Spencer method, designated as method 3, assumes parallel interslice forces making an angle of δ with the horizontal and considers the force equilibrium of each slice in the direction perpendicular to δ, the overall force equilibrium in the δ direction, and the overall moment equilibrium (Spencer 1967). If the force equilibrium of each slice in the direction perpendicular to δ is satisfied, the overall force equilibrium in the direction perpendicular to δ is automatically satisfied. Later on, Spencer (1973) improved the method by considering both the moment and force equilibrium of each slice. This improved method now generally is called the Spencer method and will be presented in Section 8.5. However, the difference in the factor of safety between these two methods is usually quite small. The original Spencer method has the advantage that it always converges, whereas the Spencer method sometimes may have convergence problems.

It should be noted that all equations of equilibrium are based on total forces, including the neutral forces, if any. When assuming the direction of interslice forces, the forces due to soil and water pressures are considered as an entity and act in the same direction. The soil pressure is separated from the water pressure only when the shear strength is to be evaluated.

8.4.1 Factors of Safety Based on Moment Equilibrium

The equation for computing the factor of safety based on overall moment equilibrium is the same as Eq. (8-7) or (8-8) used in the normal and simplified Bishop methods. However, the forces between two slices are assumed to incline at an angle δ with the horizontal. Consequently, the normal force, N', depends not only on F but also on δ. To determine the factor of safety by the original Spencer method, three different values of δ must be assumed, and the factor of safety for each δ is then determined.

Fig. 8-7 shows all the forces considered in the original Spencer method, including the two interslice forces, Z_1 and Z_2, which make an angle of δ with the horizontal, and the forces, P_1 and P_2, due to the water pressure. By considering the equilibrium of all forces on a slice in a direction perpendicular to δ,

$$N'\cos(\theta-\delta)+ub\sec\theta\cos(\theta-\delta)+\frac{c'b\sec\theta+N'\tan\phi'}{F}\sin(\theta-\delta)$$
$$+C_sW\sin\delta-W\cos\delta-L\sin(\alpha-\delta)=0 \qquad (8\text{-}20)$$

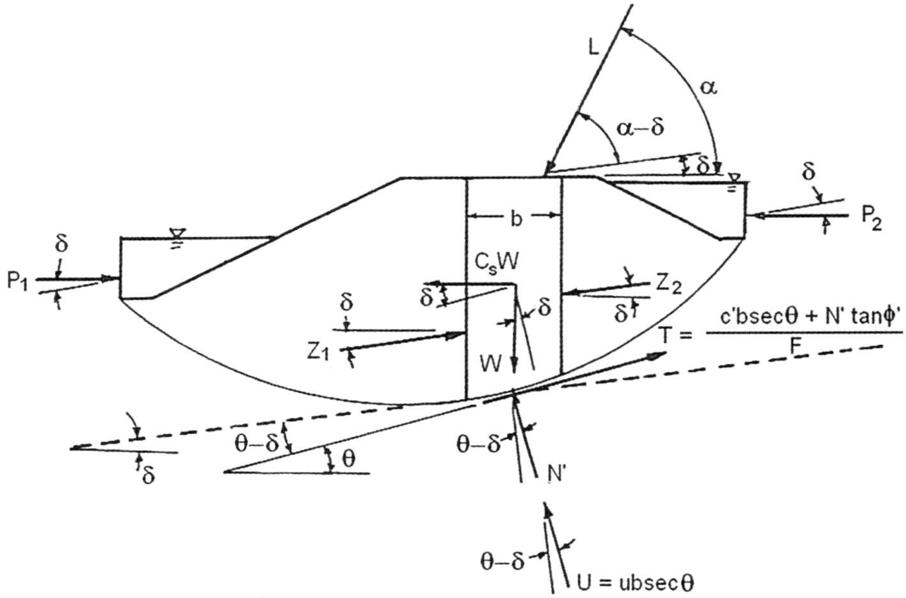

Fig. 8-7. Resolution of forces based on the original Spencer method

After simplification,

$$N' = \frac{F[W'\cos\delta - ub\tan\theta\sin\delta - C_sW\sin\delta + L\sin(\alpha-\delta)] - c'b\sec\theta\sin(\theta-\delta)}{F\cos(\theta-\delta) + \sin(\theta-\delta)\tan\phi'} \quad (8\text{-}21)$$

To determine the factor of safety with respect to moment equilibrium for a given value of δ, a factor of safety is first assumed, and the value of N' is computed by Eq. (8-21). Substituting the N' thus determined into Eq. (8-7) or (8-8), a new factor of safety is obtained. Using the new factor of safety as the assumed factor of safety, the process is repeated until the factor of safety converges. It is interesting to note that when $\delta = 0$, the factor of safety is the same as that by the simplified Bishop method.

8.4.2 Factors of Safety Based on Force Equilibrium

The factor of safety with respect to force equilibrium can be obtained by summing the forces in the δ direction:

$$P_1\cos\delta + \sum\left(\frac{c'b\sec\theta + N'\tan\phi'}{F}\right)\cos(\theta-\delta) - P_2\cos\delta - \sum W\sin\delta$$
$$-\sum C_sW\cos\delta - \sum L\cos(\alpha-\delta) - \sum(N'+U)\sin(\theta-\delta) = 0$$

Note that the interslice forces, Z_1 and Z_2, do not appear in the equation, because they are internal forces. Moving F to one side results in

$$F = \frac{\sum (c'b\sec\theta + N'\tan\phi')\cos(\theta-\delta)}{\sum W\sin\delta + \sum (N'+U)\sin(\theta-\delta) + C_s\sum W\cos\delta + \sum L\cos(\alpha-\delta) + (P_2-P_1)\cos\delta} \quad (8\text{-}22)$$

in which U can be computed by Eq. (8-3). Eq. (8-22) in conjunction with Eq. (8-21) can be used to determine the factor of safety with respect to force equilibrium. In both force and moment equilibrium, the factor of safety obtained by the normal method is used as the initial trial value. If N' obtained by Eq. (8-21) for any given slice is negative, $\tan\phi'$ in Eqs. (8-21) and (8-22) for that particular slice must be changed to 0.

8.4.3 Overall Factor of Safety

To determine the factor of safety that satisfies both moment and force equilibrium, three different values of δ, 0, 0.3, and 0.6 rad, are assumed. For each δ, the factors of safety with respect to moment and force equilibrium are determined, both by Newton's method of tangent described in Section 8.6.1. These factors of safety always converge when $\delta = 0$. However, if the actual value of δ is small, the factor of safety at δ of 0.3 or 0.6 may not converge. If the factor of safety at δ of 0.3 does not converge, then the factors of safety at 0.1125 and 0.225 are determined, so the three values of δ to be used are 0, 0.1125, and 0.225. If the factor of safety at δ of 0.6 does not converge, then the factor of safety at δ of 0.15 is determined, so the three values of δ to be used are 0, 0.15, and 0.3. If the factor of safety still does not converge even after the stated adjustment, further reduction of δ should be made until the factor of safety converges. After the factors of safety at the three values of δ are determined, a parabola relating F to δ is developed for both moment and force equilibrium. The intersection of these two parabolas, as shown in Fig. 8-8, gives the final values of F and δ desired. If the final value of δ thus determined is greater than 0.6 but smaller than 0.7, the extrapolated value is considered acceptable. If the extrapolated value of δ is greater than 0.7, δ values of 0, 0.45, and 0.9 will be used for interpolation.

Although the values of δ and F can be determined manually by the graphical method shown in Fig. 8-8, a numerical method by converting each curve into an equation and then finding their intersection is needed for machine computations. The well-known Lagrange interpolation formula, or Eq. (8-23), can be used for this purpose.

$$F = \frac{(\delta-\delta_2)(\delta-\delta_3)}{(\delta_1-\delta_2)(\delta_1-\delta_3)}F_1 + \frac{(\delta-\delta_1)(\delta-\delta_3)}{(\delta_2-\delta_1)(\delta_2-\delta_3)}F_2 + \frac{(\delta-\delta_1)(\delta-\delta_2)}{(\delta_3-\delta_1)(\delta_3-\delta_2)}F_3 \quad (8\text{-}23)$$

The curve represented by Eq. (8-23) apparently passes through the three points (δ_1, F_1), (δ_2, F_2), and (δ_3, F_3), because when $\delta = \delta_1$, $F = F_1$, whereas the other

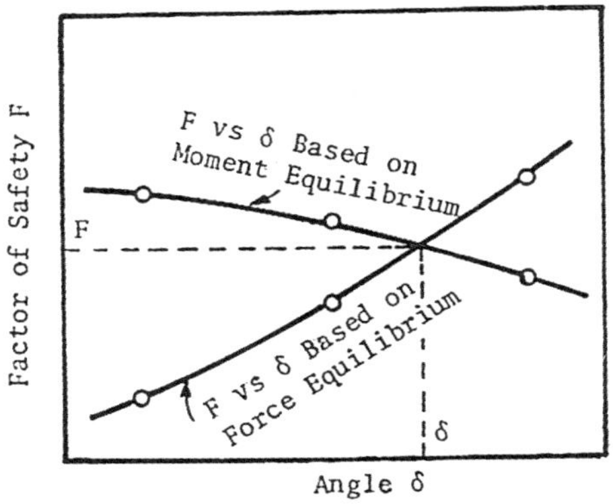

Fig. 8-8. *Factor of safety by the original Spencer method*

two terms are 0. The same is true when $\delta = \delta_2$, $F = F_2$ and when $\delta = \delta_3$, $F = F_3$. If the three points are spaced equally, Eq. (8-23) can be written as

$$(2s^2)F = A\delta^2 + B\delta + C \tag{8-24}$$

in which s = spacing between δ_1 and δ_2, which is the same as between δ_2 and δ_3,

$$A = F_1 - 2F_2 + F_3 \tag{8-25}$$

$$B = 2(\delta_1 + \delta_3)F_2 - (\delta_2 + \delta_3)F_1 - (\delta_1 + \delta_2)F_3 \tag{8-26}$$

$$C = \delta_2\delta_3 F_1 - 2\delta_1\delta_3 F_2 + \delta_1\delta_2 F_3 \tag{8-27}$$

Let A_m, B_m, and C_m be the coefficients when the three factors of safety based on moment equilibrium are substituted into Eqs. (8-25), (8-26), and (8-27). With $\delta_1 = 0$, $\delta_2 = 0.3$, and $\delta_3 = 0.6$, or $s = 0.3$, Eq. (8-24) can be written as

$$0.18F = A_m\delta^2 + B_m\delta + C_m \tag{8-28}$$

Similarly, let A_f, B_f, and C_f be the coefficients for force equilibrium, or

$$0.18F = A_f\delta^2 + B_f\delta + C_f \tag{8-29}$$

Subtracting Eq. (8-28) from Eq. (8-29), the following quadratic equation is obtained for solving δ:

$$(A_f - A_m)\delta^2 + (B_f - B_m)\delta + C_f - C_m = 0 \tag{8-30}$$

or $$\delta = \frac{-(B_f - B_m) + \sqrt{(B_f - B_m)^2 - 4(A_f - A_m)(C_f - C_m)}}{2(A_f - A_m)} \tag{8-31}$$

Substituting δ into Eq. (8-28), the factor of safety, F, can be obtained by

$$F = (A_m\delta^2 + B_m\delta + C_m)/0.18 \tag{8-32}$$

The number of equations and unknowns for the original Spencer method are tabulated as follows:

EQS.		UNKNOWNS	
Description	**No.**	**Description**	**No.**
Σ forces in δ direction = 0	1	Effective normal forces, N'	n
Σ forces on slice ⊥ to δ = 0	n	Angle of interslice force, δ	1
Overall moment equilibrium	1	Factor of safety, F	1
Total	n + 2	Total	n + 2

Example 8.4 Based on the information provided by Example 8.2, determine the factor of safety by the original Spencer method, or method 3.

Solution To determine the factor of safety by the original Spence method, six factors of safety at three different values of δ must be computed. Table 8-4 is the spreadsheet for computing these six factors of safety, three based on moment equilibrium and three based on force equilibrium, and a factor of safety of 2.127, which satisfies both moment and force equilibrium. Details about the spreadsheet are as follows:

1. Rows 5 to 9 are the input data obtained from Table 8-2. Rows 12 to 21 compute the three factors of safety for moment equilibrium and rows 24 to 33 compute those for force equilibrium. Based on the six factors of safety, rows 37 to 42 determine the factor of safety that satisfies both moment and force equilibrium.
2. For moment equilibrium with δ = 0, a factor of safety must first be assumed and then the normal force in cell B15, the driving moment in cell C15, and the resisting moment in cell D15 must be filled out. When typing an equation, be sure that the symbol for c is \$I\$5, for tan ϕ is \$I\$7, the assumed angle δ is D\$12, the assumed factor of safety is D\$13, and all input data from rows 5 to 9 (e.g., B5 and G9) are \$B5 and \$G9. Once these three cells (B15, C15, and D15) are filled, they can be copied vertically to the cells from rows 16 to 19. Row 20 sums up the driving moment and the resisting moment. Dividing the sum of the resisting moment by the sum of the driving moment gives the computed factor of safety. Using the computed factor of safety as the assumed factor of safety, the process is repeated until the factor of safety converges to 2.544. After completing δ = 0, cells in columns B to D and rows 12 to 21 can be copied as a group to cell E12 and then again to cell H12. If there are no mistakes in the spreadsheet, the three sets of data all should be identical. Then change the

Table 8-4. *Spreadsheet for Noncircular Failure Surface by Original Spencer Method*

	A	B	C	D	E	F	G	H	I	J
3	Data input									
4	Slice	b	W	θ in radian	λN	λT	λw			
5	1	40	40,000	0.0995	−16.22	61.71	−10	c =	500	psf
6	2	40	120,000	0.0995	23.98	61.71	30	ϕ =	18	deg
7	3	40	200,000	0.0995	64.18	61.71	70	tan ϕ =	0.32492	
8	4	40	220,000	0.1972	98.27	69.61	110			
9	5	40	100,000	0.7854	88.39	123.74	150			
10										
11	Factor of safety based on moment equilibrium									
12	Assumed angle δ			0			0.3			0.6
13	Assumed Factor of Safety			2.544			2.117			2.015
14	Slice	Normal N′	Driving M	Resisting M	Normal N′	Driving M	Resisting M	Normal N′	Driving M	Resisting M
15	1	38,914	231,184	2,020,590	42,242	285,162	2,087,316	47,252	366,430	2,187,777
16	2	118,299	763,180	3,612,332	122,742	656,647	3,701,410	129,789	487,651	3,842,715
17	3	197,685	1,312,585	5,204,073	203,242	955,912	5,315,503	212,327	372,879	5,497,652
18	4	217,204	2,855,392	6,332,355	215,699	3,003,292	6,298,315	216,562	2,918,408	6,317,851
19	5	115,546	4,786,894	8,145,492	93,401	6,744,313	7,255,130	78,952	8,021,415	6,674,220
20		Sum	9,949,236	25,314,842	Sum	11,645,325	24,657,673	Sum	12,166,782	24,520,215
21				2.544			2.117			2.015

Table 8-4. (Continued)

	A	B	C	D	E	F	G	H	I	J
22										
23	Factor of safety based on force equilibrium									
24	Assumed angle δ			0			0.3			0.6
25	Assumed Factor of Safety			1.992			2.133			2.278
26	Slice	Normal N′	Driving F	Resisting F	Normal N′	Driving F	Resisting F	Normal N′	Driving F	Resisting F
27	1	38,564	3,831	32,468	42,217	3,413	33,139	46,047	489	30,761
28	2	117,673	11,689	58,045	122,697	11,026	58,765	127,673	6,491	54,029
29	3	196,783	19,548	83,622	203,178	18,639	84,391	209,298	12,494	77,298
30	4	215,286	42,180	88,595	215,665	42,883	89,991	214,203	40,255	82,792
31	5	109,381	77,344	45,131	93,502	73,177	51,889	79,517	71,122	53,193
32		Sum	154,592	307,861	Sum	149,138	318,175	Sum	130,851	298,073
33				1.991			2.133			2.278
34										
35	Factor of safety and angle δ that satisfy both moment and force equilibrium									
36										
37		D1 =	0	D2 =	0.3	D3 =	0.6			
38		FM1 =	2.544	FM2 =	2.117	FM3 =	2.015			
39		FF1 –	1.991	FF2 =	2.133	FF3 =	2.278			
40		AM=	0.325	BM =	–0.354	CM =	0.458			
41		AF =	0.003	BF =	0.084	CF =	0.358			
42			Angle D or δ	0.288		Factor of safety F =	2.127			

assumed angle δ to 0.3 and 0.6 and determine the factor of safety by trial and error in the same way as when $\delta = 0$.

3. The normal force, N, in cell B15 can be obtained from Eq. (8-21) and typed as (D$13*$C5*COS(D$12)-$I$5*$B5*SIN($D5-D$12)/COS($D5))/(D$13*COS($D5-D$12) + SIN($D5-D$12)*$I$7). The driving moment in cell C15 can be obtained from the denominator of Eq. (8-8) and typed as $C5*$G5-B15*$E5. The resisting moment in cell D15 is the numerator of Eq. (8-8) and can be typed as $F5*($I$5*$B5/COS($D5) + B15*$I$7).
4. The same procedures can be applied for force equilibrium. The normal force in cell B27 is the same as that in cell C15 except that D$12 and D$13 must be changed to D$24 and D$25, respectively. The driving force in cell C27 can be obtained from the denominator of Eq. (8-22) and typed as $C5*SIN(D$24) + B27*SIN($D5-D$24). The resisting force in cell D27 is the numerator of Eq. (8-22) and can be typed as ($I$5*$B5/COS($D5) + B27*$I$7)*COS ($D5-D$24).
5. The input parameters for determining the factor of safety that satisfies both moment and force equilibrium are the three δ angles in row 37, the three factors of safety for moment equilibrium in row 38, and the three factors of safety for force equilibrium in row 39. The coefficients of Eq. (8-24) for moment equilibrium can be obtained from Eqs. (8-25), (8-26), and (8-27), with AM in cell C40 typed as C38-2*E38 + G38, BM in cell E40 as 2*(C$37 + G$37)*E38-(E$37 + G$37)*C38-(C$37 + E$37)*G38, and CM in cell G40 as E$37*G$37*C38-2*C$37*G$37*E38 + C$37*E$37*G38. The coefficients AF, BF, and CF for force equilibrium in row 41 can be obtained by copying cell C40 into C41, cell E40 into E41, and cell G40 into G41. Angle δ in cell D42 can be determined from Eq. (8-31) and typed as 0.5*(E40-E41 + SQRT((E41-E40)∧2-4*(C41-C40)*(G41-G40)))/(C41-C40). The factor of safety in cell G42 can be computed by Eq. (8-32) and typed as (C40*D42∧2 + E40*D42 + G40)/0.18. The factor of safety that satisfies both moment and force equilibrium is 2.127, which checks with the 2.126 obtained by LEAME.

8.5 Spencer Method

The Spencer method (Spencer 1973; Chugh 1981), designated as method 4, is the most refined method, because it satisfies all the equations of equilibrium. Fig. 8-9 shows the most general case with normal force, E, and shear force, S, at the right side of the slice. The differences in normal and shear forces between the left and right sides are ΔE and ΔS. The assumption of $S = E \tan\delta$ in this method is the same as in the original Spencer method but the moment is taken at the midpoint of the base of each slice. The general procedure is first to assume $S = 0$, and, based on the force equilibrium, determine the normal forces, E and N', the tangential force, T, and the factor of safety, F. Then, based on the moment equilibrium, determine the angle of inclination, δ, and a new set of the shear forces,

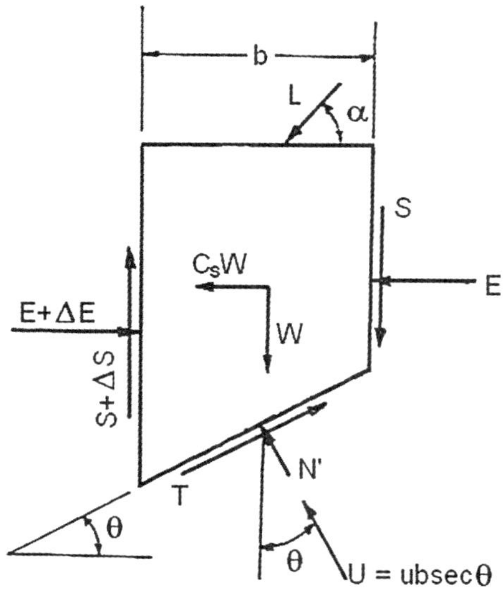

Fig. 8-9. General case of forces on a slice

S. Using the *F* and the new set of *S* thus obtained, the process is repeated until the factor of safety converges.

The Spencer method is similar to the well-known Morgenstern-Price method (Morgenstern and Price 1965), which assumes that the shear forces, *S*, between the slices not only vary with the normal forces, *E*, but also change from slice to slice according to a function $f(x)$, or

$$S = \lambda f(x) E \tag{8-33}$$

in which λ is an unknown constant to be determined, and $f(x)$ can be a constant, a linear function, a sine curve, or a numerical value at each vertical side. It can be seen that the Spencer method is a special case of Morgenstern-Price with $\lambda = \tan\delta$ and $f(x) = 1$. The major difference between the two methods is that the Morgenstern-Price method provides more flexibility in the assumptions of the inclinations of interslice forces. However, the assumptions generally appear to have very little effect on the computed factor of safety when the static equilibrium is satisfied. Although the solution techniques used in the Spencer method can also be applied to the Morgenstern-Price method, the Morgenstern-Price method is not used in LEAME.

8.5.1 Force Equilibrium of Each Slice

Based on the equilibrium of forces in the vertical direction,

$$N'\cos\theta + T\sin\theta + ub + \Delta S - W - L\sin\alpha = 0$$

$$\text{or} \quad N' = (W - ub - \Delta S)\sec\theta - T\tan\theta + L\sin\alpha\sec\theta$$

$$\text{and} \quad N' = (W' - \Delta S)\sec\theta - T\tan\theta + L\sin\alpha\sec\theta \tag{8-34}$$

in which $W' = W - ub$. Without subscript i, Eq. (8-2) can be rewritten as

$$T = \frac{c'b\sec\theta + N'\tan\phi'}{F} \tag{8-35}$$

Substituting Eq. (8-34) into (8-35),

$$T = \frac{c'b\sec\theta + [(W' - \Delta S)\sec\theta - T\tan\theta + L\sin\alpha\sec\theta]\tan\phi'}{F} \tag{8-36}$$

From Eq. (8-36), the shear force, T, can be determined by

$$T = \frac{c'b\sec\theta + [(W' - \Delta S)\sec\theta + L\sin\alpha\sec\theta]\tan\phi'}{F + \tan\theta\tan\phi'} \tag{8-37}$$

Based on the equilibrium of forces in the horizontal direction,

$$\Delta E = N'\sin\theta + ub\tan\theta + C_s W + L\cos\alpha - T\cos\theta \tag{8-38}$$

Substituting Eq. (8-34) into Eq. (8-38),

$$\Delta E = (W - \Delta S)\tan\theta - T\sec\theta + C_s W + L(\sin\alpha\tan\theta + \cos\alpha) \tag{8-39}$$

Since the overall horizontal force equilibrium must be satisfied,

$$\sum \Delta E = P_2 - P_1 \tag{8-40}$$

or

$$\sum T\sec\theta = \sum(W - \Delta S)\tan\theta + C_s\sum W + \sum L(\sin\alpha\tan\theta + \cos\alpha) - (P_2 - P_1) \tag{8-41}$$

Note that P_1 and P_2 may be zero if there is no water pressure at both ends. Combining Eqs. (8-36) and (8-41),

$$F = \frac{\sum\{c'b\sec\theta + [(W' - \Delta S)\sec\theta - T\tan\theta + L\sin\alpha\sec\theta]\tan\phi'\}\sec\theta}{\sum(W - \Delta S)\tan\theta + C_s\sum W + \sum L(\sin\alpha\tan\theta + \cos\alpha) - (P_2 - P_1)} \tag{8-42}$$

Eq. (8-37) in conjunction with Eq. (8-42) can be used in the Spencer method for determining the factor of safety. Note that both equations contain the unknown ΔS, which must be evaluated from the moment equilibrium. The value of T obtained from Eq. (8-37) must not be negative. If $T < 0$, then $\tan\phi'$ in Eqs. (8-37) and (8-42) must be assigned 0.

8.5.2 Moment Equilibrium of Each Slice

Fig. 8-10 shows the forces involved in moment equilibrium. Assuming that the side forces, Z_1 and Z_2, are applied at h_1 and h_2 above the base, and taking the moment at the midpoint of the base, the following equation is obtained:

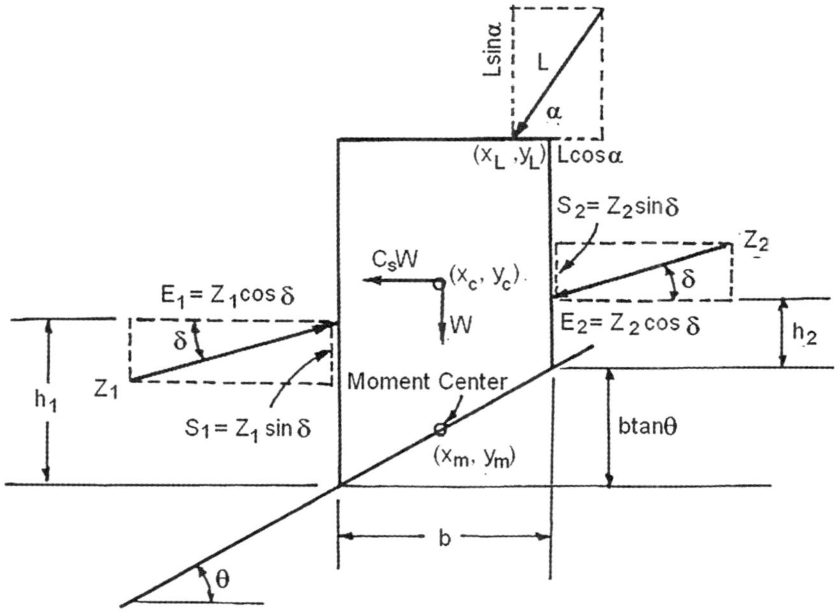

Fig. 8-10. *Forces for moment equilibrium based on the Spencer method*

$$Z_1\cos\delta\left(h_1-\frac{b}{2}\tan\theta\right)+\frac{b}{2}(Z_1\sin\delta+Z_2\sin\delta)+L\sin\alpha(x_L-x_m)$$
$$-L\cos(y_L-y_m)-C_sW(y_c-y_m)-Z_2\cos\delta\left(h_2+\frac{b}{2}\tan\theta\right)=0$$

Moving h_2 to one side and replacing Z_1 by $E_1/\cos\delta$ and Z_2 by $E_2/\cos\delta$,

$$h_2=\left(\frac{E_1}{E_2}\right)h_1+\frac{b}{2}\left(1+\frac{E_1}{E_2}\right)(\tan\delta-\tan\theta)-\frac{C_sW(y_c-y_m)}{E_2}+\frac{L[\sin\alpha(x_L-x_m)-\cos\alpha(y_L-y_m)]}{E_2} \tag{8-43}$$

Eq. (8-43) can be used for all intermediate slices to determine h_2 based on the known or computed value of h_1. For the first slice shown in Fig. 8-11(a), Eq. (8-43) should be modified to

$$h_2=\left(\frac{P_1}{E_2}\right)h_1-\frac{b}{2}\left(1+\frac{P_1}{E_2}\right)\tan\theta+\frac{b}{2}\tan\delta-\frac{C_sW(y_c-y_m)}{E_2}+\frac{L[\sin\alpha(x_L-x_m)-\cos\alpha(y_L-y_m)]}{E_2} \tag{8-44}$$

In most cases when there is no pounding of water on the slope surface, P_1 and h_1 are both zero. If water is pounded on the slope surface, $h_1=d_1/3$, where

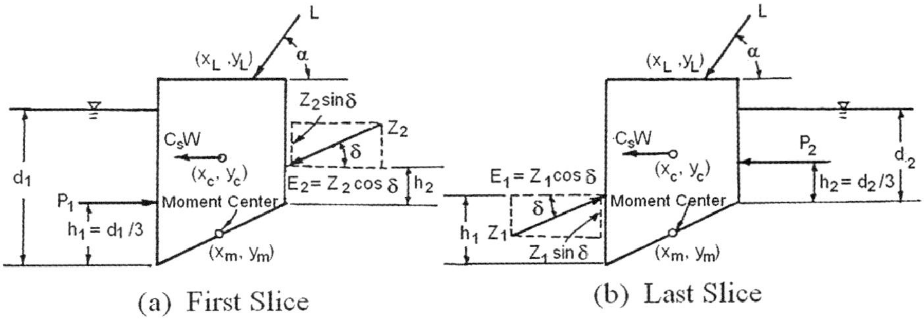

Fig. 8-11. *Forces for moment equilibrium of first and last slices*

d_1 is the water table above the failure surface for the first slice. For the last slice shown in Fig. 8-11(b),

$$h_2 = \left(\frac{E_1}{P_2}\right)h_1 - \frac{b}{2}\left(1 + \frac{E_1}{P_2}\right)\tan\theta + \frac{b}{2}\left(\frac{E_1}{P_2}\right)\tan\delta - \frac{C_s W(y_c - y_m)}{P_2} + \frac{L[\sin\alpha(x_L - x_m) - \cos\alpha(y_L - y_m)]}{P_2} \tag{8-45}$$

Starting from the first slice with given values of P_1 and h_1, Eqs. (8-44), (8-43), and (8-45) are applied successively until h_2 of the last slice is obtained. The value of δ can be adjusted gradually by trial and error until $h_2 = d_2/3$, where d_2 is the water table above the failure surface for the last slice. A subroutine called RTMI, originally developed by IBM (1970), is used in LEAME to solve a general non-linear equation of the form Function(δ) = 0 by means of Mueller's iteration scheme of successive bisection and inverse parabolic interpolation. When $P_2 = 0$, h_1 for the last slice can be determined by

$$h_1 = \frac{b}{2}\tan\theta - \frac{b}{2}\tan\delta + \frac{C_s W(y_c - y_m)}{E_1} - \frac{L[\sin(x_L - x_m) - \cos\alpha(y_L - y_m)]}{E_1} \tag{8-46}$$

The value of δ is so selected that h_2 of the next to the last slice, as obtained from Eq. (8-43), is equal to h_1 of the last slice, as obtained from Eq. (8-46). The number of equations and unknowns are tabulated as follows:

Eqs.		**Unknowns**	
Description	**No.**	**Description**	**No.**
Σ forces in horizontal direction = 0	n	Normal forces between slices, E	n − 1
Σ forces in vertical direction = 0	n	Height of forces between slices, h	n − 1
Σ moments at base of each slice = 0	n	Effective normal forces, N′	n
		Angle of interslice forces, δ	1
		Factor of safety, *F*	1
Total	3n	Total	3n

The Spencer method can be summarized as follows:

1. Based on an initial F by the normal method and $\delta = 0$, which is the same as $S = \Delta S = 0$, determine T from Eq. (8-37) and a new value of F from Eq. (8-42). Using the new F as the assumed F, repeat the process until F converges.
2. Based on $\Delta S = 0$ and the value of F obtained in step 1, compute T by Eq. (8-37) and ΔE by Eq. (8-39). Starting from the left side of the first slice where $E_1 = 0$ or P_1, compute E_2 on the right side of the first slice by $E_2 = E_1 - \Delta E$. Apply this procedure recursively, slice by slice, until the last slice is reached. Because the factor of safety is obtained through Eq. (8-40), E_2 at the right side of the last slice automatically should be equal to 0 or P_2.
3. Based on $\Delta S = 0$, the given P_1 and h_1 of the first slice, and the values of E obtained in step 2, apply Eq. (8-43) recursively to determine h_2 of the last slice. Instead of Eq. (8-43), Eq. (8-44) should be used for the first slice and Eq. (8-45) for the last slice. Vary δ until h_2 of the last slice is equal to $d_2/3$. If $P_2 = 0$, vary δ until h_2 of the next-to-last slice obtained from Eq. (8-43) is equal to h_1 of the last slice obtained from Eq. (8-46).
4. Based on the values of E obtained in step 2 and δ in step 3, compute the shear force between slices by $S = E \tan\delta$ and ΔS by difference, that is, $\Delta S = S_1 - S_2$. This completes the first cycle of iteration for $\Delta S = 0$.
5. Based on the factor of safety obtained in step 1 and the value of ΔS in step 4, repeat steps 1 to 4 and find new values of F and ΔS. This completes the second cycle of iteration.
6. Continue the cycles until F converges.

Example 8.5 Based on the information provided by Example 8.2, determine the factor of safety by the Spencer method, or method 4.

Solution The Spencer method applies the force equilibrium to evaluate the factor of safety, F, and the moment equilibrium to evaluate the angle of inclination, δ. Table 8-5 is the spreadsheet, the details of which are as follows:

1. After assuming a factor of safety in cell E12, the cells in row 14 from columns B to F should be filled out first. For the first iteration, ΔS in cell B14 should be assumed 0. The driving force in cell C14 is the denominator of Eq. (8-42) and can be expressed as $W \times \tan\theta$. The tangential force, T, in cell D14 can be computed by Eq. (8-37) and expressed as $(c \times b + W \times \tan\phi) \times \sec\theta/(F + \tan\theta \times \tan\phi')$. The resisting force in cell E14 is the numerator of Eq. (8-42) and can be expressed as $(c \times b \times \sec\theta + (W \times \sec\theta - T \times \tan\theta)\tan\phi) \times \sec\theta$. ΔE in cell F14 can be computed by Eq. (8-39) and expressed as $W \times \tan\theta - T \times \sec\theta$. Before copying row 14 from columns B to F into rows 15 to 18, be sure that c in cell G5 is typed as $G$5, $\tan\phi$ in cell G7 as $G$7, and the assumed factor of safety in cell E12 as E$12. The driving forces are summed in cell C19 and the resisting forces in cell E19. Dividing the sum of resisting forces by the sum of driving forces gives the

Table 8-5. Spreadsheet for Noncircular Failure Surface by Spencer Method

	A	B	C	D	E	F	G	H	I	J
3	Data input									
4	Slice	b	W	θ in radian						
5	1	40	40,000	0.0995		c =	500 psf			
6	2	40	120,000	0.0995		ϕ =	18 deg			
7	3	40	200,000	0.0995		tan ϕ =	0.325			
8	4	40	220,000	0.1972						
9	5	40	100,000	0.7854						
10										
11	Iteration no. 1									
12		Assumed factor of safety			1.991	Computed δ		0.291	0.292	
13	Slice	ΔS	Driving F.	T	Resisting F.	ΔE	E	h2	h2	S
14	1	0	3,993	16,388	32,791	−12,477	12,477	3.993	4.015	3,750
15	2	0	11,980	29,298	58,623	−17,465	29,941	7.322	7.362	9,000
16	3	0	19,966	42,209	84,455	−22,452	52,394	10.460	10.517	15,749
17	4	0	43,955	45,377	92,130	−2,318	54,712	13.920	14.018	16,446
18	5	0	100,000	32,054	90,255	54,669	43	*14.010	*13.988	13
19		Sum	179,894		358,255	−43	Δh	−0.090	0.029	
20		Computing factor of safety			1.991					
21										
22	Iteration no. 2									
23		Assumed factor of safety			2.153	Computed δ		0.285	0.284	
24	Slice	ΔS	Driving F.	T	Resisting F.	ΔE	E	h2	h2	S
25	1	−3,750	4,368	15,734	34,043	−11,445	11,445	3.863	3.841	3,341
26	2	−5,250	12,504	27,911	60,391	−15,546	26,991	7.139	7.099	7,878
27	3	−6,749	20,640	40,088	86,739	−19,648	46,638	10.230	10.172	13,613
28	4	−697	44,094	42,166	92,578	1,095	45,544	14.248	14.145	13,294
29	5	16,433	83,567	26,911	81,940	45,509	35	*14.141	*14.162	10
30		Sum	165,173		355,691	−35	Δh	0.107	−0.017	
31		Computing factor of safety			2.153					

Table 8-5. (Continued)

	A	B	C	D	E	F	G	H	I	J
	To save space, iterations no. 3 to no. 5 are not shown.									
	Iteration no. 6									
	Assumed factor of safety				2.127	Computed δ		0.285	0 286	
	Slice	ΔS	Driving F.	T	Resisting F.	ΔE	E	h2	h2	S
	1	−3,418	4,334	15,873	33,930	−11,618	11,618	3.863	3.885	3,416
	2	−4,665	12,445	28,159	60,191	−15,853	27,471	7.130	7.170	8,078
	3	−5,913	20,556	40,444	86,453	−20,089	47,560	10.213	10.270	13,986
	4	155	43,924	42,538	92,266	546	47,014	14.080	14.182	13,825
	5	13,836	86,164	27,683	83,272	47,014	0	*14.141	*14.119	0
		Sum	167,424		356,112	0	Δh	−0.061	0.063	
		Computing factor of safety			2.127					
	Iteration no. 7									
	Assumed factor of safety				2.127	Computed δ		0.285	0.286	
	Slice	ΔS	Driving F.	T	Resisting F.	ΔE	E	h2	h2	S
	1	−3,416	4,334	15,873	33,929	−11,617	11,617	3.863	3.885	3,416
	2	−4,662	12,445	28,158	60,190	−15,853	27,471	7.130	7.170	8,078
	3	−5,907	20,556	40,444	86,451	−20,089	47,560	10.213	10.270	13,985
	4	161	43,923	42,537	92,264	546	47,014	14.080	14.182	13,825
	5	13,825	86,175	27,685	83,279	47,022	−8	*14.141	*14.119	−2
		Sum	167,433		356,113	8	Δh	−0.061	0.063	
		Computing factor of safety			2.127					

Note: *indicates that the value is h1 of slice 5 and h2 of slice 5 is 0.

computed factor of safety. By trial and error, the factor of safety converges to 1.991. Note that the sum of ΔE is −43, which is nearly equal to 0 compared with other ΔE's, as is expected.

2. Next compute E in column G by $E_i = E_{i-1} - \Delta E$. For the first slice, E_{i-1} is 0, so $E = -\Delta E$; for the last slice, E should be 0 or negligibly small. Then assume δ in cell H12 as H$12 and compute h_2 in the first slice by Eq. (8-44) expressed as $0.5 \times b \times (\tan\delta - \tan\theta)$ and those for the intermediate slices by Eq. (8-43) expressed as $E_{i-1} \times h_{i-1}/E_i + 0.5 \times b \times (1 + E_{i-1}/E_i) \times (\tan\delta - \tan\theta)$. Because h_2 for the last slice is 0, the value shown for the last slice is actually h_1 computed by Eq. (8-46) and expressed as $0.5 \times b \times (\tan\theta - \tan\delta)$. The difference between h_2 of slice 4 and h_1 of slice 5 is Δh, as shown

on the line beneath slice 5. Finally, copy column H from rows 12 to 19 into column I. Theoretically, Δh should be equal to 0 but, practically, a value of δ can be found by trial and error that makes Δh as close to 0 as possible. For example, when $\delta = 0.291$, $\Delta h = -0.090$ and when $\delta = 0.292$, $\Delta h = 0.029$. It is apparent that $\Delta h = 0$ lies between $\delta = 0.291$ and 0.292, so the one closest to 0 is placed in column I. After δ is determined, S in column J can be computed by $E \tan \delta$.

3. Before copying iteration 1 from rows 11 to 20 to form iteration 2 from rows 22 to 31, change all variables with a row number less than 10, such as B5 and D9, to B$5 and D$9, the assumed factor of safety from E$12 to E12, and the assumed δ from H$12 and I$12 to H12 and I12. After copying, iteration 2 should be exactly the same as iteration 1. If not, something must be wrong in the spreadsheet and should be corrected. In iteration 2, change ΔS in column B from 0 to $S_{i-1} - S_i$. Then adjust the assumed factor of safety in cell E23 until it converges to 2.153 and the assumed δ in cells H23 and H24 until a value of 0.284 is obtained.
4. Iteration 2 from rows 22 to 31 can be copied repeatedly for the remaining iterations. Adjust the factor of safety and the δ angle until they converge. The factor of safety converges to 1.127 and the δ angle to 0.286 at the seventh iteration. The factor of safety obtained by LEAME is 2.125, which checks closely with the 2.127 by the spreadsheet.
5. In this example, the use of the computed ΔS as the assumed ΔS does not cause any problem on convergence. If the factor of safety does not converge, a relaxation factor to reduce the amount of change in ΔS between two iterations must be used, as described in Section 8.6.2. If $\Delta S'$ is the previously assumed value and ΔS, or $S_{i-1} - S_i$, is the newly computed value, instead of using ΔS directly as the assumed ΔS, the assumed ΔS is modified as $\Delta S = \Delta S' + R_f(\Delta S - \Delta S')$, where R_f is the relaxation factor.

8.6 Special Solution Techniques

Except for the normal method, all the methods require some kinds of iterations. To speed up convergence, Newton's method of tangent can be applied to the simplified Bishop and the original Spencer methods and a relaxation factor can be used in the Spencer method.

8.6.1 Newton's Method of Tangent

A very efficient method to solve a nonlinear equation $f(F) = 0$ is by Newton's method of tangent, as shown in Fig. 8-12. The relationship between the factor of safety, F, and the function $f(F)$ is represented by the smooth curve. The intersection of the curve with the F-axis gives the solution for $f(F) = 0$. To obtain the solution graphically, a factor of safety, F, is assumed, and the function $f(F)$ is determined, as shown by point 1 in Fig. 8-12. A tangent to the curve is drawn at

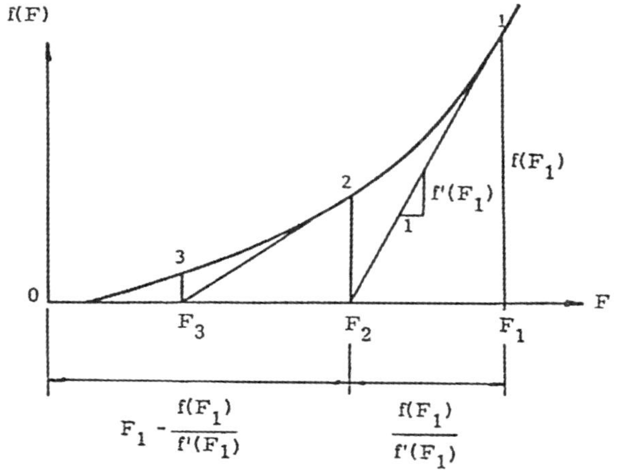

Fig. 8-12. Newton's method of tangent

point 1 and its intersection with the F-axis gives a new factor of safety, F. Repeat the process for point 2 and then point 3 until the two successive factors of safety, F_{m+1} and F_m, converge to a specified tolerance. The new F can be determined from the previous F by

$$F_{m+1} = F_m - \frac{f(F_m)}{f'(F_m)} \tag{8-47}$$

in which $f'(F_m)$ is the slope of the tangent or the first derivative of $f(F)$ at point m.

According to Eq. (8-8) for noncircular failure surfaces, the factor of safety can be reduced to the following simplified form:

$$F = \frac{\sum(a_1 + a_2 N')}{\sum(b_1 - b_2 N')} \tag{8-48}$$

in which the constants a_1, a_2, b_1, and b_2 are independent of F and can be determined directly from the geometry of each slice, the soil parameters, and the seepage and loading conditions. Eq. (8-22) for the force equilibrium in the original Spencer method can be expressed in the same form except that the negative sign before b_2 must be replaced by a positive sign. Note that the effective normal force, N', is a function of F.

Eq. (8-48) can also be applied to circular failure surfaces by assigning b_2 to 0. Eq. (8-48) can be written as

$$f(F) = F\sum(b_1 - b_2 N') - \sum(a_1 + a_2 N') \tag{8-49}$$

The first derivative of $f(F)$ is

$$f'(F) = \sum(b_1 - b_2N') - F\sum b_2\left(\frac{dN'}{dF}\right) - \sum a_2\left(\frac{dN'}{dF}\right) \tag{8-50}$$

As indicated by Eq. (8-19) for the simplified Bishop method and Eq. (8-21) for the original Spencer method, the effective normal forces can be expressed as

$$N' = \frac{a_3F - a_4}{b_3F + b_4} \tag{8-51}$$

Note that the constants a_3, a_4, b_3, and b_4 are different for each slice and, for the original Spencer method, also depend on the assumed value of δ. The first derivative of N' with respect to F is

$$\frac{dN'}{dF} = \frac{a_3b_4 + a_4b_3}{(b_3F + b_4)^2} \tag{8-52}$$

$f'(F)$ can be evaluated by substituting Eqs. (8-51) and (8-52) into Eq. (8-50).

The iterative procedure proceeds as follows. First assume $F = F_m$ and determine $f(F_m)$ from Eqs. (8-49) and (8-51), then compute $f'(F)$ from Eqs. (8-50), (8-51), and (8-52), and finally obtain a new F_{m+1} from Eq. (8-47). The process is repeated until the difference between F_{m+1} and F_m becomes negligibly small.

Because the simplified Bishop method with circular failure surfaces is employed most frequently and is repeated many times, a more concise iterative equation is developed here to save computer time. After substituting Eq. (8-19) into Eq. (8-7), the factor of safety can be computed directly by

$$F = \frac{R\sum\dfrac{c'b + (W' + L\sin\alpha)\tan\phi'}{\cos\theta + \sin\theta\tan\phi'/F}}{\sum W\lambda_w + C_s\sum W\lambda_s + \sum L\lambda_L - P_1\lambda_1 + P_2\lambda_2} \tag{8-53}$$

To make the following equations shorter, name the overturning moment in the denominator of Eq. (8-53), which is independent of F, as M_o:

$$f(F) = FM_o - R\sum\frac{c'b + (W' + L\sin\alpha)\tan\phi'}{\cos\theta + \sin\theta\tan\phi'/F} \tag{8-54}$$

$$f'(F) = M_o - R\sum\frac{[c'b + (W' + L\sin\alpha)\tan\phi']\sin\theta\tan\phi'}{(F\cos\theta + \sin\theta\tan\phi')^2} \tag{8-55}$$

From Eq. (8-47),

$$F_{m+1} = F_m\left\{1 - \frac{M_o - R\sum\dfrac{c'b + (W' + L\sin\alpha)\tan\phi'}{F_m\cos\theta + \sin\theta\tan\phi'}}{M_o - R\sum\dfrac{[c'b + (W' + L\sin\alpha)\tan\phi']\sin\theta\tan\phi'}{(F_m\cos\theta + \sin\theta\tan\phi')^2}}\right\} \tag{8-56}$$

Eq. (8-56) can be used directly to compute F_{m+1} based on the value of F_m.

Table 8-6. Spreadsheet for Computing Factor of Safety Using Newton's Method

Slice No.	W	Assumed F_m		1.618		1.750		1.693
		θ in deg	M1	M2	M1	M2	M1	M2
1	71,000	–8.9	27,818	–903	25,654	–768	26,552	–823
2	182,000	2.6	48,517	438	44,884	375	46,394	401
3	252,500	14.2	61,909	2,994	57,443	2,577	59,302	2,747
4	228,500	26.4	66,585	5,360	55,045	4,645	56,751	4,937
5	97,000	40.2	35,639	5,171	33,312	4,518	34,285	4,785
	SUM		240,468	13,060	216,340	11,347	223,283	12,047
Mo =	44,659,000	Computed F_{m+1}		1.750		1.693		1.693

Example 8.6 Based on the information provided by Example 8.1, determine the factor of safety by the simplified Bishop method using Eq. (8-56).

Solution To use the spreadsheet, Eq. (8-56) can be simplified to

$$F_{m+1} = F_m\left(1 - \frac{M_o - R\sum M_1}{M_o - R\sum M_2}\right) \tag{8-57}$$

in which $M_o = 44{,}659{,}000$ ft-lb (see Table 8-3), $R = 200$ ft, $M_1 = (c'b + W\tan\phi')/(F_m \cos\theta + \sin\theta\,\tan\phi')$, and $M_2 = [(c'b + W\tan\phi')\sin\theta\,\tan\phi']/(F_m\cos\theta + \sin\theta\,\tan\phi')^2$. Table 8-6 is the spreadsheet for computing the factor of safety.

In Table 8-6, items W and θ can be obtained from Table 8-1. An initial safety factor of 1.618 by the normal method is assumed as F_m and a new safety factor, F_{m+1}, of 1.750 is computed by Eq. (8-57). Using the new F_{m+1} as the assumed F_m, the process is repeated until the factor of safety converges to 1.693 at the third iteration, instead of the fourth iteration in Example 8.3.

8.6.2 Use of a Relaxation Factor

To enhance convergence, a relaxation factor can be used in the Spencer method. In the Spencer method, a set of ΔS is assumed and the values of F and δ are determined. Instead of using δ directly to determine the new set of ΔS, the actual δ to be used for the next iteration is

$$\delta_a = \delta' + R_f(\delta - \delta') \tag{8-58}$$

in which δ_a is the assumed angle for the next iteration, δ' is the previously assumed angle, δ is the newly computed angle, and R_f is the relaxation factor. To start iterations by LEAME, a relaxation factor of 1 initially is assumed. A relaxation factor of 1, or $\delta_a = \delta$, indicates that no relaxation factor actually is applied, because the newly computed δ is used directly as the assumed δ for the next iteration. If the factor of safety does not converge, the relaxation factor is reduced to 0.5 and the iteration is started from the very beginning. A relaxation factor of 0.5 implies that the assumed δ_a for the next iteration is the average of the previous δ', and the newly computed δ. Eq. (8-58) reduces the amount of change in δ between two iterations and prevents δ from oscillating back and forth, which may cause difficulty in convergence. If the factor of safety still does not converge, the relaxation factor will be reduced to 0.25 and then 0.125. When the relaxation factor has been reduced to 0.125 but the factor of safety still diverges, the factor of safety just before divergence is taken as the final factor of safety, and the fact that the factor of safety does not converge will be noted. If the factor of safety of the most critical failure surface diverges, LEAME will reanalyze the most critical failure surface by the original Spencer method. Because the original Spencer method usually does not have the convergence problem, the factor of safety it provides should be considered reliable. In the simplified Bishop and the original Spencer methods, Newton's method of tangent is used and no relaxation factor is applied.

Summary

1. Of the four limit equilibrium methods discussed in this chapter, the normal, simplified Bishop, and original Spencer methods consider the overall moment equilibrium by taking moments at the center of the circle, or, in the case of a noncircular failure surface, at an arbitrarily selected point. The factor of safety based on overall moment equilibrium depends on the effective normal force, N', at the bottom of each slice. The difficulty in determining N' is because of the presence of the unknown interslice forces. Each method applies a different procedure to eliminate the effect of interslice forces. The normal and simplified Bishop methods are not suitable for noncircular failure surfaces, because they do not consider the overall force equilibrium, and the location of the moment center will have some effect on the factor of safety obtained. The original Spencer method considers not only the overall moment equilibrium but also the overall force equilibrium, so any reasonable moment centers may be selected and should yield about the same factor of safety.

2. The normal method is similar to the well-known Fellenius method in which the forces between two slices are assumed to be zero. By summing the forces on each slice in a direction normal to the failure surface, N' can be determined directly and the factor of safety computed. The difference between the normal and Fellenius methods lies in the determination of pore water pressure normal to the failure surface. To avoid the occurrence of negative pore pressure on steeply inclined failure surfaces, the normal method invokes the concept of submerged weight, so the neutral force normal to the failure surface can be assumed equal to $ub\cos\theta$ instead of $ub\sec\theta$, where u is the pore pressure, b is the width of slice, and θ is the

angle of inclination of the failure surface. The normal method computes the factor of safety directly without iterations, whereas all the other methods require some kinds of iterations, or trial and error, until the factor of safety converges to a specified tolerance. The normal method usually gives a factor of safety somewhat smaller than the other methods and can be used to provide an initial factor of safety for all the other methods.

3. The simplified Bishop method is the most widely used method recognized by the engineering profession. It is recommended by LEAME as a standard method for use with circular failure surfaces, and it should yield a factor of safety very close to the more refined methods. By assuming the forces between two slices as horizontal, and considering the vertical equilibrium of each slice, the effective normal force, N', can be determined even without knowing the magnitude of the horizontal interslice forces. Unlike in the normal method, the N' thus determined depends on the factor of safety and the factor of safety depends on N', so an iteration method must be used to solve the factor of safety.

4. The original Spencer method assumes parallel interslice forces, all making an angle of δ with the horizontal. It considers the force equilibrium of each slice in the direction perpendicular to δ to determine N', the overall force equilibrium in the δ direction to determine the factor of safety with respect to force equilibrium, and the overall moment equilibrium to determine the factor of safety with respect to moment equilibrium. Because these two factors of safety are not equal and vary with the value of δ assumed, the value of δ must be adjusted by trial and error until the two factors of safety become the same. Instead of trial and error, a graphical method also can be used. The original Spencer method considers the overall force and moment equilibrium but not the force and moment equilibrium for each slice and is not as refined as the Spencer method, which satisfies all equations of equilibrium. However, the difference in the factor of safety between the two methods is usually quite small. The original Spencer method has the advantage in that it always converges, whereas the Spencer method sometimes may have convergence problems.

5. The Spencer method is a special case of the well-known Morgenstern-Price method by assigning the function $f(x) = 1$, thus avoiding the sophistication of a user-defined function, as required by the Morgenstern-Price method. Similar to the original Spencer method, it also assumes that all the interslice forces incline at an angle δ with the horizontal. It considers both the force and the moment equilibrium of each slice. If each slice is in equilibrium, the overall force and moment equilibrium automatically is satisfied. The force equilibrium equations are used to determine the factor of safety, F, and the moment equilibrium equations are used to determine the angle δ. The determination of F and δ requires two separate iterations. Because F depends on δ and δ depends on F, a third iteration is

needed to solve F and δ. The Spencer method is the most refined method and is recommended by LEAME as a standard method for use with non-circular and composite failure surfaces.

6. All the methods presented in this chapter, except for the normal method, require some kinds of iterations. To speed up convergence, Newton's method of tangent can be applied to the simplified Bishop and original Spencer methods. The application of this technique is described and an equation applicable to the simplified Bishop method with circular failure surfaces is presented. For the Spencer method, a relaxation factor should be applied to avoid oscillations and ensure convergence.

7. For a given failure surface, spreadsheets can be applied easily to compute the factor of safety. The use of a trial-and-error method by a spreadsheet to find a factor of safety is so easy that it is really not necessary to change the factor of safety gradually until it converges, as is done by LEAME. A few trials and errors will be sufficient to find a computed factor of safety that is equal or nearly equal to the assumed factor of safety. It should be noted that the spreadsheet only solves one of the hundreds of trial failure surfaces and cannot be used as a substitute for a computer program. The factors of safety obtained by the spreadsheets were compared with those by the LEAME computer program and found to be in good agreement.

Problems

8.1 Fig. P8-1 shows a 3:1 slope and the location of a circular failure surface. The failure mass is first divided evenly into three slices, as shown by the solid lines, and then subdivided at the two break points of the ground line, as shown by the dashed lines, so there are a total of five slices. If the soil has a cohesion of 500 psf, a friction angle of 10°, and a total unit weight of 125 pcf, determine the factor of safety by the normal method. [Answer: 1.445]

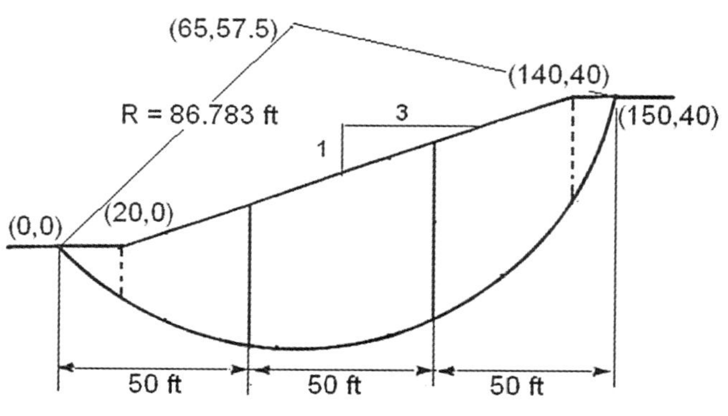

Fig. P8-1.

8.2 Same as Problem 8.1, but determine the factor of safety by the simplified Bishop method.
[Answer: 1.625]

8.3 Same as Problem 8.1, but determine the factor of safety by the original Spencer method.
[Answer: 1.624]

8.4 Same as Problem 8.1, but determine the factor of safety by the Spencer method.
[Answer: 1.654]

8.5 Fig. P8-5 shows a 2:1 slope and the location of the noncircular failure surface. The failure mass is first divided evenly into four slices, as shown by the solid lines, and then subdivided at the break point of the failure surface, as shown by the dashed lines, so there are a total of five slices. The soil has a cohesion of 400 psf, a friction angle of 15°, and a total unit weight of 125 pcf. By assuming an arbitrary moment center at (100, 200), determine the factor of safety by the normal method.
[Answer: 1.250]

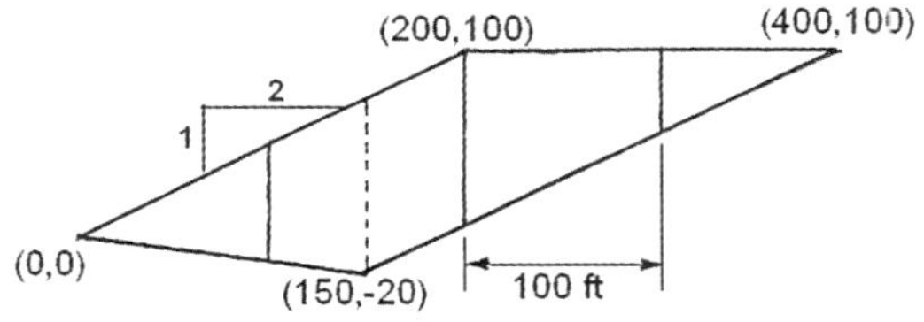

Fig. P8-5.

8.6 Same as Problem 8.5, but determine the factor of safety by the original Spencer method.
[Answer: 1.393]

8.7 Same as Problem 8.5, but determine the factor of safety by the Spencer method.
[Answer: 1.393]

8.8 Also considered in the LEAME computer program but not presented in this chapter is a special case where the center of the circle is located below the top of embankment, as shown by a vertical cut in Fig. P8-8. Because part of the circular arc is replaced by a vertical line, an additional resisting force of $c'\lambda/F$ over the vertical line segment must be taken into consideration, where c' is the effective cohesion of soil, λ is the length of the vertical line, and F is the factor of safety. First, modify Eqs. (8-7) and (8-19) for the first slice to include this additional resisting force. Then, by the use of spreadsheets, divide the failure mass into five slices and solve the following two cases: (a) if the soil has an undrained shear strength of 1,050 psf and a total unit weight of 125 pcf, determine the factors of safety by the normal method, and (b) if the soil has a cohesion of 500 psf, a friction angle of 20°, and a total unit weight of 125 pcf, determine the factor of safety by the simplified Bishop method.
[Answer: 1.397, 1.089]

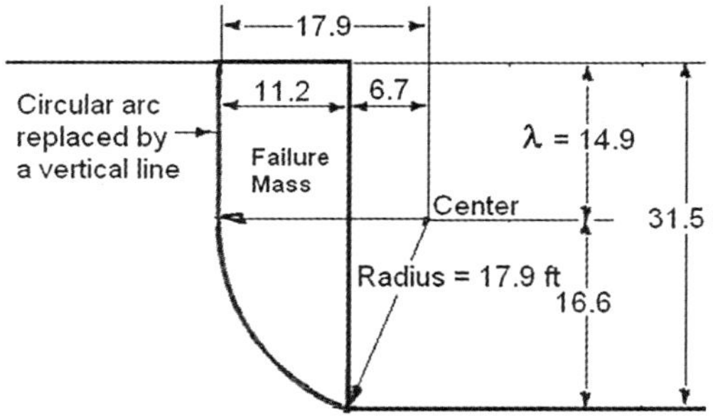

Fig. P8-8.

Chapter 9

Methods for Three-Dimensional Analysis

This chapter presents theoretical background for three-dimensional (3D) analysis and extends the method of slices to the method of columns. To simplify the analysis, only the normal, simplified Bishop, and original Spencer methods will be used. The simplified Bishop method is recommended for circular failure surfaces and the original Spencer method for noncircular and composite failure surfaces. The purpose of 3D analysis is to include the end effects, so the same cross section for 2D analysis will be used for 3D analysis. Two types of failures will be discussed, one with ellipsoidal ends and the other with planar ends. Major equations incorporated in LEAME will be derived and the use of spreadsheets to solve these equations will be demonstrated. Details on the application of these two types of failures and their practical implications are discussed in Chapter 3 of the companion volume to this book, *Slope Stability by Limited Equilibrium Analysis: LEAME Software and User's Manual.*

9.1 Failure Surfaces with Ellipsoidal Ends

This type of 3D failure surface consists of a cylinder formed by circles of constant radius at the central part and two half ellipsoids formed by circles of decreasing radius at each end. The failure mass is divided into columns and the method of slices used in 2D analysis can be extended easily to 3D analysis. Instead of the four methods for 2D analysis, only the first three methods (normal, simplified Bishop, and original Spencer) can be used for 3D analysis. As indicated in the 2D analysis, the original Spencer method always can yield a factor of safety very close to the more refined Spencer method.

9.1.1 Dimensions of Failure Mass

Fig. 9-1(b) shows a 3D failure mass and the Cartesian coordinates, x, y, and z. Because of symmetry, only one-half of the slope is shown, so some of the length is referred to as half length. The failure mass is formed by rotating the rectangle and the ellipse, as shown in Fig. 9-1(a), until they intersect with the slope below. The rectangle has a half length of λ_c and the ellipse has a half length of λ_e. Because the axis of rotation, which is the center (x_o, z_o) for 2D analysis, is located above the slope surface, the half length of the failure mass, $\lambda_c + \lambda$, is always shorter than $\lambda_c + \lambda_e$. The failure mass is divided arbitrarily into 10 slices in the x direction and 10 columns in the y direction. Note that the entire length of the cylinder, λ_c, in the y direction is considered as one column, whereas the ellipsoidal part of the failure mass is divided into nine columns. The radius of rotation is R for the cylinder and reduces gradually, as indicated by r in Fig. 9-1(a). If a weak soil layer intercepts the cylinder or ellipsoid, a composite failure surface through the bottom of the weak layer also may be formed.

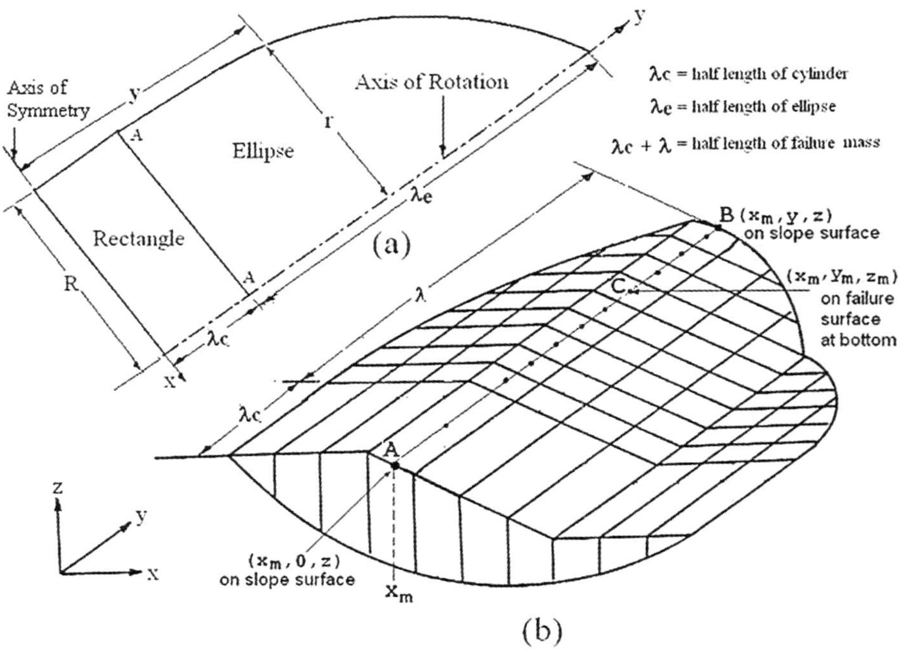

Fig. 9-1. Intersection of cylinder and ellipsoid with slope surface

In the stability analysis, λ_c and λ_e must be assumed or determined by trial and error. If λ_c is given, such as a uniform load distributed over a short distance or a steep slope with a short length, one basic assumption is that the half ellipsoidal length, λ_e, must be equal to or greater than half the height of the tallest column, as explained in Sections 3.8.1 and 3.8.7 of the companion volume. If the length of embankment is given, the lengths of λ_c and λ_e must be proportioned so that the factor of safety is minimum when the half length of the failure mass is equal to the half length of the embankment, as discussed in Sections 3.7.1, 3.8.3 and 3.8.6 of the companion volume.

The equation for an ellipse with semi-axes R and λ_e, as shown in Fig. 9-1(a), can be expressed as

$$\frac{x^2}{R^2}+\frac{(y-\lambda_c)^2}{\lambda_e^2}=1 \tag{9-1}$$

The reason why the y coordinate is replaced by $(y-\lambda_c)$ is because the origin of the y axis starts at the plane of symmetry instead of from plane A-A. For any given y, the radius of rotation, r, is

$$r=x=R\sqrt{1-\frac{(y-\lambda_c)^2}{\lambda_e^2}} \tag{9-2}$$

Knowing the radius, r, and the center of rotation, (x_o, z_o), the intersection between the ellipse of revolution and the slope surface can be calculated.

The equation for a circle with radius, r, can be written as

$$\frac{(x-x_o)^2}{r^2}+\frac{(z-z_o)^2}{r^2}=1 \tag{9-3}$$

Substituting r from Eq. (9-2) into Eq. (9-3), the equation of ellipsoid becomes

$$\frac{(x-x_o)^2}{R^2}+\frac{(y-\lambda_c)^2}{\lambda_e^2}+\frac{(z-z_o)^2}{R^2}=1 \tag{9-4}$$

in which x_o and z_o are the coordinates of the rotary center for 2D analysis, λ_e = half length of the ellipsoid in the y direction, λ_c = half length of the cylinder, and R = radius of the cylinder.

For any given values of x and z on the surface of the slope, the coordinate y where the ellipsoid intersects the slope surface can be computed by Eq. (9-4), which can be rewritten as

$$y=\lambda_c+\frac{\lambda_e}{R}\sqrt{R^2-(x-x_o)^2-(z-z_o)^2} \tag{9-5}$$

In Fig. 9-1(b), x_m is the x coordinate at the center of a slice. Point A $(x_m, 0, z)$ is located on the slope surface at the plane of symmetry, where x_m and z can be determined from the given cross section. Point B (x_m, y, z) is located on the slope surface where it intersects with the failure surface, and y can be determined by Eq. (9-5). After comparing these y coordinates at several slices, the maximum value of y, which is the half length of the failure mass, can be found. Dividing the half length of the failure mass into a number of columns, the y_m coordinate at the center of each column, such as point C (x_m, y_m, z_m), can be found, where z_m is the z coordinate at the bottom of the column and can be determined by

$$z_m = z_o - R\sqrt{1 - \frac{(x_m - x_o)^2}{R^2} - \frac{(y_m - \lambda_c)^2}{\lambda_e^2}} \tag{9-6}$$

The height of each column is the difference in z coordinates between the slope surface and the bottom of the column.

Example 9.1 Fig. 9-2 shows the cross section of the failure mass on the plane of symmetry with $x_o = 40$ ft (12.2 m), $z_o = 100$ ft (30.5 m), and $R = 107.7$ ft (32.8 m). The failure mass is divided into 10 slices in the x direction and 10 columns in the y direction, including the first column. The first column has a length λ_c, while the remaining nine columns are divided evenly. If the failure mass has a half cylindrical length, λ_c, of 100 ft (30.5 m), and a half ellipsoidal length, λ_e, of 200 ft (61.0 m), determine the half length of the failure mass. What is the height of the sixth column in the fifth slice?

Solution From Eq. (9-5), $y = 100 + 1.857\sqrt{(107.7)^2 - (x-40)^2 - (z-100)^2}$. It is estimated that the half length of the failure mass is maximum near the middle of the cross section, so the y coordinates at three different centers of slices is computed and compared as follows:

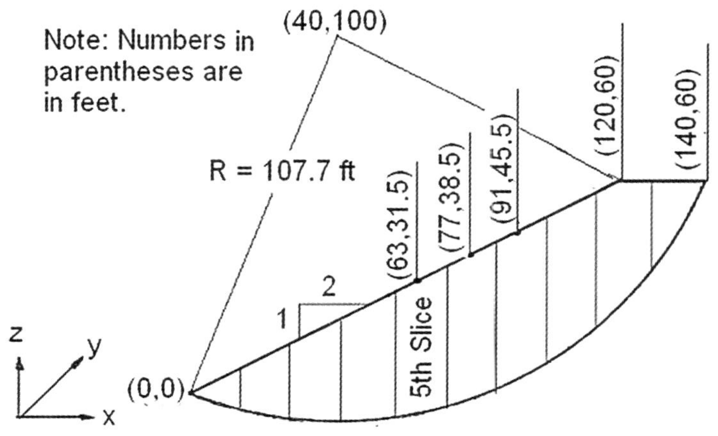

Fig. 9-2. Example 9.1

Note: 1 ft = 0.305 m

x in ft	63	77	91
z on slope surface in ft	31.5	38.5	45.5
y in ft	248.3	249.1	237.1

The maximum y occurs at $x = 77$ ft, so the half length of the failure mass is 249.1 ft. At the center of slice 5, $x = 63$ ft, at the center of column 6, $y = 100 + 4.5 \times (249.1 - 100)/9 = 174.6$ ft, and from Eq. (9-6),

$$z_m = 100 - 107.7\sqrt{1 - \frac{(63-40)^2}{(107.7)^2} - \frac{(174.6-100)^2}{(200)^2}} = 2.8 \text{ ft},$$

or Height of column = 31.5 − 2.8 = 28.7 ft.

9.1.2 Orientation and Area of Failure Surface

The failure surface at the bottom of each column is tilted in two different directions, as shown in Fig. 9-3. Given the x_o and z_o coordinates of the rotary center and the x_m and z_m coordinates at the bottom of the column, as shown in Fig. 9-3(a), the angle of inclination, θ_{xz}, on the xz plane can be determined by

$$\tan\theta_{xz} = \frac{x_m - x_o}{z_o - z_m} \tag{9-7}$$

The inclination, θ_{yz}, on the yz plane is shown in Fig. 9-3(b) and can be obtained from Eq. (9-4) by taking partial derivative of z with respect to y and assuming θ_{yz} as positive, or

$$\frac{\partial z}{\partial y} = \tan\theta_{yz} = \frac{y_m - \lambda_c}{z_o - z_m}\left(\frac{R}{\lambda_e}\right)^2 \tag{9-8}$$

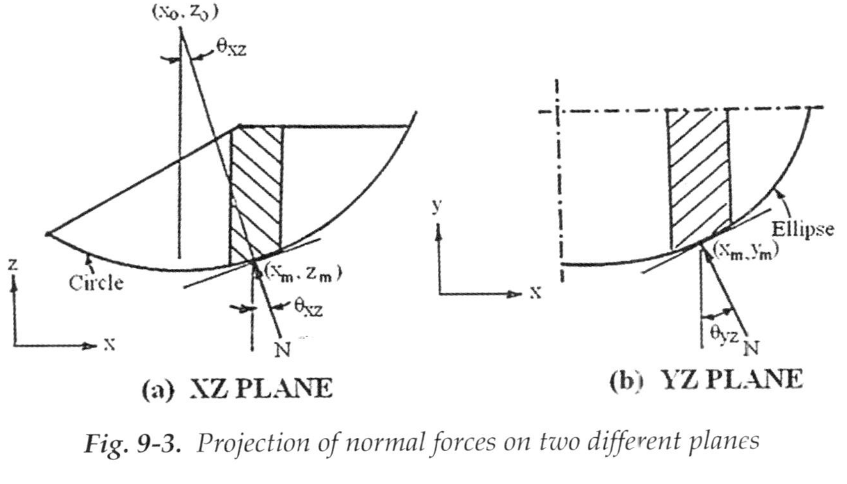

Fig. 9-3. Projection of normal forces on two different planes

In 3D analysis, the movement of the failure mass is assumed parallel to the *xz* plane, so all forces involved in force and moment equilibrium must be projected onto the *xz* plane. The only forces that are not parallel to the *xz* plane are the effective normal force and the pore water pressure, which act in a direction perpendicular to the failure surface. Therefore, it is necessary to determine the dip angle, θ_d, between the normal force and the vertical, *z*, axis so the normal force and water pressure on the failure surface can be projected to the *xz* plane. Eqs. (9-11) and (9-17) for computing the dip angle, θ_d, and the surface area, *S*, were presented originally by Hovland (1997) and is derived following.

Fig. 9-4(a) shows the 3D view of a soil column. On the horizontal, or *xy*, plane, the area of the column is a rectangle with a width Δx and a length Δy, whereas the actual surface area is a parallelogram with a width of *a*, a length of *b*, and an angle of θ between them. Also shown in the figure are the terms strike and dip, which indicate the orientation of the failure surface. According to the usual geological definition, the intersection between the failure surface and the horizontal, or *xy*, plane is called a strike, and the intersection of the failure surface with a vertical plane perpendicular to the strike is called a dip. The directions of the strike and dip also are shown in Fig. 9-4(b).

Fig. 9-4(b) shows the dimensions of the failure surface on different planes. Area ABC is a portion of the failure surface. Area ACD is the projection of the failure surface on the *xz* plane, area ABD on the *yz* plane, and area BCD on the *xy* plane. The angle between the strike and the *x* axis is θ_s and the angle between

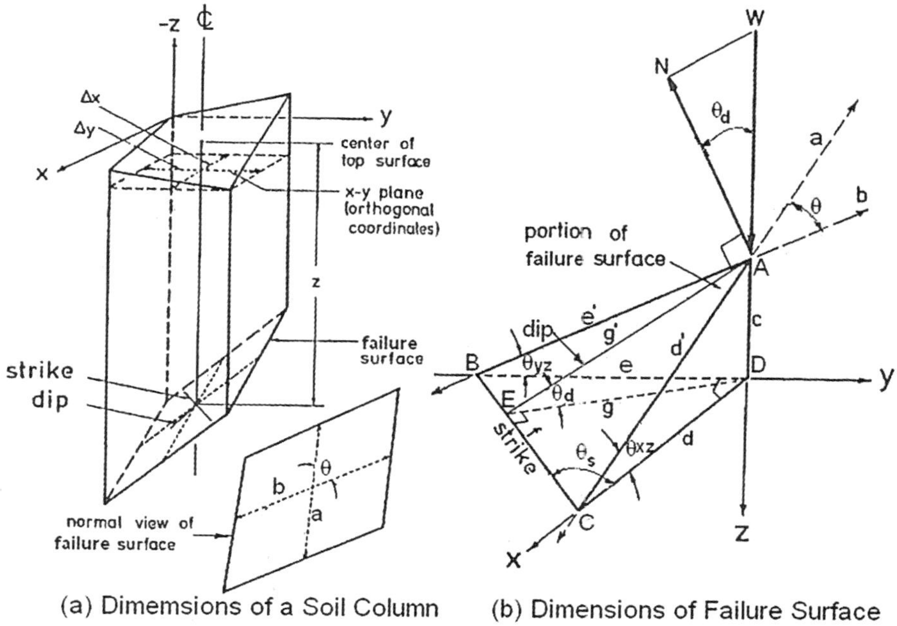

(a) Dimemsions of a Soil Column (b) Dimensions of Failure Surface

Fig. 9-4. Three-dimensional view of failure surface

the dip and the horizontal is θ_d, which is the same as the angle between the normal force, N, and the vertical axis. The dimensions d′, e′, and g′ are measured on the failure surface, while those without the prime are measured on the three orthogonal planes. In ΔBCD, ∠BDC is a right angle, so

$$\sin\theta_s = \frac{e}{f} = \frac{\dfrac{c}{\tan\theta_{yz}}}{\sqrt{e^2+d^2}} = \frac{\dfrac{c}{\tan\theta_{yz}}}{\sqrt{\left(\dfrac{c}{\tan\theta_{yz}}\right)^2+\left(\dfrac{c}{\tan\theta_{xz}}\right)^2}} \tag{9-9}$$

$$\text{or}\quad \sin\theta_s = \frac{1}{\sqrt{1+\left(\dfrac{\tan^2\theta_{yz}}{\tan^2\theta_{xz}}\right)}}$$

Similarly, a general expression can be derived for $\cos\theta_d$:

$$\cos\theta_d = \frac{g}{g'} = \frac{d\sin\theta_s}{\dfrac{c}{\sin\theta_d}} = \frac{\left(\dfrac{c}{\tan\theta_{xz}}\right)\sin\theta_s}{\dfrac{c}{\sin\theta_d}} = \frac{\sin\theta_d\sin\theta_s}{\tan\theta_{xz}} \tag{9-10}$$

$$\text{or}\quad \cos^2\theta_d = \frac{(1-\cos\theta_d)\sin^2\theta_s}{\tan^2\theta_{xz}}$$

After substituting Eq. (9-9) into Eq. (9-10) and simplifying,

$$\cos\theta_d = \frac{1}{\sqrt{1+\tan^2\theta_{yz}+\tan^2\theta_{xz}}} \tag{9-11}$$

in which θ_d = angle between the normal to the failure plane and the vertical, z, axis.

The normal force on the xz plane always makes an angle, θ_{xz}, with the z axis, but the angle between the force, N, normal to the failure plane and the z axis is θ_d. To project N onto the xz plane, N must be multiplied by $\cos(\theta_d - \theta_{xz})$. For example, when the failure surface is cylindrical, $\theta_{yz} = 0$, and, from Eq. (9-11), $\cos\theta_d = \cos\theta_{xz}$, or $\cos(\theta_d - \theta_{xz}) = 1$, so the normal force actually lies on the xz plane. When the failure surface is ellipsoidal or planar, with $\theta_{xz} = 0$, then $\cos\theta_d = \cos\theta_{yz}$, so $\cos(\theta_d - \theta_{xz}) = \cos\theta_{yz}$, or the normal force makes an angle of θ_{yz} with the xz plane and therefore should be multiplied by $\cos\theta_{yz}$ to project onto the xz plane.

From the parallelogram shown in Fig. 9-4(a), the area of the failure surface can be computed by

$$S = ab\sin\theta = \left(\frac{\Delta x}{\cos\theta_{xz}}\right)\left(\frac{\Delta y}{\cos\theta_{yz}}\right)\sin\theta \tag{9-12}$$

From ΔABC and the law of cosine,

$$f^2 = e'^2 + d'^2 - 2e'd'\cos\theta$$
$$\text{or} \quad \cos\theta = \frac{1}{2}\left(\frac{e'}{d'} + \frac{d'}{e'} - \frac{f^2}{e'd'}\right) \tag{9-13}$$

in which

$$e' = c/\sin\theta_{yz},\ d' = c/\sin\theta_{xz},\ f^2 = e^2 + d^2 = (c/\tan\theta_{yz})^2 + (c/\tan\theta_{yz})^2 \tag{9-14}$$

Substituting Eq. (9-14) into Eq. (9-13) and simplifying,

$$\cos\theta = \sin\theta_{yz}\sin\theta_{xz} \tag{9-15}$$

$$\text{or} \quad \sin\theta = \sqrt{1 - \sin^2\theta_{yz}\sin^2\theta_{xz}} \tag{9-16}$$

From Eqs. (9-12) and (9-16),

$$S = \left(\frac{\Delta x}{\cos\theta_{xz}}\right)\left(\frac{\Delta y}{\cos\theta_{yz}}\right)\sqrt{1 - \sin^2\theta_{yz}\sin^2\theta_{xz}} \tag{9-17}$$

For cylindrical failure surfaces, $\theta_{yz} = 0$, so the surface area at the bottom of the cylinder is $S = \Delta x \Delta y / \cos\theta_{xz}$, which is as expected.

Example 9.2 In Example 9.1, determine $\cos(\theta_d - \theta_{xz})$ for column 6 in slice 5. If the soil has a unit weight of 125 pcf (19.7 kN/m³), determine the weight of the column. What is the surface area at the bottom of this column?

Solution From Example 9.1, the coordinates at the bottom of column 6 in slice 5 are $x = 63$ ft, $y = 174.6$ ft, $z = 2.8$ ft. From Eq. (9-7), $\tan\theta_{xz} = (63 - 40)/(100 - 2.8) = 0.237$, or $\theta_{xz} = 13.2°$. From Eq. (9-8),

$$\tan\theta_{yz} = \frac{174.6 - 100}{100 - 2.8}\left(\frac{107.7}{200}\right)^2 = 0.223,$$
$$\text{or} \quad \theta_{yz} = 12.6°$$

From Eq. (9-11),

$$\cos\theta_d = (1 + \tan^2 12.6° + \tan^2 13.2°)^{-1/2} = 0.951, \text{ so } \theta_d = 18.0°,$$
$$\text{or} \quad \cos(\theta_d - \theta_{xz}) = \cos(18.0° - 13.2°) = 0.996$$

From Example 9.1, $\Delta x = 14$ ft, $\Delta y = (249.1 - 100)/9 = 16.6$ ft, height of the column = 28.7 ft, so the weight of the column = 14 × 16.6 × 28.7 × 125 = 834,000 lb. From Eq. (9-17), the surface area at the bottom is

$$S = \left(\frac{14}{\cos 13.2°}\right)\left(\frac{16.6}{\cos 12.6°}\right)\sqrt{1 - \sin^2 12.6° \sin^2 13.2°} = 244.3 \text{ ft}^2$$

9.2 Failure Surfaces with Planar Ends

Instead of ellipsoids, this type of failure surface is cut short by two symmetrical planes called the end planes, each with a given slope and oriented at a given direction. A simple way to visualize this type of failure surface is an embankment between two banks, and the surfaces of the bank are considered as the end planes. These two end planes may not be parallel and can rotate at a given angle apart. The failure surface between the end planes can be circular, composite, or noncircular. When the failure surface intersects the end plane, it will follow the end plane until the top of the fill is reached. In other words, the end planes also form part of the failure surfaces.

Fig. 9-5 shows a 3D failure mass with an end plane. Due to symmetry, only one-half of the failure mass is shown in (a) and (d). Section A-A at the central part near the plane of symmetry, as shown in (b), is usually the most critical section for 2D analysis. In 2D analysis, this section is assumed to be infinitely long, whereas in 3D analysis the length is limited at each end by an end plane. The intersection of the end plane and the failure surface is indicated by the dashed line in (a). For any given cross section, such as B-B, the failure surface inside the dashed line is the same as Section A-A, and that outside the dashed line has a uniform slope, because the failure surface is on the end plane. Depending on the inclination of the end plane, this uniform slope may become quite steep. Consequently, Section B-B may be more critical than Section A-A, thus resulting in a lower factor of safety based on 3D analysis.

Fig. 9-6 shows how the end plane is defined and how the dimensions of the failure mass are determined. The location of the end plane depends on the half width of fill, W_t, at a given point A, the angle of end plane, α, and the slope of end plane, g. Note that g is a dimensionless parameter; for example, $g = 0.5$ if the slope is 2:1. First, a horizontal plane is passed through point A with coordinates (x_t, w_t, z_t). Next, on this horizontal plane, a line with an angle α with the x axis is drawn through point A. Finally, using this line as the axis, the horizontal plane

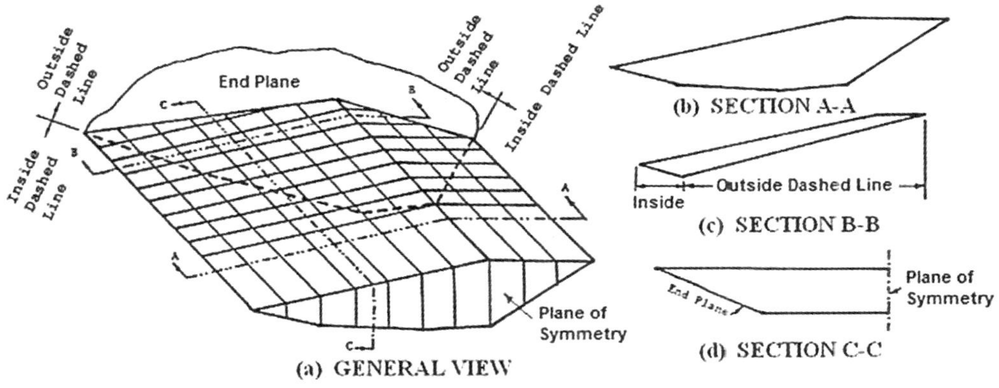

Fig. 9-5. Geometry of 3D failure surfaces with planar ends

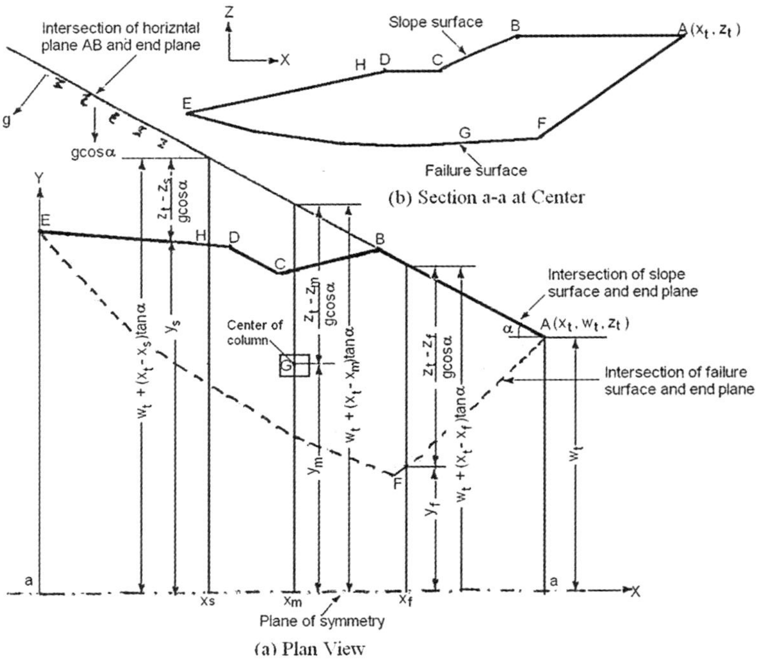

Fig. 9-6. Analysis of 3D failure surfaces with planar ends

is rotated to incline at a slope of g, measured in a direction perpendicular to the axis of rotation. This inclined plane is the end plane for 3D analysis.

In Fig. 9-6(a), ABCDE is the intersection of the slope surface with the end plane, where A to E are the breaking points on the slope surface. Because these breaking points are not located on the centerline of each slice, they are not used to define y_s for the length of the failure mass. Note also that the slope of the end plane in the y direction is $g \cos\alpha$, so the half length of the failure mass, y_s, at the centerline of each slice with coordinates (x_s, z_s) on the slope surface can be determined by

$$y_s = w_t + (x_t - x_s)\tan\alpha - \frac{z_t - z_s}{g\cos\alpha} \tag{9-18}$$

The half length of the failure mass is the maximum y_s among all the slices. This maximum should occur on a slice next to one of the breaking points on the slope surface.

The distance, y_f, to the end plane at the centerline of each slice with coordinates (x_f, z_f) on the failure surface can be determined by

$$y_f = w_t + (x_t - x_f)\tan\alpha - \frac{z_t - z_f}{g\cos\alpha} \tag{9-19}$$

The minimum y_f among all the slices is used as the length of the first row of columns, similar to the half length of the cylinder, λ_c, in the ellipsoidal case. The minimum y_f is not located at the breaking point, F, but should occur on a slice next to one of the breaking points on the failure surface.

Based on the half length of the failure mass, y_s, and the half length of the first row of columns, y_f, the remaining failure mass can be divided evenly into a number of columns, and the y coordinate at the center of each column, y_m, can be determined. If $y_m < y_f$, the cross section shown in Fig. 9-6(b) applies, so $\theta_{yz} = 0$. If $y_m > y_f$, the failure surface is on the end plane and the z coordinate at the center of the column can be determined as follows.

Similar to Eq. (9-19), y_m can be expressed as

$$y_m = w_t + (x_t - x_m)\tan\alpha - \frac{z_t - z_m}{g\cos\alpha} \tag{9-20}$$

or $$z_m = z_t - g\cos\alpha\left[w_t + (x_t - x_m)\tan\alpha - y_m\right] \tag{9-21}$$

The orientation at the bottom of each column can be computed by

$$\theta_{xz} = \tan^{-1}(g\sin\alpha) \tag{9-22}$$

$$\theta_{yz} = \tan^{-1}(g\cos\alpha) \tag{9-23}$$

Example 9.3 Fig. 9-7(a) shows the upper half of the plan view for a 3D noncircular failure surface intercepted by an end plane. The end plane has an angle, α, of 30° and a slope, g, of 0.5. The cross section of the fill is shown in Fig. 9-6(b). The half top width, w_t, is 150 ft (45.8 m). The cross section is divided into 10 slices in the x direction and then subdivided into 10 columns in the y direction, including the first column. (1) Compute the half length of the failure mass, y_s, at each breaking point of the slope surface, and determine the half length of the failure mass, which gives the largest y_s measured along the centerline of the slice. (2) Compute y_f, where the failure surface intersects with the end plane at each breaking point of the failure surface, and determine the half length of the first column. (3) Determine z_m, θ_{xz}, and θ_{yz} at the center of columns 2 and 4, both located in slice 4. (4) Plot the intersection of the end plane with the slope surface and also that with the failure surface, and divide the failure mass into columns.

Solution (1) Use Eq. (9-18) to determine y_s at the four breaking points of the slope surface:

When $x_s = 0$, $z_s = 0$, $y_s = 150 + (400 - 0)\tan 30° - (120 - 0)/(0.5\cos 30°) = 103.8$ ft.
When $x_s = 200$ ft, $z_s = 80$ ft, $y_s = 150 + (400 - 200)\tan 30° - (120 - 80)/(0.5\cos 30°) = 173.1$ ft.
When $x_s = 240$ ft, $z_s = 80$ ft, $y_s = 150 + (400 - 240)\tan 30° - (120 - 80)/(0.5\cos 30°) = 150.0$ ft.
When $x_s = 360$ ft, $z_s = 120$ ft, $y_s = 150 + (400 - 360)\tan 30° - (120 - 120)/(0.5\cos 30°) = 173.1$ ft.

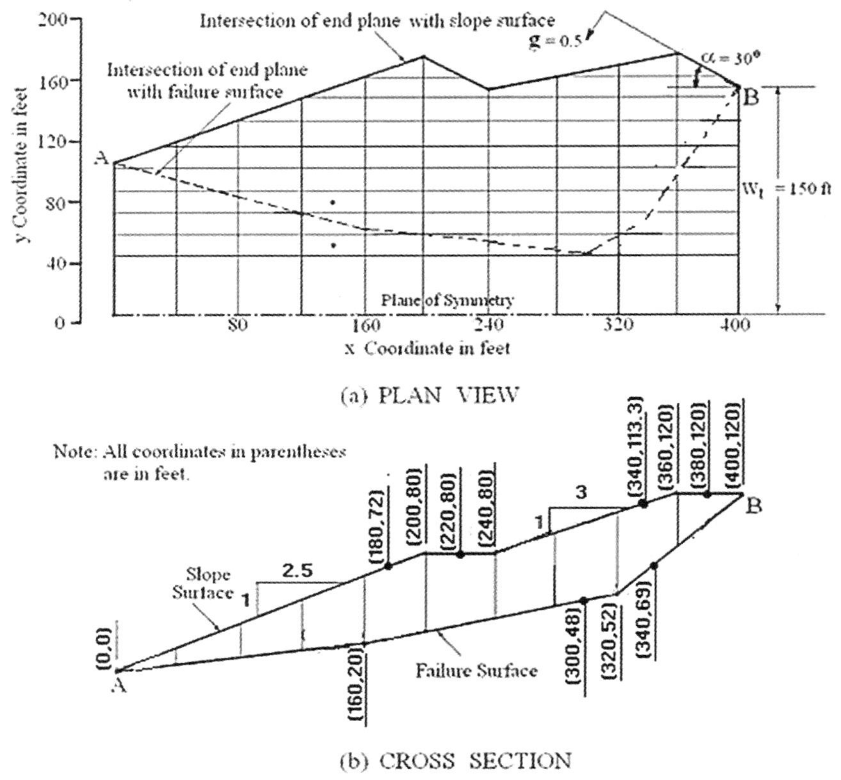

Fig. 9-7. Example 9.3

The maximum is 173.1 ft, and it occurs when $x_s = 200$ ft and $x_s = 360$ ft. To compute the area of the end columns, the half length of the failure mass is measured along the centerline of the slices next to the breaking point, instead of at the breaking point, so the following four cases should be investigated:

When $x_s = 180$ ft, $z_s = 72$ ft, $y_s = 150 + (400 - 180)\tan 30° - (120 - 72)/(0.5 \cos 30°) = 166.2$ ft.

When $x_s = 220$ ft, $z_s = 80$ ft, $y_s = 150 + (400 - 220)\tan 30° - (120 - 80)/(0.5 \cos 30°) = 161.5$ ft.

When $x_s = 340$ ft, $z_s = 113.3$ ft, $y_s = 150 + (400 - 340)\tan 30° - (120 - 113.3)/(0.5 \cos 30°) = 169.2$ ft.

When $x_s = 380$ ft, $z_s = 120$ ft, $y_s = 150 + (400 - 380)\tan 30° - (120 - 120)/(0.5 \cos 30°) = 161.5$ ft.

The maximum is 169.2 so the half length of the failure mass is 169.2 ft.

(2) Use Eq. (9-19) to determine y_f at the four breaking points of the failure surface.

At point A, the coordinates of the failure and slope surfaces are the same, so $y_f = 103.8$ ft.

When $x_f = 160$ ft, $z_f = 40$ ft, $y_f = 150 + (400 - 160)\tan 30° - (120 - 20)/(0.5 \cos 30°) = 57.6$ ft.

When $x_f = 320$ ft, $z_f = 52$ ft, $y_f = 150 + (400 - 320)\tan 30° - (120 - 52)/(0.5 \cos 30°) = 39.1$ ft.

At point B, the coordinates of the failure and slope surfaces are the same, so $y_f = 150$ ft.

The minimum is 39.1 ft, and it occurs when $x_f = 320$ ft, so the following two cases should be investigated:

When $x_f = 300$ ft, $z_f = 48$ ft, $y_f = 150 + (400 - 300)\tan 30° - (120 - 48)/(0.5 \cos 30°) = 41.5$ ft.

When $x_f = 340$ ft, $z_f = 69$ ft, $y_f = 150 + (400 - 340)\tan 30° - (120 - 69)/(0.5 \cos 30°) = 66.9$ ft.

The minimum is 41.5, so the half length of the first column is 41.5 ft.

(3) At the fourth slice with $x_f = 140$ ft and $z_f = 17.5$ ft, from Eq. (9-19), $y_f = 150 + (400 - 140)\tan 30° - (120 - 17.5)/(0.5\cos 30°) = 63.4$ ft. The y coordinate at the center of the second column is $y_m = 41.5 + 0.5 \times (169.2 - 41.5)/9 = 48.6$ ft, which is smaller than y_f of 63.4 ft, so z_m, θ_{xz}, and θ_{yz} are the same as column 1, or $z_m = 17.5$ ft, $\theta_{xz} = \tan^{-1}(20/160) = 7.1°$, and $\theta_{yz} = 0$.

At the fourth column, $y_m = 41.5 + 2.5 \times (169.2 - 41.5)/9 = 77.0$ ft, which is greater than y_f of 63.4 ft, so the bottom is on the end plane. From Eq. (9-21), $z_m = 120 - 0.5 \cos 30°[150 + (400 - 140)\tan 30° - 77.0] = 23.4$ ft. From Eq. (9-22), $\theta_{xz} = \tan^{-1}(0.5 \sin 30°) = 14.0°$. From Eq. (9-23), $\theta_{yz} = \tan^{-1}(0.5 \cos 30°) = 23.4°$.

(4) The lines of intersection and the division into columns are shown in Fig. 9-7(a).

9.3 Equation for Overall Moment Equilibrium

The equation for overall moment equilibrium is used in all three methods and is presented in this section. The equation for determining the 3D factor of safety is similar to that in 2D except that the summation of resisting and driving moments must be extended to all of the slices in the x direction and all of the columns in the y direction. Similar to Eq. (8-7) for the 2D case, the factor of safety for circular failure surfaces can be expressed as

$$F = \frac{\sum_{j=1}^{m}\left[R_j \sum_{i=1}^{n}\left(c'_{ij}S_{ij} + N'_{ij}\tan\phi'_{ij}\right)\right]}{\sum_{j=1}^{m}\left[R_j \sum_{i=1}^{n} W_{ij}\sin\theta_{ij}\right] + C_s \sum_{j=1}^{m}\sum_{i=1}^{n} W_{ij}\lambda_{si} + \sum_{j=1}^{m}\Delta y_j \sum_{i=1}^{n} L_i\lambda_{Li}} \tag{9-24}$$

in which n = number of slices in the x direction, m = number of columns in the y direction, and S = surface area computed by Eq. (9-17). For circular and composite failure surfaces with ellipsoidal ends, the last term in the denominator of Eq. (8-7) does not appear in Eq. (9-24), because the water pond is considered a part of the failure mass, with a cohesion and friction angle equal to 0 and a unit weight equal to 62.4 pcf (9.8 kN/m^3), so there are no water pressures P_1 and P_2 at both ends. The problem in using water pressures is that P_1 and P_2 change from the column to the column and are more difficult to evaluate.

In Eq. (9-24), all variables with a subscript ij indicate that they vary with both x and y coordinates. The radius of rotation, R, has a subscript j, because it varies with the y coordinate only, whereas the line loads, L and λ_L, have a subscript i, because they vary with the x coordinate only. If there is no line load on a column, L should be assigned 0. If there is more than one line load on a column, they should be superimposed to obtain the total effect. Note that the actual number of slices or columns, n or m, in the x or y direction is not a fixed entity but varies with the x or y coordinate.

For noncircular failure surfaces, similar to Eq. (8-8), the factor of safety can be written as

$$F = \frac{\sum\sum \lambda_T (c'S + N' \tan\phi')}{\sum\sum [W\lambda_w + C_s W\lambda_s + \Delta y L\lambda_L - \lambda_N \cos(\theta_d - \theta_{xz})(N' + U)] - \sum \Delta y (P_2\lambda_2 - P_1\lambda_1)} \tag{9-25}$$

Note that the subscript ij is not shown in Eq. (9-25) to save space. For a composite failure surface, Eq. (9-24) can be used for the circular part and Eq. (9-25) for the noncircular part. Eq. (9-25) is applicable to the simplified and original Spencer methods and will be slightly modified for the normal method, as will be discussed in the next section. In Eq. (9-25),

$$U = uS \tag{9-26}$$

in which u = pore pressure normal to the failure surface with a surface area S. One expression for the overturning moment due to weight is $W\lambda_w = W(\lambda_T \sin\theta_{xz} + \lambda_N \cos\theta_{xz})$. However, this expression is not as simple and direct as $W\lambda_w = W(x_m - x_o)$. In 3D analysis, it is assumed that the movement of the failure mass is parallel to the xz plane. Because the effective normal force, N', and water pressure, U, are the only two forces not parallel to the xz plane, they must be projected onto the xz plane by multiplying with $\cos(\theta_d - \theta_{xz})$.

9.4 Normal Method

Fig. 9-8 shows the forces on a slice. The solid lines with arrows are the directions of actual forces, and the dashed lines with arrows are the components normal to the failure surface. The normal method applies the concept of submerged weight

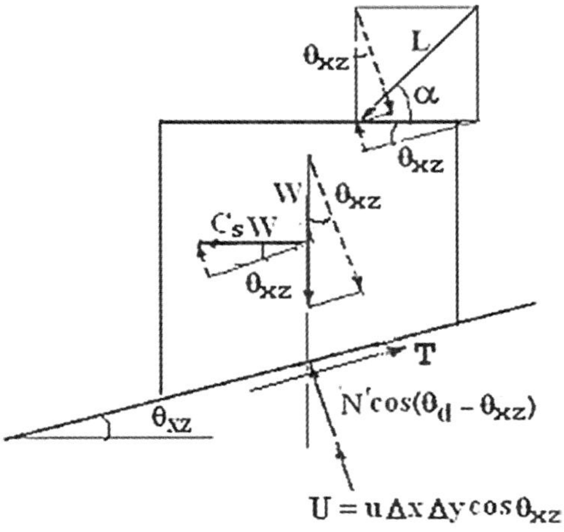

Fig. 9-8. Components of forces normal to failure surface

to determine the effective stress normal to the failure surface. Because the submerged weight is applied in the vertical direction, no projection of the water pressure onto the xz plane is required. Similar to the 2D case in Fig. 8-3, the pore pressure, U, on the failure surface is expressed as $u\Delta x\Delta y\cos\theta_{xz}$, instead of $u\Delta x\Delta y\sec\theta_{xz}$.

Consider the equilibrium of the forces in the normal direction:

$$\begin{aligned} N'\cos(\theta_d - \theta_{xz}) &= W\cos\theta_{xz} - u\Delta x\Delta y\cos\theta_{xz} - C_s W\sin\theta_{xz} \\ &\quad + \Delta y L(\sin\alpha\cos\theta_{xz} - \cos\alpha\sin\theta_{xz}) \\ N' &= \frac{W'\cos\theta_{xz} - C_s W\sin\theta_{xz} + \Delta y L(\sin\alpha\cos\theta_{xz} - \cos\alpha\sin\theta_{xz})}{\cos(\theta_d - \theta_{xz})} \end{aligned} \tag{9-27}$$

in which W' = submerged weight, or

$$W' = W - u\Delta x\Delta y \tag{9-28}$$

Eq. (9-27) is very similar to Eq. (8-9) except that θ is replaced by θ_{xz}, L is replaced by ΔyL, and the expression in the right-hand side is divided by $\cos(\theta_d - \theta_{xz})$. Substituting Eq. (9-27) into Eq. (9-24) or (9-25), the factors of safety for circular or noncircular failure surfaces can be obtained.

In view of the fact that the normal method is used in both the simplified Bishop and the original Spencer methods for determining the initial factor of safety, three examples, each using a different type of failure surface, are given to illustrate how the failure mass is divided into rectangular columns. All calculations were made by hand, and they checked well with the results obtained by LEAME, thus validating the correctness of the LEAME computer program.

Example 9.4 Fig. 9-9(a) shows the central cross section and Fig. 9-9(b) the plan view of a 3D circular failure surface. The failure mass consists of a half cylinder with $\lambda_c = 90$ ft (27.5 m) and a half ellipsoid with $\lambda_e = 200$ ft (61 m). The soil has a cohesion of 200 psf (`9.6 kN/m^2), a friction angle of 30°, and a total unit weight of 125 pcf (19.7 kN/m^3). If the failure mass is divided into five slices in the x direction and three columns in the y direction, determine the factor of safety by the normal method.

Solution The division of the failure mass into five slices and three columns is not sufficient but is used only to demonstrate how the failure mass is divided, especially how to handle the columns of irregular shape adjacent to the boundary.

Fig. 9-9(a) shows the coordinates at the centerline of each slice. Given the x coordinates, the z coordinates on the surface of the slope are equal to $0.5x$ and those on the failure surface can be determined by $z = 165 - \sqrt{(167.7)^2 - (x-30)^2}$.

First, consider the dimensions in the y direction. Given the x and z coordinates of points 1, 2, and 3, as shown in Fig. 9-9(a), the corresponding y coordinate of the ellipsoid can be computed by Eq. (9-5), or

$$y = 90 + \frac{200}{167.7}\sqrt{(167.7)^2 - (x-30)^2 - (z-165)^2} \tag{9-29}$$

Table 9-1 presents the y coordinates of the failure surface computed by Eq. (9-29) and the length of each column. The maximum y is 210 ft. Because the length

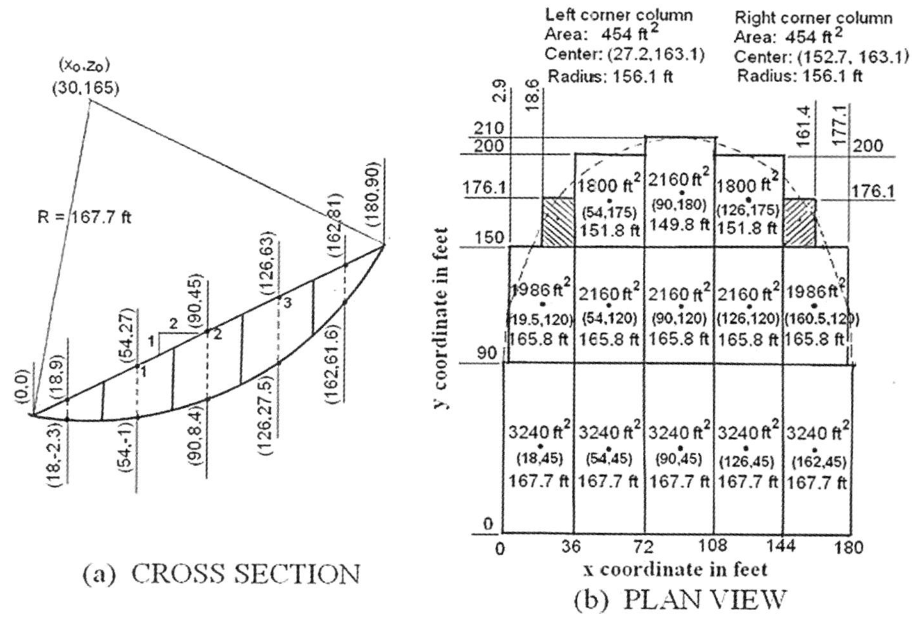

Fig. 9-9. Example 9.4

Table 9-1. Length of Each Column in y Direction

Location	1	2	3
x	54	90	126
z	27	45	63
y	200	210	200
1st column length	90	90	90
2nd column length	60	60	60
3rd column length	50	60	50
y coordinate at center of 3rd column	175	180	175

Note: All dimensions are in feet.

of the first column should be equal to λ_c, or 90 ft, each of the remaining two columns should be $(210 - 90)/2 = 60$ ft. The length of column 3 for slices 2 and 4 is $200 - 150 = 50$ ft, and that for slice 3 is $210 - 150 = 60$ ft.

Next, consider the dimensions in the x direction. For any given y coordinate at the center of a column, the radius of circle, r, can be computed by Eq. (9-2), or

$$r = 167.7\sqrt{1 - \frac{(y-90)^2}{40{,}000}} \tag{9-30}$$

The intersection of the circle with the slope surface can be determined as follows.

The equation of the circle is $(x - 30)^2 + (z - 165)^2 = r^2$. Replacing z by $0.5x$ ($z = 0.5x$ is the equation for slope surface), the equation becomes $1.25x^2 - 225x + 28125 - r^2 = 0$, or

$$x = \frac{225 \pm \sqrt{5r^2 - 90{,}000}}{2.5} \tag{9-31}$$

Two values of x are obtained, depending on the sign before the square root. The small x, designated as x_1, is applicable to the first slice and the larger x, designated as x_2, is applicable to the last slice. For the second column with $y = 120$ ft, it can be found from Eqs. (9-30) and (9-31) that $r = 165.8$ ft, $x_1 = 2.9$ ft, and $x_2 = 177.1$ ft. Except for the two shaded columns at the corner (to be discussed later), the area, the center, and the radius of each column for computing the factor of safety are shown in Fig. 9-9(b). Also shown but in dashed lines are the actual boundaries of the failure mass. The factor of safety is based on the information provided by each rectangular column, instead of the actual boundaries.

To determine the dimensions of the shaded columns, it is necessary to know the x_m and y_m at its center. Because x_m depends on y_m and both are not known a priori, an iteration process must be used. First, assume x_m, and the y coordinate of the boundary can be computed by Eq. (9-29), so y_m can be determined. Knowing

Table 9-2. Iteration Method to Determine Dimensions of Left Corner Column

Iteration	Assumed x_m	Boundary, y	Center, y_m	Radius, r	Boundary, x	Computed x_m
1	27.0	175.7	162.9	156.2	18.5	27.2
2	27.2	176.0	163.0	156.1	18.6	27.3
3	27.3	176.1	163.1	156.1	18.6	27.3
Note: All dimensions are in feet.						

y_m, the x coordinates of the boundary can be computed by Eqs. (9-30) and (9-31), so x_m can be determined. Using the new x_m as the assumed x_m, the process is repeated until x_m converges. Table 9-2 illustrates the iteration process for determining the dimensions of the column at the left upper corner.

Similar to the LEAME program, the width of the corner column is assumed as half of the regular width, so x_m is initially assumed as 27 ft. After three iterations, the computed x_m is 27.3 ft, which is equal to the assumed x_m. The results show that the x coordinate of the left boundary is 18.6 ft and that of the top boundary is 176.1 ft, so the area of the column on the xy plane is $(36 - 18.6) \times (176.1 - 150) = 454.1\ \text{ft}^2$, and its center is located at (27.3, 163.1).

The same procedure can be applied to the column at the right corner except that x is taken as x_2, instead of x_1. Because of symmetry, the area of the right corner column is the same as the left corner column, or $454.1\ \text{ft}^2$, and its center is located at (152.7, 163.1).

Table 9-3 shows the information on each column and how the factor of safety can be calculated. The factor of safety is a ratio between the resisting moment and the overturning moment, or $F = 9{,}353{,}195/6{,}387{,}903 = 1.464$, which checks with the 1.469 obtained by LEAME. A slight discrepancy between the two is expected because the manual method carries the length to one decimal point and the angle to two decimal points, whereas the LEAME computer program carries the numbers to many more decimal points.

Each column of Table 9-3 is explained as follows:

1. The failure mass is divided into five slices in the x direction.
2. The failure mass is divided into three columns in the y direction.
3. The radius corresponding to each column center is computed by Eq. (9-30).
4. The area is obtained by multiplying the slice width with the column length. This is the area projected on a horizontal plane, not the surface area for computing the cohesion resistance.
5. z_m is the z coordinate at the center of the column bottom. These coordinates at the cylindrical part are shown in Fig. 9-9(a) and those at the ellipsoidal part can be computed by Eq. (9-6), or

$$z_m = 165 - 167.7\sqrt{1 - \frac{(x_m - 30)^2}{28{,}123} - \frac{(y_m - 90)^2}{40{,}000}}$$

Table 9-3. *Analysis of 3D Failure Surface with Ellipsoidal Ends by Normal Method*[a]

Sl. (1)	Co. (2)	Radius ft (3)	Area ft² (4)	z_m ft (5)	Weight 10^3 lb (6)	$\sin\theta_{xz}$ (7)	θ_{xz} deg (8)	θ_{yz} deg (9)	$\cos(\theta_d - \theta_{xz})$ (10)	S ft² (11)	N' 10^3 lb (12)	M_r 10^3 ft-lb (13)	M_o 10^3 ft-lb (14)
1	1	167.7	3,240	−2.3	4,576	−0.072	−4.13	0.00	1.000	3,248	4,564	550,831	−55,252
1	2	165.8	1,986	−0.5	2,545	−0.064	−3.67	7.25	0.997	2,006	2,547	310,329	−26,584
1	3	156.1	454	8.9	266	−0.017	−0.97	18.19	0.955	478	278	39,978	−706
2	1	167.7	3,240	−1.0	11,340	0.143	8.22	0.00	1.000	3,274	11,223	1,196,439	271,945
2	2	165.8	2,160	0.9	7,047	0.145	8.33	7.25	0.999	2,200	6,980	741,110	169,417
2	3	151.8	1,800	15.1	2,678	0.158	9.09	21.72	0.970	1,959	2,726	298,387	64,230
3	1	167.7	3,240	8.4	14,823	0.358	20.98	0.00	1.000	3,470	13,840	1,456,395	889,923
3	2	165.8	2,160	10.4	9,342	0.362	21.22	7.69	1.000	2,335	8,709	911,095	560,703
3	3	149.8	2,160	27.8	4,644	0.401	23.64	24.77	0.988	2,560	3,944	417,803	278,964
4	1	167.7	3,240	27.5	14,378	0.572	34.89	0.00	1.000	3,950	11,794	1,274,397	1,379,201
4	2	165.8	2,160	29.8	8,964	0.579	35.38	8.89	1.000	2,671	7,309	788,222	860,144
4	3	151.8	1,800	47.7	3,510	0.632	39.20	26.87	0.997	2,495	2,728	314,835	336,741
5	1	167.7	3,240	61.6	7,857	0.787	51.91	0.00	1.000	5,252	4,847	645,447	1,036,966
5	2	165.8	1,986	62.7	4,357	0.787	51.91	11.76	1.000	3,246	2,626	359,010	567,608
5	3	156.1	454	68.5	445	0.786	51.81	27.99	0.999	773	275	48,917	54,599
								Sum				9,353,195	6,387,903

Note: Sl. = Slice; Co. = Column; $\cos(\theta_d - \theta_{xz})$ = Projection from normal direction to xz plane; S = Surface area at the bottom of each column; N' = Normal force at the bottom of each column; M_r = Resisting moment; M_o = Overturning Moment.

[a]Factor of safety by normal method = 9,353,195/6,387,903 = 1.464, which is about the same as 1.469 by LEAME.

The x_m and y_m coordinates at each center are shown in Fig. 9-9(b). The dashed lines in Fig. 9-9(b) indicate the actual boundaries of the failure mass, which has nothing to do with the computation in Table 9-3.

6. The weight of each column = 125 × Area × $(0.5x_m - z_m)$.
7. From Eq. (9-7), $\tan\theta_{xz} = (x_m - 30)/(165 - z_m)$, so $\sin\theta_{xz}$ can be determined.
8. $\theta_{xz} = \tan^{-1}(\tan\theta_{xz})$.
9. For the first column, $\theta_{yz} = 0$; for the other columns, from Eq. (9-8),

$$\tan\theta_{yz} = 0.703(y_m - 90)/(165 - z_m), \text{ so } \theta_{yz} = \tan^{-1}(\tan\theta_{yz})$$

10. From Eq. (9-11), $\theta_d = \cos^{-1}[(1 + \tan^2\theta_{xz} + \tan^2\theta_{yz})^{-0.5}]$, so

$$\cos(\theta_d - \theta_{xz}) = \cos\{\cos^{-1}[(1 + \tan^2\theta_{xz} + \tan^2\theta_{yz})^{-0.5}] - \theta_{xz}]\}$$

11. From Eq. (9-17), surface area

$$S = \text{Area} \times \sqrt{1 - \sin^2\theta_{xz} \times \sin^2\theta_{yz}} \,/ (\cos\theta_{xz} \times \cos\theta_{yz})$$

12. From Eq. (9-27), normal force $N' = \text{Weight} \times \cos\theta_{xz}/\cos(\theta_d - \theta_{xz})$
13. The resisting moment is the numerator of Eq. (9-24),

$$M_r = \text{Radius} \times (0.2 \times S + N' \times 0.57735)$$

14. The overturning moment is the denominator of Eq. (9-24), or

$$M_o = \text{Radius} \times \text{Weight} \times \sin\theta_{xz}$$

The above method for treating the irregular columns at the boundary is used in LEAME and can be summarized as follows:

1. The failure mass is divided into a number of slices. The y coordinates through the centerline of each slice are computed by Eq. (9-5) and compared. The one with the largest y is used to divide the failure mass into a number of columns, and all the other y's are used to define the boundary of the last column in each slice and the y coordinate of its center.
2. Based on the y coordinate at the center of column, the radius of the circle can be determined from Eq. (9-2) and the x coordinates x_1 and x_2, where the circle intersects with the slope surface and forms the left and right column boundaries, can be computed. Knowing the column boundaries, the x coordinate of the column center can be relocated as shown in Fig. 9-9(b).
3. The stated procedures for relocating the column center are applicable to those columns where one of the dimensions, either x or y, is fixed, and the other is to be determined. For the two corner columns where both dimensions are not known, a trial-and-error method must be used to determine the column boundaries. First, the x coordinate, x_m, at the center of the

column is assumed, and the y coordinate can be computed by Eq. (9-5), thus the y coordinate, y_m, at the center of the column can be determined. From y_m, the radius of the circle, r, can be obtained by Eq. (9-2), so the x coordinate of the column boundary and a new x_m can be computed. The process is repeated until x_m converges.

Example 9.5 Fig. 9-10(a) shows the central cross section, and 9-10(b) shows the plan view of an embankment with a 3D composite failure surface. The embankment is placed on a soft soil layer, designated as soil 1, with a thickness of 5 ft, a cohesion of 100 psf (4.8 kN/m^2), and a friction angle of 20°. The soil in the embankment, designated as soil 2, has a cohesion of 200 psf (9.6 kN/m^2) and a friction angle of 30°. To simplify the calculation, both soils are assumed to have the

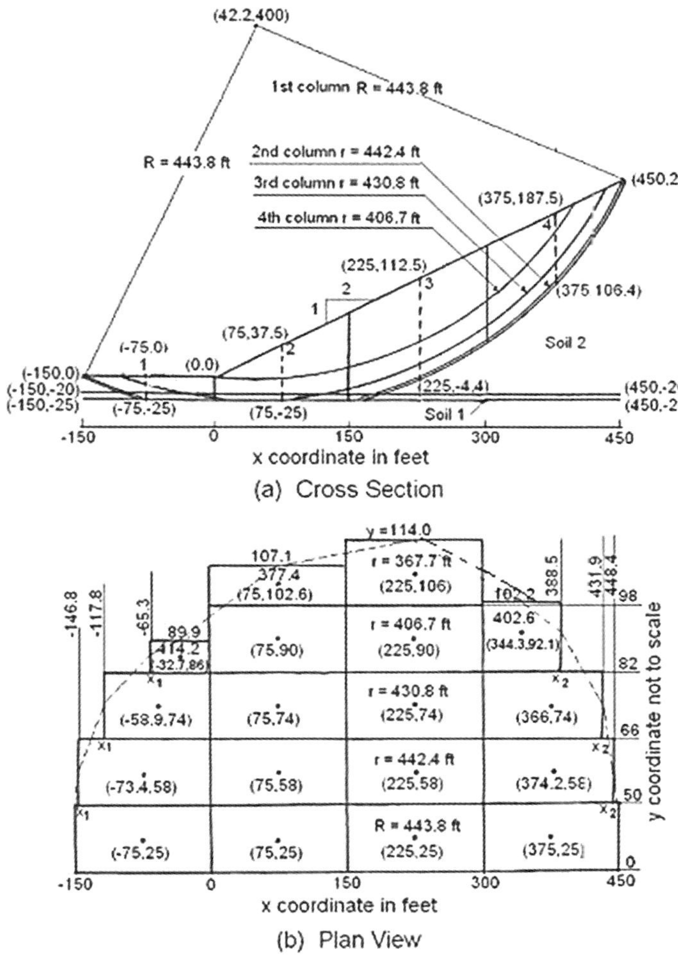

Fig. 9-10. Example 9.5

same total unit weight of 125 pcf (19.7 kN/m^3). Because a center at (42.2, 400) and a radius of 443.8 ft (135.4 m) are given, there is no need to generate other composite failure surfaces by placing an imaginary rock line below the weak soil layer. The failure mass consists of a half cylinder with $\lambda_c = 50$ ft (15.3 m) and a half ellipsoid with $\lambda_e = 100$ ft (30.5 m). If the failure mass is divided into four slices in the x direction and five columns in the y direction, determine the factor of safety by the normal method.

Solution Fig. 9-10(a) shows the dimensions of the slope and the coordinates at the centerline of each slice. Given the x and z coordinates of points 1, 2, 3, and 4 on the slope surface, as shown in Fig. 9-10(a), the corresponding y coordinate of the ellipsoid can be computed by Eq. (9-5).

For point 1 with $z = 0$,

$$y = 50 + (100/443.8)\sqrt{(443.8)^2 - (x - 42.2)^2 - (0 - 400)^2},$$
$$\text{or} \quad y = 50 + 0.225\sqrt{35{,}178 + 84.4x - x^2} \tag{9-32}$$

For the other three points with $z = 0.5x$,

$$y = 50 + (100/443.8)\sqrt{196{,}958 - (x - 42.2)^2 - (0.5x - 400)^2},$$
$$\text{or} \quad y = 50 + 0.225\sqrt{35{,}177 + 484.4x - 1.25x^2} \tag{9-33}$$

Table 9-4 presents the y coordinates of the failure surface and the division of the failure mass into columns. The maximum y is 114 ft. Because the length of the first column should be equal to λ_c, or 50 ft, each of the remaining four columns should be (114 − 50)/4 = 16 ft. Next, consider the dimensions in the x direction. For any given y_m coordinate at the center of a column, the radius of circle, r, can be computed by Eq. (9-2), or

$$r = 443.8\sqrt{1 - \frac{(y - 50)^2}{10{,}000}} \tag{9-34}$$

Table 9-4. *Division of Failure Mass into Columns in y Direction*

Location	1	2	3	4
x	−75	75	225	375
z	0	37.5	112.5	187.5
y	84.3	107.1	114.0	95.6
Number of columns	4	5	5	4
Length of Column 5	—	9.1	16.0	—

Note: All dimensions are in feet; length of column 2 to 4 is 16 ft.

The intersection of the circle with the slope surface can be determined as follows:

The equation of circle is $(x - 42.2)^2 + (z - 400)^2 = r^2$. At the left boundary $x = x_1$ and $z = 0$, so

$$x_1 = 42.2 - \sqrt{r^2 - 160{,}000} \tag{9-35}$$

At the right boundary, $x = x_2$ and $z = 0.5x_2$, so $(x_2 - 42.2)^2 + (0.5x_2 - 400)^2 = r^2$, or $1.25x_2^2 - 484.8x_2 + 161{,}780 - r^2 = 0$. The solution of this quadratic equation is

$$x_2 = \frac{484.8 + \sqrt{5r^2 - 573{,}869}}{2.5} \tag{9-36}$$

The intersection of the circle with the bottom of the weak layer can be determined as follows:

The equation of circle is $(x - 42.2)^2 + (z - 400)^2 = r^2$. With $z = -25$ ft (7.6 m),

$$x = \frac{84.4 \pm \sqrt{4r^2 - 722{,}500}}{2} \tag{9-37}$$

The use of the negative sign before the square root gives the x_a coordinate of the left intersecting point, and the positive sign gives the x_b coordinate of the right point.

Table 9-5 shows the radius, r, and the boundaries, x_1 and x_2, of the first three columns. The procedures for determining the dimensions of the two corner columns are described in the previous example and are presented in Table 9-6.

The dimensions of each column and its radius and the coordinates at each center are shown in Fig 9-10(b). For columns with the same y coordinate at the center, the radius is the same as that shown in slice 3 and is not repeated in the figure. The number of columns in slice 4 is based on the center at column 1 with the x coordinate of 375 ft. When the center of column 4 is moved to

Table 9-5. *Radius and Boundaries of First Three Columns*

Column	1	2	3
y_m	25.0	58.0	74.0
r	443.8	442.4	430.8
x_1	–150.0	–146.8	–117.8
x_2	450.0	448.4	431.9
x_a	–85.6	–80.7	–28.3
x_b	170.0	165.1	112.7
Note: All dimensions are in feet.			

Table 9-6. Iteration Method to Determine Dimensions of Corner Columns

Iteration	Assumed x_m	Boundary, y	Center, y_m	Radius, r	Boundary, x	Computed x_m
	Left Corner Column (Slice 1)				x_1	
1	−37.5	89.4	85.7	414.6	−66.9	−33.5
2	−33.5	89.8	85.9	414.2	−65.3	−32.7
3	−32.7	89.8	85.9	414.2	−65.3	−32.7
	Right Corner Column (Slice 4)				x_2	
1	337.5	103.4	92.7	401.3	386.3	343.2
2	343.2	102.4	92.2	402.3	388.0	344.0
3	344.0	102.2	92.1	402.6	388.5	344.2
4	344.2	102.2	92.1	402.6	388.5	344.2

Note: All dimensions are in feet.

an x coordinate of 344.3 ft, there are actually five columns in slice 4, so columns 4 and 5 are combined to form a single column.

Table 9-7 shows the information on each column and how the factor of safety by the normal method can be calculated. Each column is explained as follows.

1. The failure mass is divided into four slices in the x direction.
2. The failure mass is divided into five columns in the y direction.
3. The area is obtained by multiplying the slice width with the column length. This is the area projected on a horizontal plane, not the surface area for computing the cohesion resistance.
4. z_m is the z coordinate at the center of the column on the failure surface. These coordinates at the cylindrical part are shown in Fig. 9-10(a), and those at the ellipsoidal part can be computed by Eq. (9-6), or

$$z_m = 400 - 443.8\sqrt{1 - \frac{(x_m - 42.2)^2}{196{,}958} - \frac{(y_m - 50)^2}{10{,}000}}$$

   The x_m and y_m coordinates at each center are shown in Fig. 9-10(b). If the computed z_m is smaller than −25 ft, which is the elevation at the bottom of soil 1, the failure surface is along boundary line 1 with $z_m = -25$ ft.
5. For the columns in the first slice, Weight = 125 × Area × $(0 - z_m)$, and for those in the other slices, Weight = 125 × Area × $(0.5x_m - z_m)$.
6. If $z_m = -25$ ft, $\theta_{xz} = 0$. Otherwise, from Eq. (9-7), $\theta_{xz} = \tan^{-1}[(x_m - 42.2)/(400 - z_m)]$, which applies to both the cylindrical and the ellipsoidal surfaces.
7. If $z_m = -25$ ft, $\theta_{yz} = 0$. Otherwise, from Eq. (9-8), $\theta_{yz} = \tan^{-1}[19.79(y_m - 50)/(400 - z_m)]$.
8. From Eq. (9-11), $\theta_d = \cos^{-1}[(1 + \tan^2\theta_{xz} + \tan^2\theta_{yz})^{-0.5}]$, so $\cos(\theta_d - \theta_{xz}) = \cos\{\cos^{-1}[(1 + \tan^2\theta_{xz} + \tan^2\theta_{yz})^{-0.5}] - \theta_{xz}]\}$.

Table 9-7. Analysis of 3D Composite Failure Surface by Normal Method[a]

Sl. (1)	Co. (2)	Area ft² (3)	z_m ft (4)	Weight 10^3 lb (5)	θ_{xz} deg (6)	θ_{yz} deg (7)	cos $(\theta_d - \theta_{xz})$ (8)	S ft² (9)	N' 10^3 lb (10)	λ_N ft (11)	λ_T ft (12)	M_r 10^3 ft-lb (13)	M_o 10^3 ft-lb (14)
1	1	7,500	−25.0	23,438	0.00	0.00	1.000	7,500	23,438	−117.2	425.0	3,944,312	0
1	2	2,349	−25.0	7,340	0.00	0.00	1.000	2,349	7,340	−115.6	425.0	1,235,238	0
1	3	1,885	−18.8	4,429	−13.53	48.47	0.813	2,879	5,297	0	430.8	1,565,538	−446,388
1	4	516	−7.2	464	−10.43	60.59	0.639	1,055	714	0	414.2	258,141	−34,793
2	1	7,500	−25.0	58,594	0.00	0.00	1.000	7,500	58,594	32.8	425.0	9,382,501	0
2	2	2,400	−25.0	18,750	0.00	0.00	1.000	2,400	18,750	32.8	425.0	3,002,388	0
2	3	2,400	−25.0	18,750	0.00	0.00	1.000	2,400	18,750	32.8	425.0	3,002,388	0
2	4	2,400	−5.4	12,870	4.65	62.74	0.528	5,244	24,295	0	406.7	6,131,216	424,332
2	5	1,365	24.0	2,303	4.99	70.06	0.421	4,004	5,450	0	377.4	1,489,733	75,600
3	1	7,500	−4.4	109,594	24.33	0.00	1.000	8,231	99,861	0	443.8	26,317,773	20,038,368
3	2	2,400	−2.8	34,590	24.39	21.41	0.993	2,798	31,725	0	442.4	8,350,759	6,319,146
3	3	2,400	9.9	30,780	25.09	50.43	0.889	3,932	31,359	0	430.8	8,138,470	5,622,799
3	4	2,400	36.6	22,770	26.68	65.23	0.776	5,345	26,219	0	406.7	6,591,203	4,158,057
3	5	2,400	81.0	9,450	29.87	73.86	0.717	8,742	11,429	0	367.7	3,069,168	1,730,550
4	1	7,500	106.4	76,031	48.59	0.00	1.000	11,339	50,289	0	443.8	13,891,902	25,306,771
4	2	2,374	107.9	23,503	48.68	28.36	0.999	3,817	15,534	0	442.4	4,305,419	7,809,044
4	3	2,110	115.8	17,724	48.76	59.00	0.966	4,751	12,059	0	430.8	3,408,691	5,741,551
4	4	1,788	133.9	8,549	48.59	72.18	0.909	6,184	6,221	0	402.6	1,943,953	2,581,356
							Sum					106,028,793	79,326,393

Note: Sl. = Slice; Co. = Column; cos $(\theta_d - \theta_{xz})$ = Projection from normal direction to xz plane; S = Surface area at the bottom of each column; N' = Normal force at the bottom of each column; M_r = Resisting moment; M_o = Overturning Moment.

[a]The factor of safety by normal method = 106,028,793/79,326,393 = 1.337, which checks with the 1.338 obtained by LEAME.

9. From Eq. (9-17), surface area $S = \text{Area} \times \sqrt{1 - \sin^2\theta_{xz} \times \sin^2\theta_{yz}} / (\cos\theta_{xz} \times \cos\theta_{yz})$
10. From Eq. (9-27), normal force $N' = \text{Weight} \times \cos\theta_{xz}/\cos(\theta_d - \theta_{xz})$.
11. When the failure surface is along boundary line 1, or $z_m = -25$ ft, $\lambda_N = x_m - 42.2$ ft. Otherwise, $\lambda_N = 0$.
12. When $z_m = -25$ ft, $\lambda_T = 425$ ft. Otherwise, λ_T can be computed by Eq. (9-41).
13. The resisting moment is the numerator of Eq. (9-25), or $M_r = \lambda_T (cS + N'\tan\phi)$. Note that $c = 0.1$ ksf and $\phi = 20°$ when $z_m = -25$ ft, and that $c = 0.2$ ksf and $\phi = 30°$ when otherwise.
14. When $z_m = -25$ ft, or the failure surface is horizontal, the overturning moment, M_o, is 0, because, according to the denominator of Eq. (9-25), $M_o = \lambda_w W - \lambda_N \cos(\theta_d - \theta_{xz})N'$, where $\cos(\theta_d - \theta_{xz}) = 1$ and $N' = W$. If z_m is greater than -25 ft, or the failure surface is circular with $R = \lambda_T$, from the denominator of Eq. (9-24), $M_o = \lambda_T \times \text{Weight} \times \sin\theta_{xz}$.

Example 9.6 Fig. 9-11(a) shows the cross section and 9-11(b) shows the plan view of a 3D noncircular failure surface. Because of symmetry, only one-half of the failure mass is shown on the plan view. The fill has a width, w_t, of 200 ft (61 m) at the top. The end plane has an angle, α, of 20° and a slope, g, of 0.75. The soil has a cohesion of 200 psf (9.6 kN/m²), a friction angle of 30°, and a total unit weight of 125 pcf (19.7 kN/m³). If the failure mass is divided into three slices in the x direction and three columns in the y direction, determine the factor of safety by the normal method.

Solution To plot the boundary of the failure mass, it is only necessary to compute the y_s coordinate at the toe, because the intersection of the slope surface and the end plane is a straight line. With $x_s = 0$, $z_s = 0$, $x_t = 300$ ft, $z_t = 150$ ft, $w_t = 200$ ft, $\alpha = 20°$, and $g = 0.75$, from Eq. (9-18), $y_s = 200 + 300 \times \tan 20° - 150/(0.75 \times \cos 20°) = 96.4$ ft.

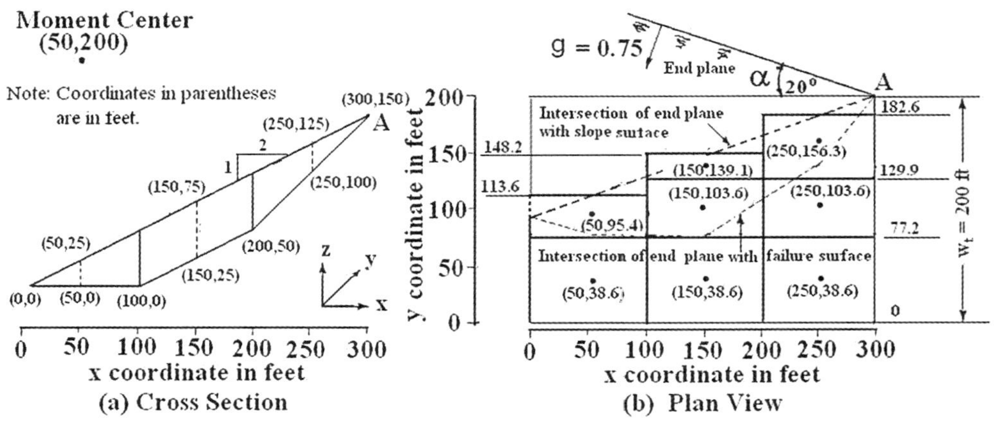

Fig. 9-11. Example 9.6

To determine the area, weight, and center of the columns requires the values of y_s through the centerline of each slice. These values for the three slices are 113.6 ft, 148.2 ft, and 182.6 ft and, because of a straight-line boundary, can be computed simply by $y_s = 96.4 + 0.345x_m$. The (x_m, y_m) coordinates at the center of each column are noted in Fig. 9-11(b). The maximum y_s is 182.6 ft, which is the half length of the failure mass.

To determine the length of the first column, the intersection of the failure surface and the end plane at the center of each slice must be found. From Eq. (9-20),

When $x_m = 50$, $z_m = 0$, $y_f = 200 + (300 - 50)\tan 20° - (150 - 0)/(0.75 \times \cos 20°) = 78.2$ ft.

When $x_m = 150$, $z_m = 25$, $y_f = 200 + (300 - 150)\tan 20° - (150 - 25)/(0.75 \times \cos 20°) = 77.2$ ft.

When $x_m = 250$, $z_m = 100$, $y_f = 200 + (300 - 250)\tan 20° - (150 - 100)/(0.75 \times \cos 20°) = 147.3$ ft.

The minimum y_f is 77.2 ft, which is the length of the first column. The length of the other columns is $(182.6 - 77.2)/2 = 52.7$ ft. The division into three slices and three columns is shown in Fig. 9-11(b). The intersection of the failure surface with the end plane is indicated in dashed lines. Any column with a center inside the dashed lines, hereafter referred to as case 1, has the cross section shown in Fig. 9-11(a), whereas that outside the dashed lines, referred to as case 2, has a base resting on the end plane.

The determination of moment arms λ_T and λ_N is similar to the 2D case. The point of intersection (x, y) of two perpendicular lines, one tangent to the failure surface and the other passing through the moment center can be obtained from Eqs. (8-15) and (8-14), or

$$x = \frac{200 - z_m + x_m \tan\theta_{xz} + \dfrac{50}{\tan\theta_{xz}}}{\tan\theta_{xz} + \dfrac{1}{\tan\theta_{xz}}} \tag{9-38}$$

$$z = 200 + \frac{50 - x}{\tan\theta_{xz}} \tag{9-39}$$

Knowing the point of intersection (x, z), λ_N and λ_T can be computed by Eqs. (8-16) and (8-17), or

$$\lambda_N = \pm\sqrt{(x - x_m)^2 + (z - z_m)^2} \tag{9-40}$$

λ_N is positive if $x < x_m$ and negative if $x > x_m$.

$$\lambda_T = \sqrt{(x - x_o)^2 + (z - z_o)^2} \tag{9-41}$$

In case 1, when the cross section in Fig. 9-11(a) applies, z_m at column 1 can be obtained by averaging the z coordinates at the two end points, and $\tan\theta_{xz}$ is the slope of the failure surface and can be computed from the coordinates of the two end points. In case 2, when the failure surface is along the end plane, from Eq. (9-21), $z_m = 150 - 0.75\cos 20° [200 + (300 - x_m)\tan 20° - y_m] = 0.257 x_m + 0.705 y_m - 67.9$; from Eq. (9-22), $\theta_{xz} = \tan^{-1}(0.75 \times \sin 20°) = 14.39°$.

The computations of moment arms, λ_N and λ_T, for each column are shown in Table 9-8.

Table 9-9 shows the information on each column and how the factor of safety can be calculated. The factor of safety by the normal method is F = 17,413,418/12,083,234 = 1.441, which checks with the 1.442 obtained by LEAME. Each column in Table 9-9 is explained as follows.

1. The failure mass is divided into three slices in the x direction, each 100 ft wide.
2. The failure mass is divided into three columns in the y direction. However, only two columns exist in slice 1, because the third column is cut off by the end plane.
3. The area is the product of the slice width and the column length. All of the slices have the same width of 100 ft. The column lengths can be found from the dimensions shown in Fig. 9-11(b).
4. The vertical coordinate, z_m, at the center of the column bottom can be obtained from Table 9-8.
5. The weight = $(0.5\,x_m - z_m) \times \text{Area} \times 125$.
6. The values of $\tan\theta_{xz}$ can be found in Table 9-8, so $\theta_{xz} = \tan^{-1}(\tan\theta_{xz})$.
7. In case 1, $\theta_{yz} = 0$. In case 2, from Eq. (9-23), $\theta_{yz} = \tan^{-1}(0.75 \times \cos 20°) = 35.18°$.
8. From Eq. (9-11), $\theta_d = \cos^{-1}[(1 + \tan^2\theta_{xz} + \tan^2\theta_{yz})^{-0.5}]$, so $\cos(\theta_d - \theta_{xz}) = \cos\{\cos^{-1}[(1 + \tan^2\theta_{xz} + \tan^2\theta_{yz})^{-0.5}] - \theta_{xz}]\}$.
9. From Eq. (9-17),

$$S = \text{Area} \times \sqrt{1 + \tan^2\theta_{xz} + \tan^2\theta_{yz}} \,/ (\cos\theta_{xz} \times \cos\theta_{yz})$$

Table 9-8. Computations of Moment Arms for Noncircular Failure Surface

Sl.	Co.	x_m (ft)	y_m (ft)	z_m (ft)	$\tan\theta_{xz}$	x (ft)	z (ft)	λ_N (ft)	λ_T (ft)
1	1	50	38.6	0	0	50.0	0.0	0.0	200.0
1	2	50	95.4	12.2	0.257	95.3	23.7	–46.7	181.9
2	1	150	38.6	25.0	0.500	140.0	20.0	11.2	201.2
2	2	150	103.6	43.7	0.257	93.9	29.2	57.9	176.4
2	3	150	139.1	68.7	0.257	87.9	52.5	64.2	152.3
3	1	250	38.6	100.0	1.000	200.0	50.0	70.7	212.1
3	2	250	103.6	100.0	1.000	200.0	50.0	70.7	212.1
3	3	250	156.3	106.5	0.257	84.9	64.2	170.4	140.2

Table 9-9. Analysis of 3D Failure Surface with Planar Ends by Normal Method[a]

Sl. (1)	Co. (2)	Area (ft^2) (3)	z_m (ft) (4)	Weight (10^3 lb) (5)	θ_{xz} (deg) (6)	θ_{yz} (deg) (7)	cos $(\theta_d - \theta_{xz})$ (8)	S (ft^2) (9)	N' (10^3 lb) (10)	λ_N (ft) (11)	λ_T (ft) (12)	M_r (10^3 ft-lb) (13)	M_o (10^3 ft-lb) (14)
1	1	7,720	0.0	24,125	0.00	0.00	1.000	7,720	24,125	0.0	200.0	3,094,515	0
1	2	3,640	12.2	5,824	14.39	35.18	0.924	4,551	6,105	–46.7	181.9	806,713	265,377
2	1	7,720	25.0	48,250	26.57	0.00	1.000	8,632	43,181	11.2	201.2	5,363,381	4,341,373
2	2	5,270	43.7	20,619	14.39	35.18	0.924	6,588	21,615	57.9	176.4	2,433,795	905,506
2	3	1,830	68.7	1,441	14.39	35.18	0.924	2,288	1,511	64.2	152.3	202,555	54,466
3	1	7,720	100.0	24,125	45.00	0.00	1.000	10,917	17,059	70.7	212.1	2,552,076	3,618,929
3	2	5,270	100.0	16,469	45.00	0.00	1.000	7,453	11,645	70.7	212.1	1,742,156	2,470,499
3	3	5,270	106.5	12,187	14.39	35.18	0.924	6,588	12,768	170.4	140.2	1,218,227	427,084
									Sum			17,413,418	12,083,234

Note: Sl. = Slice; Co. = Column; cos $(\theta_d - \theta_{xz})$ = Projection from normal direction to xz plane; S – Surface area at the bottom of each column; N' = Normal force at the bottom of each column; M_r = Resisting moment; M_o = Overturning Moment.

[a]Factor of safety by Normal method = 17,413,418/12,083,234 = 1.441, which checks with the 1.442 obtained by LEAME.

10. From Eq. (9-27), normal force $N' = \text{Weight} \times \cos\theta_{xz}/\cos(\theta_d - \theta_{xz})$.
11. Values of λ_N can be obtained from Table 9-8.
12. Values of λ_T can be obtained from Table 9-8.
13. The resisting moment is the numerator of Eq. (9-25), or $M_r = \lambda_T(0.2S + N'\tan 30°)$.
14. The overturning moment is the denominator of Eq. (9-25), or

$$M_o = \text{Weight} \times (x_m - 50) - \lambda_N \times N \times \cos(\theta_d - \theta_{xz})$$

9.5 Simplified Bishop Method

Similar to Eq. (8-18) and summing the forces in the vertical direction to zero,

$$\begin{aligned} W - \left(\frac{c'S + N'\tan\varphi'}{F}\right)\sin\theta_{xz} + \Delta y L \sin\alpha - N'\cos(\theta_d - \theta_{xz})\cos\theta_{xz} \\ - uS\cos(\theta_d - \theta_{xz})\cos\theta_{xz} = 0 \end{aligned} \quad (9\text{-}42)$$

Note that the first N' in parentheses is not multiplied by $\cos(\theta_d - \theta_{xz})$, because the terms in parentheses represent the shear force, which is parallel to the xz plane. Only the second N' term, which represents the normal force, has to be multiplied by $\cos(\theta_d - \theta_{xz})$. Eq. (9-42) can be simplified to

$$N' = \frac{F[W + \Delta y L \sin\alpha - uS\cos(\theta_d - \theta_{xz})\cos\theta_{xz}] - c'S\sin\theta_{xz}}{F\cos(\theta_d - \theta_{xz})\cos\theta_{xz} + \sin\theta_{xz}\tan\phi'} \quad (9\text{-}43)$$

By substituting Eq. (9-43) into Eq. (9-24) or (9-25), the factor of safety for circular or noncircular failure surfaces can be determined. Since N' depends on F and F depends on N', an iteration method must be applied.

Example 9.7 Same as Example 9.4, determine the factor of safety by the simplified Bishop method using the column information presented in Table 9-3.

Solution Table 9-10 is a spreadsheet for computing the factor of safety by the simplified Bishop method. Some explanations on the spreadsheet are in order:

1. Data on radius, r, surface area, S, weight, W, θ_{xz}, and $\cos(\theta_d - \theta_{xz})$ can be obtained from Table 9-3. The normal force, N, can be obtained from Eq. (9-43) and expressed as $(F \times W - c \times S \times \sin\theta_{xz})/[F \times \cos(\theta_d - \theta_{xz}) \times \cos\theta_{xz} + \sin\theta_{xz} \times \tan\phi]$. Because the unit of force is in kip, or 1,000 pounds, the cohesion, c, should be 0.2 ksf. To start the iterations, a safety factor, F, of 1.464 obtained by the normal method is assumed.
2. The resisting moment, M_r, can be obtained from the numerator of Eq. (9-24) and expressed as R × ($c \times S + N' \times \tan\phi$). The sum of the driving moment, M_o, is shown at the bottom of the table and can be obtained from

Table 9-10. *Spreadsheet for Ellipsoidal Ends Using Simplified Bishop Method*[a]

ASSUMED FACTOR OF SAFETY								1.464		1.579
SI.	Co.	R ft	S ft^2	W kip	θ_{xz} radian	cos (θ_d-θ_{xz})	N kip	M_r ft-kip	N kip	M_r ft-kip
1	1	167.7	3,248	4,576	−0.072	1.000	4,755	569,338	4,743	568,127
1	2	165.8	2,006	2,545	−0.063	0.997	2,641	319,325	2,635	318,731
1	3	156.1	478	266	−0.018	0.955	282	40,330	282	40,308
2	1	167.7	3,274	11,340	0.143	1.000	10,780	1,153,587	10,827	1,158,119
2	2	165.8	2,200	7,047	0.144	0.999	6,701	714,369	6,730	717,197
2	3	151.8	1,959	2,678	0.158	0.970	2,584	285,967	2,599	287,239
3	1	167.7	3,470	14,823	0.358	1.000	13,637	1,436,744	13,777	1,450,325
3	2	165.8	2,335	9,342	0.362	1.000	8,586	899,351	8,676	907,943
3	3	148.8	2,560	4,644	0.401	0.988	4,238	440,281	4,293	444,971
4	1	167.7	3,950	14,378	0.572	1.000	13,360	1,426,066	13,581	1,447,411
4	2	165.8	2,671	8,964	0.579	1.000	8,325	885,456	8,465	898,875
4	3	151.8	2,495	3,510	0.632	0.997	3,189	355,269	3,257	361,176
5	1	167.7	5,252	7,857	0.787	1.000	7,459	898,311	7,654	917,249
5	2	165.8	3,246	4,357	0.787	1.000	4,103	500,429	4,214	510,978
5	3	156.1	773	445	0.786	0.999	376	58,013	389	59,235
From Table 9.3, M_o				6,387,903		Sum of M_r		9,982,835		10,087,884
Computed factor of safety								1.563		1.579

[a]Factor of safety by simplified method is $F = 10,087,884/6,387,903 = 1.579$, which checks with the 1.595 obtained by LEAME.

Table 9-3. The sum of the resisting moment divided by the sum of the driving moment gives the computed factor of safety.

3. Using the computed F as the assumed F, the process is repeated until the assumed F is equal to the computed F. It was found that the factor of safety converges to 1.579 at the fourth iteration. The factor of safety obtained by LEAME is 1.595. When a spreadsheet is used, it is not necessary to show all the intermediate results. By trial and error, an assumed factor of safety exactly equal to the computed factor of safety, or sometimes with a difference of only 0.001, can be obtained.

Similar to 2D analysis, Newton's method of tangent, as described in Section 8.6.1, can be used to solve F and speed up convergence. By substituting Eq. (9-43) into (9-24) and simplifying, the factor of safety for a circular failure surface can be obtained by

$$F = \frac{R\sum\sum \dfrac{c'S\cos(\theta_d - \theta_{xz})\cos\theta_{xz} + [W + \Delta y \sin\alpha - uS\cos(\theta_d - \theta_{xz})\cos\theta_{xz}]\tan\phi'}{\cos(\theta_d - \theta_{xz})\cos\theta_{xz} + \sin\theta_{xz}\tan\phi'/F}}{\sum\sum[RW\sin\theta_{xz} + \lambda_s C_s W + \lambda_L \Delta y L]} \tag{9-44}$$

By naming the overturning moment in the denominator of Eq. (9-44) as M_o and $W - uS\cos(\theta_d - \theta_{xz})\cos\theta_{xz}$ as W_o, the following equation similar to Eq. (8-57) can be obtained:

$$F_{m+1} = F_m \left\{ 1 - \frac{M_o - R\sum\sum \dfrac{c'S\cos(\theta_d - \theta_{xz})\cos\theta_{xz} + [W_o + \Delta y L\sin\alpha]\tan\phi'}{F_m\cos(\theta_d - \theta_{xz})\cos\theta_{xz} + \sin\theta_{xz}\tan\phi'}}{M_o - R\sum\sum \dfrac{[c'S\cos(\theta_d - \theta_{xz})\cos\theta_{xz} + (W_o + \Delta y L\sin\alpha)\tan\phi']\sin\theta_{xz}\tan\phi'}{[F_m\cos(\theta_d - \theta_{xz})\cos\theta_{xz} + \sin\theta_{xz}\tan\phi']^2}} \right\} \tag{9-45}$$

Example 9.8 Same as Example 9.7, but determine the factor of safety by simplified Bishop method using Newton's method of tangent or Eq. (9-45).

Solution To use the spreadsheet, Eq. (9-45) can be expressed as

$$F_{m+1} = F_m \left(1 - \frac{M_o - \sum M_1}{M_o - \sum M_2} \right) \tag{9-46}$$

in which M_o = driving moment, $M_1 = r \times [c' \times S \times \cos(\theta_d - \theta_{xz}) \times \cos\theta_{xz} + W \times \tan\phi']/[F_m \times \cos(\theta_d - \theta_{xz}) \times \cos\theta_{xz} + \sin\theta_{xz} \times \tan\phi]$, and $M_2 = r \times [c' \times S \times \cos(\theta_d - \theta_{xz}) \times \cos\theta_{xz} + W \times \tan\phi'] \times \sin\theta_{xz} \times \tan\phi/[F_m \times \cos(\theta_d - \theta_{xz}) \times \cos\theta_{xz} + \sin\theta_{xz} \times \tan\phi]^2$. Table 9-11 is the spreadsheet for computing the factor of safety.

Table 9-11. Spreadsheet for Example 9.7 Using Newton's Method of Tangent

ASSUMED FACTOR OF SAFETY								1.464		1.579
SI.	Co.	R ft	S ft^2	W kip	θ_{xz} radian	cos (θ_d-θ_{xz})	M_1 ft-kip	M_2 ft-kip	M_1 ft-kip	M_2 ft-kip
1	1	167.7	3,248	4,576	−0.072	1.000	388,892	−11,385	359,802	−9,746
1	2	165.8	2,006	2,545	−0.063	0.997	218,118	−5,582	201,856	−4,781
1	3	156.1	478	266	−0.018	0.955	27,548	−206	25,527	−177
2	1	167.7	3,274	11,340	0.143	1.000	787,969	42,338	733,451	36,682
2	2	165.8	2,200	7,047	0.144	0.999	487,957	26,419	454,210	22,891
2	3	151.8	1,959	2,678	0.158	0.970	195,332	11,883	181,912	10,306
3	1	167.7	3,470	14,823	0.358	1.000	981,382	126,177	918,508	110,528
3	2	165.8	2,335	9,342	0.362	1.000	614,311	79,821	575,012	69,935
3	3	148.8	2,560	4,644	0.401	0.988	300,738	43,528	281,806	38,220
4	1	167.7	3,950	14,378	0.572	1.000	974,089	197,235	916,663	174,665
4	2	165.8	2,671	8,964	0.579	1.000	604,820	123,969	569,268	109,824
4	3	151.8	2,495	3,510	0.632	0.997	242,670	54,498	228,737	48,419
5	1	167.7	5,252	7,857	0.787	1.000	613,601	173,942	580,905	155,899
5	2	165.8	3,246	4,357	0.787	1.000	341,823	96,899	323,609	86,848
5	3	156.1	773	445	0.786	0.999	39,626	11,225	37,514	10,060
From Table 9.3, Mo =				6,387,903			6,818,877	970,762	6,388,780	859,575
Computed factor of safety								1.580		1.579

In Table 9-11, the first five items from r to $\cos(\theta_d - \theta_{xz})$ can be obtained from Table 9-10, and the overturning moment, M_o, from Table 9-3. First, an initial safety factor of 1.464 by the normal method is assumed as F_m, and a new safety factor, F_{m+1}, of 1.580 is computed by Eq. (9-45). Using the new 1.580 as the assumed F_m, the factor of safety rapidly converges to 1.579 at the second iteration. It can be seen that Newton's method is very efficient and reduces the number of iterations from four to two.

9.6 Original Spencer Method

The original Spencer method satisfies the overall force equilibrium in two perpendicular directions and is particularly useful for noncircular failure surfaces, because any reasonable moment center can be selected and nearly the same factor of safety can be obtained. This is not true when the normal or simplified

Bishop method is used for noncircular failure surfaces, because the location of the moment center has a significant effect on the factor of safety obtained.

Summing all forces in a direction perpendicular to the side force, the following equation similar to Eq. (8-20) is obtained:

$$N'\cos(\theta_d-\theta_{xz})\cos(\theta_{xz}-\delta)+uS\cos(\theta_d-\theta_{xz})\cos(\theta_{xz}-\delta)$$
$$+\frac{c'S+N'\tan\phi'}{F}\sin(\theta_{xz}-\delta)+C_sW\sin\delta-W\cos\delta-\Delta yL\sin(\alpha-\delta)=0$$

After simplification,

$$N'=\frac{F[W\cos\delta-uS\cos(\theta_d-\theta_{xz})\cos(\theta_{xz}-\delta)-C_sW\sin\delta+\Delta yL\sin(\alpha-\delta)]-c'S\sin(\theta_{xz}-\delta)}{F\cos(\theta_d-\theta_{xz})\cos(\theta_{xz}-\delta)+\sin(\theta_{xz}-\delta)\tan\phi'} \tag{9-47}$$

To determine the factor of safety with respect to the moment equilibrium under a given value of δ, the factor of safety is assumed first, and N' is computed by Eq. (9-43). Substituting the N' thus determined into Eq. (9-24) or (9-25), a new factor of safety is obtained. Using the new factor of safety as the assumed factor of safety, the process is repeated until the factor of safety converges.

Similar to Eq. (8-22), the factor of safety with respect to force equilibrium can be determined by

$$F=\frac{\sum\sum(c'S+N'\tan\phi')\cos(\theta_{xz}-\delta)}{\sum\sum[W\sin\delta+(N'+U)\cos(\theta_d-\theta_{xz})\sin(\theta_{xz}-\delta)+C_sW\cos\delta+\Delta yL\cos(\alpha-\delta)]} \tag{9-48}$$

in which U can be computed by Eq. (9-26). Eq. (9-48) in conjunction with Eq. (9-47) can be used to determine the factor of safety with respect to force equilibrium. After the factors of safety for both moment and force equilibrium at three different values of δ are determined by Newton's method of tangent, as described in Section 8.6.1, the factor of safety that satisfies both moment and force equilibrium can be found, as shown in Fig. 8-8 and described in Section 8.4.3.

Example 9.9 Same as Example 9.5, but determine the factor of safety of a composite failure surface by the original Spencer method using the column information presented in Table 9-7.

Solution Table 9-12 is a spreadsheet for computing the factor of safety of a composite failure surface by the original Spencer method. The factor of safety by the original Spencer method is 1.485, which checks with the 1.486 obtained by LEAME.

Most of the comments presented in Table 8-4 for 2D analysis are also applicable to 3D analysis and will not be repeated here. Some additional comments are:

Table 9-12. Spreadsheet for Composite Failure Surface by Original Spencer Method

DATA INPUT										
Slice	Col.	S ft^2	Weight kip	θ_{xz} rad	cos $(\theta_d\text{-}\theta_{xz})$	λ_N ft	λ_T ft	λ_W ft	θ_{xz} deg	
1	1	7,500	23,438	0.000	1.000	−117.2	425.0	−117.2	0.00	
1	2	2,349	7,340	0.000	1.000	−115.6	425.0	−115.6	0.00	
1	3	2,879	4,429	−0.236	0.813	0.0	430.8	−101.1	−13.53	
1	4	1,055	464	−0.182	0.639	0.0	414.2	−74.9	−10.43	
2	1	7,500	58,594	0.000	1.000	32.8	425.0	32.8	0.00	
2	2	2,400	18,750	0.000	1.000	32.8	425.0	32.8	0.00	
2	3	2,400	18,750	0.000	1.000	32.8	425.0	32.8	0.00	
2	4	5,244	12,870	0.081	0.528	0.0	406.7	32.8	4.65	
2	5	4,004	2,303	0.087	0.421	0.0	377.4	32.8	4.99	
3	1	8,231	109,594	0.425	1.000	0.0	443.8	182.8	24.33	
3	2	2,798	34,590	0.426	0.993	0.0	442.4	182.8	24.39	
3	3	3,932	30,780	0.438	0.889	0.0	430.8	182.8	25.09	
3	4	5,345	22,770	0.466	0.776	0.0	406.7	182.8	26.68	
3	5	8,742	9,450	0.521	0.717	0.0	367.7	182.8	29.87	
4	1	11,339	76,031	0.848	1.000	0.0	443.8	332.8	48.59	
4	2	3,817	23,503	0.850	0.999	0.0	442.4	332.0	48.68	
4	3	4,751	17,724	0.851	0.966	0.0	430.8	323.8	48.76	
4	4	6,184	8,549	0.848	0.909	0.0	402.6	302.1	48.59	

Table 9-12. (*Continued*)

Factor of safety based on moment equilibrium										
Assumed angle δ				0			0.3			0.6
Assumed Factor of safety				1.491			1.483			1.526
Slice	Col.	N	Driving M.	Resisting M.	N	Driving M.	Resisting M.	N	Driving M.	Resisting M.
		kip	ft-kip	ft-kip	kip	ft-kip	ft-kip	kip	ft-Kip	ft-kip
1	1	23,438	0	3,944,310	25,533	245,522	4,268,363	28,410	582,725	4,713,423
1	2	7,340	0	1,235,237	7,996	75,840	1,336,721	8,897	180,000	1,476,100
1	3	6,458	−447,772	1,854,216	8,858	−447,772	2,451,258	14,894	−447,772	3,952,575
1	4	877	−34,754	297,101	1,320	−34,754	403,121	2,573	−34,754	702,627
2	1	58,594	0	9,382,495	63,577	−163,449	10,153,334	70,421	−387,933	11,212,017
2	2	18,750	0	3,002,386	20,345	−52,304	3,249,054	22,535	−124,138	3,587,831
2	3	18,750	0	3,002,386	20,345	−52,304	3,249,054	22,535	−124,138	3,587,831
2	4	22,977	422,136	5,821,636	28,891	422,136	7,210,364	40,468	422,136	9,928,795
2	5	4,980	75,538	1,387,283	7,029	75,538	1,833,834	11,913	75,538	2,897,963
3	1	101,931	20,033,783	26,848,038	100,478	20,033,783	26,475,774	98,666	20,033,783	26,011,534
	2	32,355	6,323,052	8,512,298	31,914	6,323,052	8,399,536	31,361	6,323,052	8,258,420
3	3	31,524	5,626,584	8,179,417	31,402	5,626,584	8,149,283	31,223	5,626,584	8,104,668
3	4	25,884	4,162,356	6,512,630	26,079	4,162,356	6,558,241	26,292	4,162,356	6,608,455
3	5	10,884	1,727,460	2,953,431	11,171	1,727,460	3,014,441	11,517	1,727,460	3,087,774
4	1	78,678	25,303,117	21,165,932	68,004	25,303,117	18,431,029	58,728	25,303,117	16,054,331
4	2	24,324	7,802,996	6,550,668	21,023	7,802,996	5,707,474	18,158	7,802,996	4,975,523
4	3	18,583	5,739,031	5,031,411	16,163	5,739,031	4,429,434	14,056	5,739,031	3,905,407
4	4	8,890	2,582,653	2,564,324	7,901	2,582,653	2,334,418	7,039	2,582,653	2,134,171
Sum			79,316,181	118,245,198		79,369,486	117,654,734		79,442,696	121,199,444
Computed Factor of safety				1.491			1.482			1.526

Table 9-12. *(Continued)*

Factor of safety based on force equilibrium										
Assumed angle δ				0			0.3			0.6
Assumed Factor of safety				1.349			1.465			1.642
Slice	**Col.**	N	**Driving F.**	**Resisting F.**	N	**Driving F.**	**Resisting F.**	**N in kip**	**Driving F.**	**Resisting F.**
		kip	**kip**	**kip**	**kip**	**kip**	**kip**	**kip**	**kip**	**kip**
1	1	23,438	0	9,281	25,561	–627	9,604	27,996	–2,574	9,029
1	2	7,340	0	2,906	8,005	–196	3,008	8,767	–806	2,828
1	3	6,561	–1,248	4,242	8,906	–2,390	4,916	13,785	–5,815	5,721
1	4	894	–103	715	1,330	–257	867	2,303	–775	1,093
2	1	58,594	0	22,076	63,644	–1,492	22,846	69,436	–6,122	21,477
2	2	18,750	0	7,064	20,366	–478	7,311	22,220	–1,959	6,873
2	3	18,750	0	7,064	20,366	–478	7,311	22,220	–1,959	6,873
2	4	22,831	977	14,183	28,965	483	17,348	38,496	–2,812	20,211
2	5	4,930	181	3,634	7,055	53	4,764	11,015	–975	6,239
3	1	100,313	41,328	54,272	100,419	44,871	59,160	98,153	44,757	57,421
3	2	31,836	13,055	17,250	31,895	14,192	18,824	31,198	14,158	18,290
3	3	30,948	11,666	16,894	31,379	12,931	18,724	31,052	12,925	18,469
3	4	25,311	8,819	14,013	26,052	10,063	15,890	26,153	10,139	16,023
3	5	10,546	3,766	6,796	11,152	4,548	7,987	11,472	4,689	8,346
4	1	76,107	57,080	30,564	67,835	57,813	35,364	59,118	57,445	35,285
4	2	23,524	17,650	9,472	20,970	17,889	10,975	18,279	17,782	10,966
4	3	17,936	13,029	7,453	16,120	13,390	8,739	14,158	13,405	8,838
4	4	8,524	5,811	4,073	7,875	6,256	4,937	7,102	6,412	5,174
Sum			172,011	231,953		176,571	258,575		157,914	259,156
Computed Factor of safety				1.348			1.464			1.641

Table 9-12. *(Continued)*

Factor of safety and angle δ that satisfy both moment and force equilibrium									
D1 =	0	D2 =	0.3	D3 =	0.6				
FM1 =	1.491	FM2 =	1.482	FM3 =	1.526				
FF1 =	1.348	FF2 =	1.464	FF3 =	1.641				
AM =	0.053	BM =	−0.021	CM =	0.268				
AF =	0.061	BF =	0.051	CF =	0.243				
Angle D or δ =	0.342		Factor of safety F =		1.485				

1. All the necessary input data, except λ_w, can be obtained from Table 9-7. The moment arm, λ_w, is the horizontal distance between the moment and the column centers, or $x_o - x_m$, and can be determined from Fig. 9-10.
2. For moment equilibrium, the normal force, N', can be computed by Eq. (9-47) and expressed as $(F \times W \times \cos\delta - c' \times S \times \sin(\theta_{xz} - \delta))/(F \times \cos(\theta_d - \theta_{xz}) \times \cos(\theta_{xz} - \delta) + \sin(\theta_{xz} - \delta) \times \tan\phi')$. As explained in Example 9.5, the resisting moment can be computed by the numerator of Eq. (9-24)). Note that $c = 0.1$ ksf and $\phi = 20°$ when the failure surface is horizontal and that $c = 0.2$ ksf and $\phi = 30°$ when the failure surface is circular. The driving moment is 0 when the failure surface is horizontal, with $\theta_{xz} = 0$, and can be computed by the denominator of Eq. (9-24) when the failure surface is circular.
3. For force equilibrium, the normal force is the same as that for moment equilibrium; the resisting force is the numerator of Eq. (9-48) expressed as $(c' \times S + N' \times \tan\phi') \times \cos(\theta_{xz} - \delta)$. The driving force is the denominator of Eq. (9-48) expressed as $W \times \sin\delta + N' \times \cos(\theta_d - \theta_{xz}) \times \sin(\theta_{xz} - \delta)$.
4. The last stage of the spreadsheet for determining the factor of safety was explained in Example 8.4. The factor of safety is 1.485, which checks with the 1.486 obtained by LEAME.

Example 9.10 Same as Example 9.6, but determine the factor of safety by the original Spencer method using the column information presented in Table 9-9.

Solution Table 9-13 is the spreadsheet for determining the factors of safety by the original Spencer method. The procedures are similar to Example 9.9. All the necessary input data, except x_m, can be obtained from Table 9-9. The moment arm, λ_w, is the horizontal distance between the moment and the column centers, or $x_o - x_m$, and can be determined from Fig. 9-11. For moment equilibrium, the normal force, N', can be computed by Eq. (9-47) and expressed as $[F \times W \times \cos\delta - c' \times S \times \sin(\theta_{xz} - \delta)]/[F \times \cos(\theta_d - \delta) + \sin(\theta_{xz} - \delta) \times \tan\phi']$. The driving moment is the denominator of Eq. (9-25) expressed as $W \times \lambda_w - N' \times \lambda_N \times \cos(\theta_d - \theta_{xz})$, and the resisting moment is the numerator of Eq. (9-25) expressed as $\lambda_T \times (c \times S + N' \times \tan\phi)$. For force equilibrium, the normal force is the same as that for moment equilibrium, the driving force is the denominator of Eq. (9-48) expressed as $W \times \sin\delta + N' \times \cos(\theta_d - \theta_{xz}) \times \sin(\theta_{xz} - \delta)$, and the resisting force is the numerator of Eq. (9-48) expressed as $(c' \times S + N' \times \tan\phi') \times \cos(\theta_{xz} - \delta)$. The factor of safety obtained from the spreadsheet is 1.623 compared with 1.625 obtained by LEAME.

Summary

1. In 2D analysis, the slope is infinitely long and is represented by a single cross section, or the most critical cross section. In 3D analysis, the slope

Table 9-13. *Spreadsheet for Noncircular Failure by Original Spencer Method*

INPUT DATA										
Slice	Col.	S ft^2	W kip	θ_{xz} deg.	cos $(\theta_d$-$\theta_{xz})$	λ_N ft	λ_T ft	x_m ft		
1	1	7,720	24,125	0.00	1.000	0	200.0	50.0		
1	2	4,551	5,824	14.39	0.924	–46.7	181.9	50.0		
2	1	8,632	48,250	26.57	1.000	11.2	201.2	150.0		
2	2	6,588	20,619	14.39	0.924	57.9	176.4	150.0		
2	3	2,288	1,441	14.39	0.924	64.2	152.3	150.0		
3	1	10,917	24,125	45.00	1.000	70.7	212.1	250.0		
3	2	7,453	16,469	45.00	1.000	70.7	212.1	250.0		
3	3	6,588	12,187	14.39	0.924	170.4	140.2	250.0		

Table 9-13. (Continued)

Factor of safety based on moment equilibrium										
Assumed angle δ				0.0			0.3			0.6
Assumed factor of safety				1.687			1.630			1.629
Slice	**Col.**	**N in kip**	**D.M.in ft-kip**	**R.M.in ft-kip**	**N in kip**	**D.M.in ft-kip**	**R.M.in ft-kip**	**N in kip**	**D.M.in ft-kip**	**R.M.in ft-kip**
1	1	24,125	0	3,092,825	27,421	0	3,473,138	32,697	0	4,081,993
1	2	5,806	250,534	774,944	6,174	266,415	813,569	6,688	288,605	867,543
2	1	45,630	4,313,943	5,644,658	43,974	4,332,492	5,452,391	42,402	4,350,102	5,269,855
2	2	20,842	946,882	2,353,743	21,795	895,867	2,450,798	23,144	823,710	2,588,077
2	3	1,402	60,955	192,863	1,535	53,025	204,610	1,720	42,062	220,851
3	1	24,457	3,095,916	3,456,143	21,362	3,314,728	3,077,380	18,761	3,498,575	2,759,142
3	2	16,695	2,113,439	2,359,361	14,583	2,262,810	2,100,799	12,807	2,388,313	1,883,555
3	3	12,238	510,595	1,174,693	12,900	406,296	1,228,280	13,831	259,763	1,303,567
Sum			11,292,264	19,049,229		11,531,632	18,800,966		11,651,128	18,974,583
Computed Factor of safety				1.687			1.630			1.629

Table 9-13. (Continued)

Factor of safety based on force equilibrium										
Assumed angle δ				0.0			0.3			0.6
Assumed factor of safety				1.487			1.581			1.705
Slice	Col.	N in kip	D.F. in kip	R.F. in kip	N in kip	D.F. in kip	R.F. in kip	N in kip	D.F. in kip	R.F. in kip
1	1	24,125	0	15,464	27,536	−1,008	16,654	32,199	−4,559	16,608
1	2	5,721	1,313	4,079	6,179	1,442	4,470	6,629	1,195	4,450
2	1	44,693	19,987	24,610	43,894	21,410	26,692	42,298	21,494	25,890
2	2	20,575	4,724	12,776	21,809	5,108	13,885	22,962	4,389	13,689
2	3	1,376	316	1,212	1,537	356	1,343	1,702	276	1,353
3	1	23,521	16,629	11,142	21,240	17,036	12,772	18,824	17,089	12,822
3	2	16,057	11,352	7,606	14,499	11,630	8,719	12,850	11,666	8,753
3	3	12,070	2,771	8,022	12,909	3,019	8,756	13,715	2,549	8,675
Sum			57,093	84,912		58,993	93,289		54,099	92,240
Computed Factor of safety				1.487			1.581			1.705

Factor of safety and angle δ that satisfy both moment and force equilibrium					
D1 =	0	D2 =	0.3	D3 =	0.6
FM1 =	1.687	FM2 =	1.630	FM3 =	1.629
FF1 =	1.487	FF2 =	1.581	FF3 =	1.705
AM =	0.056	BM =	−0.051	CM =	0.304
AF =	0.030	BF =	0.047	CF =	0.268
Angle D or δ = 0.410		Factor of safety F = 1.623			

with the given cross section is cut short at each end, either by a half ellipsoid or by an end plane oriented at a certain direction. Within the ellipsoid or end plane, the cross section changes from one location to the other, so the slices must be subdivided into a number of columns, each representing a different cross section.

2. For failure surfaces with ellipsoidal ends, in addition to the coordinates (x_o, z_o) of the moment center and the radius, R, of the cylindrical failure surface, as required in 2D analysis, the half length of the cylinder, λ_c, and the half length of the ellipsoid, λ_e, also must be specified or obtained by trial and error. The y coordinates along the centers of the slices, where the ellipsoid intersects the slope surface, can be computed by Eq. (9-5), and the largest value of y is the half length of the failure mass. The length of the first column is λ_c, and the length of all other columns is equal to (half length of the failure mass $- \lambda_c)/(m-1)$, where m is the number of columns specified. Based on the y coordinates computed here for each slice, the length of the last column in each slice can be determined. Also, based on the intersection of the slope surface with the circle at the center of each column, the width of the column adjacent to the boundary of the failure mass can be computed. Knowing the width and length of each column, the x and y coordinates at its center can be located. Because the width and length at the two corner columns are interdependent, a trial-and-error process must be used to locate their centers.

3. For failure surfaces with planar ends, the location of the end plane is defined by the half top width, w_t, the angle of the end plane, α, and the slope of the end plane, g. To determine the area, weight, and center of the columns requires the values of y_s through the center of each slice. These y_s coordinates can be computed by Eq. (9-18), and the largest y_s is the half length of the failure mass. The coordinates, y_f, for the intersection of the failure surface and the end plane through the center of each slice can be computed by Eq. (9-19), and the smallest y_f is the length of the first column. The length of all other columns is (maximum y_s $-$ minimum y_f)/ $(m-1)$. However, the length of the last column in each slice must be reduced based on the y_s through the center of each slice.

4. In 3D analysis, it is assumed that the movement of the failure mass is in a direction parallel to the vertical, or xz, plane. In considering the force or moment equilibrium, all forces, if not on the xz plane, must be projected onto the xz plane. Except for the effective normal force, N', and the neutral force, U, which are normal to the failure surface, all other forces, such as weight, seismic force, and external or internal loads, are parallel to the xz plane. Therefore, all equations used for 2D analysis can be easily extended to 3D analysis by multiplying N' and U with $\cos(\theta_d - \theta_{xz})$. However, the N' and U used to compute the shear resistance should not be multiplied by $\cos(\theta_d - \theta_{xz})$, because the failure surface moves in the x direction on the xz plane. Other changes in notation include the

replacements of θ by θ_{xz} and the single Σ by double $\Sigma\ \Sigma$ to cover both slices and columns.

5. Three examples, each with a different type of failure surface (circular, composite, or noncircular), are presented to demonstrate the use of the normal method, especially the treatment of the partial columns adjacent to the boundary. The normal method assumes the absence of forces between two columns and uses the submerged weight to determine the combined effect of N' and U. Although the normal method is not recommended for 3D analysis, the information provided by these examples can be used in other examples involving the simplified Bishop and original Spencer methods.

6. The simplified Bishop method is recommended for circular failure surfaces with a cylinder at the center and a half ellipsoid at each end. The effective normal force, N', is obtained by assuming the side forces between the columns as horizontal and considering the force equilibrium of each column in the vertical direction. The method requires iterations, and Eq. (9-45), based on Newton's method of tangent, can be used to reduce the number of iterations required.

7. The original Spencer method is recommended for noncircular or composite failure surfaces. The side forces between two slices in the x direction are assumed to make an angle, δ, with the horizontal and the forces between two columns in the y direction ignored. The effective normal force, N', is obtained by considering the force equilibrium in a direction perpendicular to the side forces. Two factors of safety are computed, one based on the overall moment equilibrium and the other on the force equilibrium in the δ direction. Because N' depends on the factor of safety and the factor of safety depends on N', an iteration procedure is required to solve the factor of safety. Similar to 2D analysis, three different values of δ must be assumed, and the factors of safety are plotted against δ. Thus, two sets of curves are obtained, one based on force equilibrium and the other based on moment equilibrium. The intersection of these two curves gives an overall factor of safety which satisfies both force and moment equilibrium.

8. The use of spreadsheets to calculate the factors of safety based on the simplified Bishop and the original Spencer methods are demonstrated. The use of the trial-and-error method by a spreadsheet to find a factor of safety is so easy that it is really not necessary to change the factor of safety gradually until it converges, as is done by LEAME. A few trials and errors will be sufficient to find a computed factor of safety that is equal or nearly equal to the assumed factor of safety. It should be noted that the spreadsheet only solves one of the hundreds of trial failure surfaces and cannot be used as a substitute for a computer program. However, the close agreement in solutions between LEAME and the spreadsheets is a further confirmation on the correctness of the LEAME computer program.

Problems

9.1 Same as Example 9.4 except that the failure mass is divided into three slices and three columns. Determine the factor of safety by the normal method.
[Answer: 1.449]

9.2 Same as Problem 9.1, but determine the factor of safety by the simplified Bishop method.
[Answer: 1.581]

9.3 Same as Problem 9.1, but determine the factor of safety by the simplified Bishop method using Newton's method of tangent, or Eq. (9-45).
[Answer: 1.580]

9.4 A 3D failure surface consists of a cylinder with a half length, λ_c, of 100 ft, and a half ellipsoid with a half length, λ_e, of 50 ft. The cross section for the cylindrical part and the division of the failure mass into slices are the same as Problem 8.1 and are reproduced in Fig. P9-4. As in Problem 8.1, the soil is assumed to have a cohesion of 500 psf, a friction angle of 10°, and a total unit weight of 125 pcf. If the cylindrical part is considered as one column and the ellipsoidal part is divided evenly into two columns, plot a plan view of the failure mass showing the x and y coordinates at the center of each column and the horizontal area of each column.

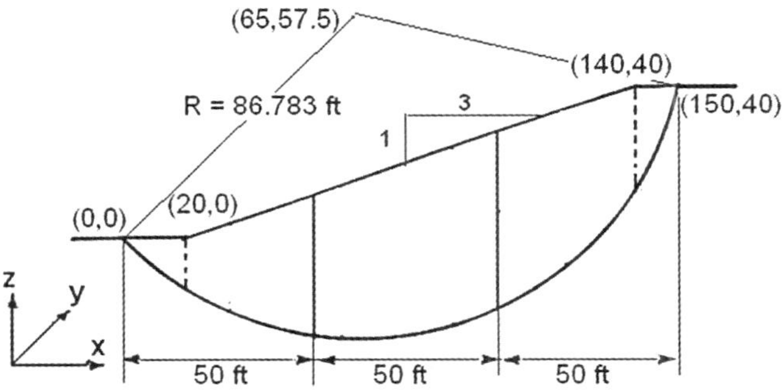

Fig. P9-4.

9.5 Same as Problem 9.4, but determine the factor of safety by the normal method.
[Answer: 1.507]

9.6 Same as Problem 9.4, but determine the factor of safety by the simplified Bishop method.
[Answer: 1.683]

9.7 Same as Problem 9.4, but determine the factor of safety by the original Spencer method.
[Answer: 1.683]

9.8 An embankment is placed between two parallel banks ($\alpha = 0°$) with a slope of 2:1 ($g = 0.5$). The distance between the two banks, measured at the top of the embankment, is 600 ft ($W_t = 300$ ft). The cross section of the noncircular failure surface and the division of the failure mass into slices are the same as Problem 8.5 and are reproduced in Fig. P9-8. As in Problem 8.5, the soil is assumed to have a cohesion of 400 psf, a friction angle of 15°, and a total unit weight of 125 pcf. If the central part of the embankment above the noncircular failure surface is considered as one column and that above the surface of the bank is divided evenly into two columns, plot a plan view of the failure mass showing the x and y coordinates at the center of each column and the horizontal area of each column.

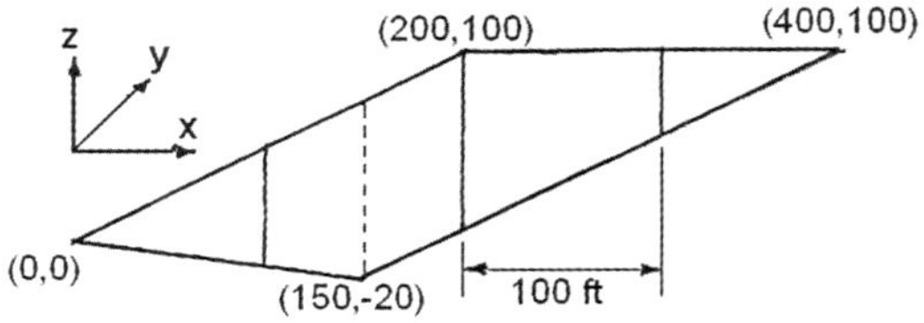

Fig. P9-8.

9.9 Assuming an arbitrary moment center at (100, 200), determine the safety factor of Problem 9.8 by the normal method.
[Answer: 1.382]

9.10 Same as Problem 9.9, but determine the safety factor by the simplified Bishop method.
[Answer: 1.558]

9.11 Same as Problem 9.9, but determine the safety factor by the original Spencer method.
[Answer: 1.571]

Chapter 10

Reliability

Two methods can be used for stability analysis: deterministic and probabilistic. In the deterministic method, each input parameter has a fixed value. If the parameter varies in time or location, the designer will select a more conservative value and use it for stability analysis. To ensure that the design is safe, a factor of safety much greater than 1, usually as high as 1.5, is required. A more realistic approach is the probabilistic method, in which each input parameter has a mean and a coefficient of variation, obtained from field or laboratory tests or based on field measurements or past experience. If these mean values are used and the factor of safety is equal to 1, the probability of failure is 50%, which is unacceptable. The design must be revised to increase the factor of safety so that the probability of failure is smaller than a specified value, say less than 1%, or a reliability of more than 99%. This chapter presents some statistical concepts and equations, which are incorporated in LEAME to determine the reliability.

10.1 Some Statistical Terms

Two of the most useful properties of a random variable are its expectation, or mean, and its variance. From these two properties, the standard deviation and the coefficient of variation can be computed. If two variables are not independent, the correlation coefficient also must be given so the covariance can be determined.

In LEAME, each input parameter must be given a mean and a coefficient of variation. If two parameters are correlated, their correlation coefficient also must be specified. Although typical coefficients of variation for shear strength, seepage, and other sources are suggested, how they are determined is outside the scope of this book. Readers interested in this subject may refer to the book by Baecher and Christian (2003) and the papers by Christian et al. (1994), Christian and Baecher (2011), Oka and Wu (1990), and Wu et al. (1989).

10.1.1 Expectation

The expectation of a random variable x is defined as

$$E[x] = \sum_{all\ x_i} x_i f(x_i) \tag{10-1}$$

in which $f(x_i)$ is the probability function of x_i. One requirement for the probability function is that the sum of $f(x_i)$ over all x_i must be equal to 1. If n independent observations of x are taken, each with the same probability $1/n$, the mean of the observations, $\bar{x}$, can be obtained from Eq. (10-1), or

$$\bar{x} = \frac{\sum_{i=1}^{n} x_i}{n} \tag{10-2}$$

This value of $\bar{x}$ is called the sample mean and is the best estimate of the true or population mean. From the definition in Eq. (10-1), it can be proved easily that, if c is a constant, then $E[c] = c$ and $E[cx] = cE[x]$.

10.1.2 Variance

The variance of a random variable x is defined as the expected value of the square of the deviation from its expectation, or

$$V[x] = E[(x - E[x])^2] \tag{10-3}$$

$$\text{or} \quad V[x] = \sum_{\text{all}\ x_i} (x_i - E[x])^2 f(x_i) \tag{10-4}$$

If n independent observations of x are taken, the variance of x is determined from

$$V[x] = \sum_{i=1}^{n} \frac{(x_i - \bar{x})^2}{n-1} \tag{10-5}$$

The sum is divided by $n - 1$ rather than n, because $\bar{x}$ is the sample mean rather than the true mean, so the degree of freedom is $n - 1$. If a, b, and c are constants, it can be proved easily that $V[c] = 0$, $V[cx] = c^2 V[x]$, and $V[a + bx] = b^2 V[x]$.

10.1.3 Standard Deviation

The standard deviation, s, of a random variable x is defined as the square root of its variance, or

$$s = \sqrt{V[x]} \tag{10-6}$$

10.1.4 Coefficient of Variation

The coefficient of variation generally is used in percentile form, but, for convenience, it can be expressed as a decimal by

$$C[x] = \frac{s}{\bar{x}} \tag{10-7}$$

Example 10.1 The undrained shear strengths, x, obtained from 10 cone penetration tests are: 2,250, 2,140, 2,030, 1,920, 1,840, 1,630, 1,480, 1,270, 1,050, and 810 psf. Compute mean, variance, standard deviation, and coefficient of variation.

Solution From Eq. (10-2), $\bar{x} = (2{,}250 + 2{,}140 + 2{,}030 + 1{,}920 + 1{,}840 + 1{,}630 + 1{,}480 + 1{,}270 + 1{,}050 + 810)/10 = 1{,}642$ psf. From Eq. (10-5), $V[x] = [(2{,}250 - 1{,}642)^2 + (2{,}140 - 1{,}642)^2 + (2{,}030 - 1{,}642)^2 + (1{,}920 - 1{,}642)^2 + (1{,}840 - 1{,}642)^2 + (1{,}630 - 1{,}642)^2 + (1{,}480 - 1{,}642)^2 + (1{,}270 - 1{,}642)^2 + (1{,}050 - 1{,}642)^2 + (810 - 1{,}642)^2]/(10 - 1) = 232{,}462$. From Eq. (10-6), $s = \sqrt{232{,}462} = 482$]. From Eq. (10-7), $C[x] = 482/1{,}642 = 0.294$, or 29.4%.

Table 10-1 shows the mean, standard deviation, and coefficient of variation of shear strength parameters from different sources as reported by Harr (1977). It can be seen that the variation of the friction angle in sand and gravel is much smaller than the variation of unconfined compressive strength in clay.

10.1.5 Covariance

The covariance of two random variables x and y is defined as the expected value of the product of the deviations of x and y from their expected values, or

$$\text{Cov}[x, y] = E[(x - E[x])(y - E[y])] \tag{10-8}$$

$$\text{or} \quad \text{Cov}[x, y] = \sum_{all\ x_i, y_i} (x_i - E[x])(y_i - E[y]) f(x_i, y_i) \tag{10-9}$$

Table 10-1. *Variability of Shear Strength Parameters*

Material	PARAMETER: Frictional Angle, Degrees	PARAMETER: Tangent of Frictional Angle	PARAMETER: Unconfined Compression Strength, ton/ft^2	Number of Samples	Mean	Standard Deviation	Coefficient of Variation, %	Source
Gravel	x			38	36.22	2.16	6.0	Private communication from Prof. R. D. Holtz of Purdue University
Sand	x			73	38.80	2.80	7.0	
Sand	x			136	36.40	4.05	11.0	
Sand	x			30	40.52	4.56	11.0	
Gravelly sand	x			81	37.33	1.97	5.3	Schultze (1972)
Sand		x		81	0.762	0.056	7.3	Schultze (1972)
Sand		x		50	0.717	0.093	13.0	Schultze (1975)
Sand: loose	x						14.0	Singh (1972)
dense	x						12.0	Singh (1972)
Silty sand		x		82	0.692	0.096	13.8	Lumb (1966)
Clay: depth, ft								
5			x	279	2.08	1.02	49.1	Fredlund and Dahlman (1972)
10			x	295	1.68	0.69	40.9	Fredlund and Dahlman (1972)
15			x	187	1.49	0.59	39.6	Fredlund and Dahlman (1972)
20			x	53	1.30	0.62	47.7	Fredlund and Dahlman (1972)

Table 10-1 (*Continued*)

Material	PARAMETER Frictional Angle, Degrees	Tangent of Frictional Angle	Unconfined Compression Strength, ton/ft^2	Number of Samples	Mean	Standard Deviation	Coefficient of Variation, %	Source
Clay			x	231	0.97	0.26	29.0	Matsuo and Kuroda (1974)
Clay			x	97			30.0–40.0	Ladd, et al. (1972)
Clay shale†			x				37.0–51.0	Lumb (1972)
till†			x				60.0–85.0	Lumb (1972)
till			x		3.24	1.17	36.1	Morse (1972)

†Lumb notes these two materials are extremely variable and believes that these results are probably close to the upper possible limits of variability for any natural soils.

(Harr 1977). Reproduced with permission from McGraw-Hill, Inc.

Note: 1 tsf = 95.8 kPa

From Eq. (10-8),

$$\text{Cov}[x, y] = E[xy - yE[x] - xE[y] + E[x]E[y]] = E[xy] - E[x]E[y] \quad (10\text{-}10)$$

For actual computation of covariance, the use of Eq. (10-10) is more convenient than Eq. (10-8).

If large positive deviations of x are associated with large positive deviations of y, then the covariance will be positive. If positive deviations of x are associated with negative deviations of y, and vice versa, the covariance will be negative. Conversely, if positive and negative deviations of x occur as frequently as positive and negative deviations of y, then the covariance will tend to 0. Therefore, the covariance is a measure of correlation between two random variables. It should be noted that variance is a special case of covariance of a random variable with itself:

$$\text{Cov}[x, x] = E[(x - E[x])(x - E[x])] = E[(x - E[x])^2] = V[x] \quad (10\text{-}11)$$

If x and y are correlated, the variance of $x + y$ can be expressed as

$$\begin{aligned} V[x+y] &= E[(x + y - E[x+y])^2] = E[(x - E[x] + y - E[y])^2] \\ &= E[(x - E[x])^2] + E[(y - E[y])^2] + 2E[(x - E[x])(y - E[y])] \quad (10\text{-}12) \\ &= V[x] + V[y] + 2\text{Cov}[x, y] \end{aligned}$$

10.1.6 Correlation Coefficient

The correlation coefficient between random variables x and y is defined as

$$\rho(x, y) = \frac{Cov[x, y]}{\sqrt{V[x]V[y]}} \quad (10\text{-}13)$$

It can be shown that $-1 \le \rho \le 1$ and that $\rho = 1$ when $y = a + bx$, and $\rho = -1$ when $y = a - bx$, where a and b are constants. For partially saturated soil under undrained conditions, a correlation coefficient may exist between c and ϕ. However, for effective stress analysis, c' is small and the correlation coefficient between c' and ϕ' can be assumed 0.

Example 10.2 Five sets of undrained triaxial tests were made on a partially saturated soil. The pairs of cohesion and friction angle (c in psf, ϕ in deg) determined from the tests are (5,120, 10.3), (4,370, 12.2), (3,200, 15.5), (2,050, 17.8), and (1,140, 20.3). Determine the coefficients of variation for c and ϕ and the correlation coefficient between them.

Solution An inspection of the pairs clearly indicates that there is a strong correlation between c and ϕ. Because a larger c always is associated with a smaller ϕ, a negative correlation coefficient is expected.

From Eq. (10-2), $E[c] = (5{,}120 + 4{,}370 + 3{,}200 + 2{,}050 + 1{,}140)/5 = 3{,}176$ psf, and $E[\phi] = (10.3 + 12.2 + 15.5 + 17.8 + 20.3)/5 = 15.22°$.

From Eq. (10-5), $V[c] = [(5{,}120 - 3{,}176)^2 + (4{,}370 - 3{,}176)^2 + (3{,}200 - 3{,}176)^2 + (2{,}050 - 3{,}176)^2 + (1{,}140 - 3{,}176)^2]/(5 - 1) = 2{,}654{,}630$ and $V[\phi] = [(10.3 - 15.22)^2 + (12.2 - 15.22)^2 + (15.5 - 15.22)^2 + (17.8 - 15.22)^2 + (20.3 - 15.22)^2]/(5 - 1) = 16.47$.

From Eq. (10-10), $\text{Cov}[c, \phi] = (5{,}120 \times 10.3 + 4{,}370 \times 12.2 + 3{,}200 \times 15.5 + 2{,}050 \times 17.8 + 1{,}140 \times 20.3)/5 - 3{,}176 \times 15.22 = -5{,}282$.

From Eq. (10-7), $C[c] = \sqrt{2{,}654{,}630}/3{,}176 = 0.51$ and $C[\phi] = \sqrt{16.47}/15.22 = 0.27$.

From Eq. (10-13), $\rho[c, \phi] = -5282/\sqrt{2{,}654{,}630 \times 16.47} = -0.80$

10.2 Taylor's Expansion

Taylor's expansion for a function $f(x, y)$ about the point (a, b) can be expressed as

$$f(x, y) = f(a, b) + f_x(a, b)(x - a) + f_y(a, b)(y - b) + \frac{1}{2}[f_{xx}(a, b)(x - a)^2 + 2f_{xy}(a, b)(x - a)(y - b) + f_{yy}(a, b)(y - b)^2] + \ldots \quad (10\text{-}14)$$

in which the subscripts x and y indicate the partial differentiation with respect to x and y. As a first approximation, the second-order terms in Eq. (10-14) may be neglected:

$$f(x, y) = f(a, b) + f_x(a, b)(x - a) + f_y(a, b)(y - b) \quad (10\text{-}15)$$

10.2.1 Mean

If a and b are taken as the means of x and y, the expectation of $f(x, y)$ can be obtained from Eq. (10-15) by taking the expectation on both sides:

$$E[f(x, y)] = E[f(a, b)] + f_x(a, b)(E[x] - a) + f_y(a, b)(E[y] - b) = E[f(a, b)] \quad (10\text{-}16)$$

If $g(x_1, x_2, \ldots x_n)$ is a function of n random variables x_i and μ_i is the mean value for each of these random variable, then from Eq. (10-16),

$$E[g] = g(\mu_i) \quad (10\text{-}17)$$

Eq. (10-17) indicates that the mean value of g can be obtained simply by substituting the mean value of each random variable into the function.

10.2.2 Variance

The variance of $f(x, y)$ can be obtained by taking the variance on both sides of Eq. (10-15) and applying Eq. (10-12), or

$$V[f(x,y)] = \{f_x(a,b)\}^2 V[x] + \{f_y(a,b)\}^2 V[y] + 2\{f_x(a,b)\}\{f_y(a,b)\}\mathrm{Cov}[x,y] \quad (10\text{-}18)$$

Eq. (10-18) can be extended to $g(x_1, x_2, \ldots x_n)$ by

$$V[g] = \sum_{i=1}^{n}\sum_{j=1}^{n}\left(\frac{\partial g}{\partial x_i}\right)_{\mu_i}\left(\frac{\partial g}{\partial x_j}\right)_{\mu_j}\mathrm{Cov}[x_i, x_j] \quad (10\text{-}19)$$

If $x_1, x_2, \ldots x_n$ are independent, there are no cross-product terms, and Eq. (10-19) becomes

$$V[g] = \sum_{i=1}^{n}\left(\frac{\partial g}{\partial x_i}\right)_{\mu_i}^2 V[x_i] \quad (10\text{-}20)$$

Example 10.3 A partially saturated soil is placed on an infinite slope with an angle, β, of 22° and a depth, *d*, of 18 ft (5.5 m). Both β and *d* are considered as fixed quantities with no variation. The cohesion, friction angle, and unit weight of the soil are considered as random variables with properties shown in the following table:

Variables	c	ϕ	γ
Mean	800 psf	12°	120 pcf
Coefficient of variation	0.5	0.3	0.1
Correlation coefficient between c and $\phi = -0.4$			

Determine the mean and variance of the factor of safety by Taylor's expansion.

Solution The static factor of safety for an infinite slope with no seepage can be obtained from Eq. (6-3) by setting C_s and r_u to zero:

$$F = \frac{\left(\frac{c}{\gamma d}\right)\sec\beta + \cos\beta\tan\phi}{\sin\beta} \quad (10\text{-}21)$$

From Eq. (10-17),

$$E[F] = \frac{\left(\frac{800}{120\times 18}\right)\sec 22° + \cos 22°\tan 12°}{\sin 22°} = 1.592$$

To determine $V[F]$, take partial derivative of F with respect to each of the variables in Eq. (10-21), or

$$\frac{\partial F}{\partial c} = \frac{1}{\gamma d\cos\beta\sin\beta} = \frac{1}{120\times 18\cos 22°\sin 22°} = 0.001333$$

$$\frac{\partial F}{\partial \phi} = \frac{\cos\beta \sec^2\phi}{\sin\beta} = \frac{\cos 22^\circ}{\sin 22^\circ \cos^2 12^\circ} = 2.5869$$

$$\frac{\partial F}{\partial(\gamma)} = -\frac{c}{(\gamma)^2 d \sin\beta\cos\beta} = -\frac{800}{(120)^2 18 \sin 22^\circ \cos 22^\circ} = -0.008886$$

From Eq. (10-19),

$$\begin{aligned} V[F] &= (0.001333 \times 0.5 \times 800)^2 + (2.5869 \times 0.3 \times 12 \times \pi/180)^2 \\ &\quad + (0.008886 \times 0.1 \times 120)^2 \\ &\quad - 2(0.001333 \times 2.5869)(0.5 \times 800)(0.3 \times 12 \times \pi/180) \times 0.4 \\ &= 0.2843 + 0.0264 + 0.0114 - 0.0693 \\ &= 0.2528 \end{aligned}$$

10.3 Mean-Value First Order Second Moment Method

Taylor's expansion can be used for simple cases where the factor of safety can be expressed in a close-form formula. For complex cases involving computer programming, the mean-value first order second moment method (MFOSM) can be used (Cornell 1971). This simplified method is particularly useful in geotechnical engineering, where the coefficient of variation for each variable is based on past experience or engineering judgment rather than on site-specific testing or measurements.

In this method, two factors of safety are computed for each variable, one at one standard deviation above the mean and the other at one standard deviation below the mean, whereas the other variables do not change but remain at the same mean values. If the factor of safety is a function of n variables from x_1 to x_n, from Eq. (10-20), the variance of F can be written as

$$V[F] = \left(\frac{\partial F}{\partial x_1}\right)^2 V[x_1] + \left(\frac{\partial F}{\partial x_2}\right)^2 V[x_2] + \ldots\ldots\ldots\ldots + \left(\frac{\partial F}{\partial x_n}\right)^2 V[x_n]$$

$$= \left(\frac{F_{x1+} - F_{x1-}}{2\sqrt{V[x_1]}}\right)^2 V[x_1] + \left(\frac{F_{x2+} - F_{x2-}}{2\sqrt{V[x_2]}}\right)^2 V[x_2] + \ldots\ldots + \left(\frac{F_{xn+} - F_{xn-}}{2\sqrt{V[x_n]}}\right)^2 V[x_n]$$

$$\text{or} \quad V[F] = \left(\frac{F_{x1+} - F_{x1-}}{2}\right)^2 + \left(\frac{F_{x2+} - F_{x2-}}{2}\right)^2 + \ldots\ldots\ldots\ldots + \left(\frac{F_{xn+} - F_{xn-}}{2}\right)^2 \qquad (10\text{-}22)$$

in which F_{xn+} and F_{xn-} are the values of F evaluated at one standard deviation above and below the mean, respectively. If any two variables, say x_1 and x_2, are correlated, an additional term $2[(F_{x1+} - F_{x1-})/2][(F_{x2+} - F_{x2-})/2]\rho(x_1, x_2)$ must be added to Eq. (10-22), as indicated by Eq. (10-19). If there are n random variables, the factor of safety must be evaluated repeatedly for $2n$ times to determine its variance.

Example 10.4 Same as Example 10.3, but determine the mean and variance of factor of safety by the MFOSM method.

Solution The factor of safety is computed by

$$F = \frac{\left(\frac{c}{18\gamma}\right)\sec 22^\circ + \cos 22^\circ \tan\phi}{\sin 22^\circ} = 0.15995\left(\frac{c}{\gamma}\right) + 2.475\tan\phi$$

The three variables, c, ϕ, and γ have the following values:

Variables	c	ϕ	γ
Mean	800 psf	12°	120 pcf
Coefficient of variation	0.5	0.3	0.1
Standard deviation, s	400 psf	3.6°	12 pcf
Mean + s	1,200 psf	15.6	132 pcf
Mean – s	400 psf	8.4	108 pcf
Correlation coefficient between c and $\phi = -0.4$			

Table 10-2 shows the computation of variance by the MFOSM method. To compute the factor of safety, each variable is evaluated twice: one at mean plus one standard deviation and the other at mean minus one standard deviation, whereas the other variables keep the same mean values.

The factor of safety has a mean of 1.597 and a variance of 0.2533, which check closely with the 1.592 and 0.2528 computed by Taylor's expansion in Example 10.3.

Table 10-2. *Computation of Mean and Variance by MFOSM Method*

Variable	Values	Factor of safety	$F_+ - F_-$	$[(F_+ - F_-)/2]^2$
c at mean + s	1,200	2.126		
at mean – s	400	1.059	1.067	0.2846
ϕ at mean + s	15.6	1.757		
at mean – s	8.4	1.432	0.325	0.0264
γ at mean + s	132	1.495		
at mean – s	108	1.711	–0.216	0.0117
Correlation between c and $\phi = 2(1.067/2)(0.325/2)(-0.4)$				–0.0694
	Mean	1.597	Variance	0.2533
Note: Reliability = 1 – probability of failure				

10.4 Normal Distribution

The distribution function used most frequently as a probabilistic model is called the normal or Gaussian distribution. Although this symmetrical and bell-shaped distribution is very important, it is not the only type of distribution to be used in the probabilistic method.

The mathematical equation of a normal distribution expressing the frequency of occurrence of the random variable x is

$$f(x) = \frac{1}{s\sqrt{2\pi}} \exp\left[-\frac{1}{2}\left(\frac{x-\mu}{s}\right)^2\right] \tag{10-23}$$

in which s = standard deviation and μ = mean.

Fig. 10-1 shows a plot of a normal distribution with $s = 1$ and $\mu = 0$ and 4, respectively. Note that the two curves are similar, except that the x coordinate is displaced by a constant distance. If the x at the peak is not equal to 0, it can be set to zero by a simple displacement.

The cumulative distribution function for a normally distributed random variable x can be expressed as

$$\varphi(x) = \int_{-\infty}^{x} \frac{1}{s\sqrt{2\pi}} \exp\left[-\frac{1}{2}\left(\frac{x-\mu}{s}\right)^2\right] dx \tag{10-24}$$

A simple way to eliminate μ and s in Eq. (10-24) is to introduce the normal deviate defined as

$$u = \frac{x-\mu}{s} \tag{10-25}$$

Replacing x in Eq. (10-24) by u gives

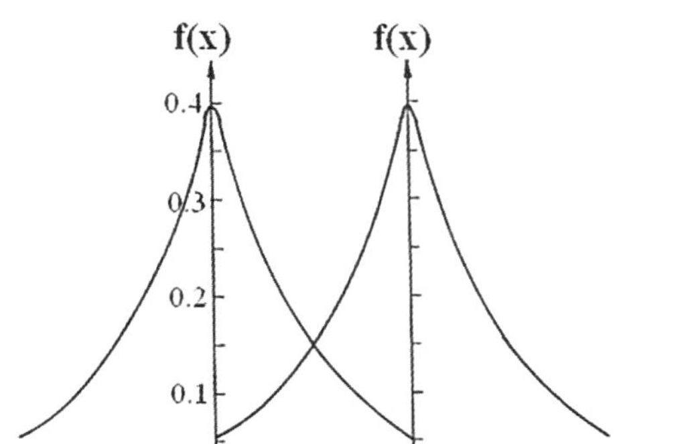

Fig. 10-1. Normal distribution

$$\varphi(z) = \frac{1}{\sqrt{2\pi}} \int_0^z \exp\left(-\frac{u^2}{2}\right) du \tag{10-26}$$

in which $\varphi(z)$ is the area under the standard normal distribution curve $f(u)$ between 0 and z, and z is the value of u computed by Eq. (10-25), as shown in Fig. 10-2 and tabulated in Table 10-3.

Using Eq. (10-26) and recognizing that the area under half of the standardized normal curve is 0.5, the probability associated with the value of the random variable being less than any specified value can be determined.

When applying Eq. (10-25) to slope stability, x is equal to 1, because it is the factor of safety when a slope fails; μ is the expectation of F, $E[F]$; and s is the standard deviation of F, $s[F]$. Because the value of z shown in Fig. 10-2 is negative, a positive term commonly used in slope stability analysis is the reliability index defined by

$$\beta = \frac{E[F]-1}{s[F]} \tag{10-27}$$

in which F is the computed factor of safety. Given β, which is equivalent to z in Fig. 10-2, the area under the normal curve, $\varphi(\beta)$, as indicated by the hatched lines in Fig. 10-3, can be determined from Table 10-3. The probability of failure is: Probability $(F < 1) = 0.5 - \varphi(\beta)$ and reliability = 1 – probability of failure.

Example 10.5 Based on the mean and variance of the factor of safety in Example 10.4 and assuming the factor of safety as a normal distribution, determine the probability of failure.

Solution From Example 10.4, $E[F] = 1.597$ and $V[F] = 0.2533$. $s = \sqrt{0.2533} = 0.5033$. From Eq. (10-27), $\beta = (1.597 - 1)/0.5033 = 1.186$. From Table 10-3, $\varphi(\beta) = 0.382$, so the probability of failure is: Probability $(F < 1) = 0.5 - 0.382 = 0.118$, or 11.8%.

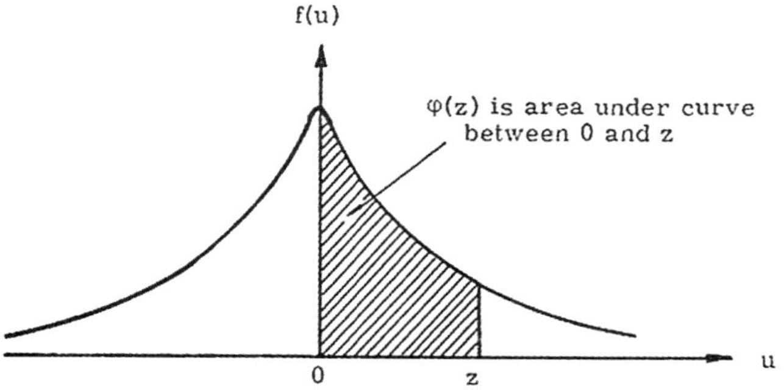

Fig. 10-2. Area under normal curve for given z

Table 10-3. *Area φ(β) under Normal Curve*

z or β	.00	.01	.02	.03	.04	.05	.06	.07	.08	.09
0	0	.003969	.007978	.011966	.015953	.019939	.023922	.027903	.031881	.035856
.1	.039828	.043795	.047758	.051717	.055670	.059618	.063559	.067495	.071424	.075345
.2	.079260	.083166	.087064	.090954	.094835	.098706	.102568	.106420	.110251	.114092
.3	.117911	.121720	.125516	.129300	.133072	.136831	.140576	.144309	.148027	.151732
.4	.155422	.159097	.162757	.166402	.170031	.173645	.177242	.180822	.184386	.187933
.5	.191462	.194974	.198466	.201944	.205401	.208840	.212260	.215661	.219043	.222405
.6	.225747	.229069	.232371	.235653	.234914	.242154	.245373	.248571	.251748	.254903
.7	.258036	.261148	.264238	.257305	.270350	.273373	.276373	.279350	.282305	.285236
.8	.288145	.291030	.293892	.296731	.299546	.302337	.305105	.307850	.310570	.313267
.9	.315940	.318589	.321214	.323814	.326391	.328944	.331472	.333977	.336457	.338913
1.0	.341345	.343752	.346136	.348495	.350830	.353141	.355428	.357690	.359929	.362143
1.1	.364334	.366500	.368643	.370762	.372857	.374928	.376976	.379000	.381000	.382977
1.2	.384930	.386861	.388768	.390651	.392512	.393350	.396165	.397958	.399727	.401475
1.3	.403200	.404902	.406582	.408241	.409877	.411492	.413085	.414657	.416207	.417736
1.4	.419243	.420730	.422196	.423641	.425066	.426471	.427855	.429219	.430563	.431888
1.5	.433193	.434476	.435745	.436992	.438220	.439429	.440620	.441792	.442947	.444083
1.6	.445201	.446301	.447384	.448449	.449497	.450529	.451543	.452540	.453521	.454486
1.7	.455435	.456367	.457284	.458185	.459070	.459941	.460796	.461636	.462462	.463273
1.8	.464070	.464852	.465620	.466375	.467116	.467843	.468557	.469258	.469946	.470621
1.9	.471283	.471933	.472571	.473197	.473610	.474412	.475002	.475581	.476148	.476705

Table 10-3 (*Continued*)

z or β	.00	.01	.02	.03	.04	.05	.06	.07	.08	.09
2.0	.477250	.477784	.478308	.478822	.479325	.479818	.480301	.480774	.481237	.481691
2.1	.482136	.482571	.482997	.483414	.483823	.484222	.484614	.484997	.485371	.485738
2.2	.486097	.486447	.486791	.487126	.487455	.487776	.488089	.488396	.488696	.488989
2.3	.489276	.489556	.489830	.490097	.490358	.490613	.490863	.491106	.491344	.491576
2.4	.491802	.492024	.492240	.492451	.492656	.492857	.493053	.493244	.493431	493613
2.5	.493790	.493963	.494132	.494297	.494457	.494614	.494766	.494915	.495060	.495201
2.6	.495339	.495473	.495604	.495731	.495855	.495975	.496093	.496207	.496319	.496427
2.7	.496533	.496636	.496736	.496833	.496928	.497020	.497110	.497197	.497282	.497365
2.8	.497445	.497523	.497599	.497673	.497744	.497814	.497882	.497948	.498012	.498074
2.9	.498134	.498193	.498250	.498305	.498359	.498411	.498462	.498511	.498559	.498605
3.0	.498650	.498694	.498736	.498777	.498817	.498856	.598893	.498930	.498965	.498999
3.1	.499032	.499065	.499096	.499126	.499155	.499184	.499211	.499238	.499264	.499289
3.2	.499313	.499336	.499359	.499381	.499402	.499423	.499443	.499462	.499481	.499499
3.3	.499517	.499534	.499550	.499566	.499581	.499596	.499610	.499624	.499638	.499651
3.4	.499663	.499675	.499687	.499698	.499709	.499720	.499730	.499740	.499749	.499758
3.5	.499767	.499776	.499784	.499792	.499800	.499807	.499815	.499822	.499828	.499835
3.6	.499841	.499847	.499853	.499858	.499864	.499869	.499874	.499879	.499883	.499888
3.7	.499892	.499896	.499900	.400904	.499908	.499912	.499915	.499918	.499922	.499925
3.8	.499928	.499931	.499933	.499936	.499938	.499941	.499943	.499946	.499948	.499950
3.9	.499952	.499954	.499956	.499958	.499959	.499961	.499963	.499964	.499966	.499967

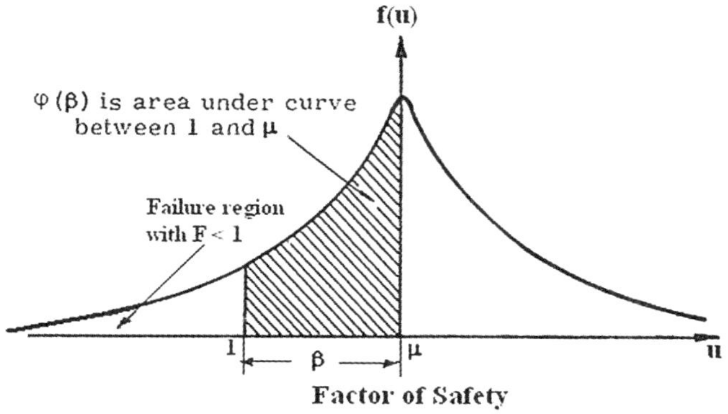

Fig. 10-3. Area under normal curve for given β

10.5 Lognormal Distribution

Another distribution used widely in geotechnical engineering is the lognormal distribution with a reliability index defined as

$$\beta_{\ln} = \frac{E[\ln F]}{s[\ln F]} \tag{10-28}$$

in which $s[\ln F]$ is the standard deviation based on the lognormal distribution of F. From Eq. (10-20),

$$V[\ln F] = \frac{V[F]}{F^2} \tag{10-29}$$

$$s[\ln F] = \frac{s[F]}{F} = C[F] \tag{10-30}$$

Eq. (10-30) indicates that the standard deviation of ln F is the coefficient of variation of F. Knowing $E[\ln F]$ and $s[\ln F]$, $\beta_{\ln}$ can be determined from Eq. (10-28) and the reliability from Table 10-3. In the stated derivations, the distribution is assumed to be natural log with base e. The same reliability will be obtained if the base 10 log is used.

Example 10.6 Same as Example 10.5, but assuming the factor of safety as a lognormal distribution.

Solution Given $F = 1.597$ and $s[F] = 0.5033$, from Eq. (10-30), $s[\ln F] = 0.5033/1.597 = 0.315$. From Eq. (10-28,) $\beta_{\ln} = \ln(1.597)/0.315 = 1.486$. From Table 10-3, $\varphi(\beta) = 0.431$, so the probability of failure is: Probability $(F < 1) = 0.5 - 0.431 = 0.069$, or 6.9%. It can be seen that the probability of failure based on a normal distribution is greater than that based on a lognormal distribution, so the use of a normal distribution gives a smaller reliability and is on the safe side.

Spreadsheets can be used conveniently to determine the reliability, as illustrated by the following example.

Example 10.7 A triangular fill is the same as that in Example 6.2, with $H = 40$ ft (12.2 m), $\alpha = 20°$, and $\beta = 36°$. After substituting these values of H, α, and β into Eq. (6-8), the following equation is obtained to solve the factor of safety, F:

$$F = 0.312\left(\frac{c'}{\gamma}\right) + 2.747(1 - r_u)\tan\phi' \tag{10-31}$$

Values of c', ϕ', γ, and r_u and their coefficients of variation can be found in the first two rows of Table 10-4. Determine the reliability of the design based on both normal and lognormal distributions.

Solution Table 10-4 is the spreadsheet for computing the factor of safety and the reliability. Details about the spreadsheet are as follows:

1. In cell B4, type Eq. (10-31) as 0.312*$B$1/$F$1 + 2.747*(1-$H$1)*TAN (RADIANS ($D$1)). Dollar signs are used, so when the equation is copied from one cell to another, these input parameters never will be changed. Copy this equation to B5 through B7.
2. In column B, "F_+" is the factor of safety when one of the variables in Column A has a value one standard deviation greater than the given value. Therefore, in cell B4, c' must be multiplied by (1 + $C[c']$), so the equation must be changed to 0.312*$B$1*(1 + B2)/$F$1 + 2.747(1-$H$1)*TAN (RADIANS($D$1)). The same should be applied to other variables, i.e., the value of the variable must be multiplied by (1 + CV), where CV is the coefficient of variation of the variable.
3. In column C, "F_-" is the factor of safety when one of the variables in Column A has a value one standard deviation smaller than the given value. Copy rows 4 to 7 from column B to column C and change (1 + CV) to (1 – CV).

***Table 10-4.** Spreadsheet for Computing Factor of Safety and Reliability*

	A	B	C	D	E	F	G	H
1	c' in psf =	160	ϕ' in deg =	24	γ in pcf =	125	ru =	0.05
2	$C[c']$ =	0.4	$C[\phi']$ =	0.15	$C[\gamma]$ =	0.1	C[ru] =	0.2
3	Variable	F_+	F_-	Variance		Normal distribution		
4	c'	1.72100	1.40151	0.02552	E[F]	s[F]	β	Reliability
5	ϕ'	1.76365	1.36988	0.03876	1.56364	0.25702	2.19298	98.58%
6	γ	1.52495	1.60562	0.00163	E[lnF]	s[lnF]	βln	
7	ru	1.54902	1.57348	0.00015	0.44702	0.16437	2.71952	99.67%
8	Sum	6.55861	5.95049	0.06606				

4. In column D, variance is computed by Eq. (10-22). In cell D4, the equation is 0.25*(B4-C4)^2. Copy the equation from D4 to D5 through D7.
5. For a normal distribution, $E[F]$ is the average of all F's and is expressed as (B8 + C8)/8; $s[F]$ can be obtained from Eq. (10-6) and expressed as SQRT(D8); β can be obtained from Eq. (10-27) and expressed as ((E5-1)/ F5; and reliability can be obtained from Table 10-4.
6. For a lognormal distribution, $E[\ln F]$ is expressed as ln(E5); $s[\ln F]$ can be obtained from Eq. (10-30) and expressed as F5/E5; $\beta_{\ln}$ can be obtained from Eq. (10-28) and expressed as E7/F7, and reliability can be obtained from Table 10-4.

Summary

1. Because of the large variations in site conditions, the probabilistic method of stability analysis has gained popularity in recent years. The major disadvantage of the method is that a large number of measurements or tests is required to ascertain the variability of those parameters that affect the factor of safety. It is only after the variability of each parameter is evaluated or assumed that the probability of failure can be determined.

2. The probabilistic method is a complement to the conventional deterministic methods by considering some input parameters as random variables with a mean and a variance. The requirement for the deterministic method is the factor of safety that the design must exceed, whereas the requirement for the probabilistic method is the reliability that the design can accept. It is more reasonable to use the probabilistic method, because the required factor of safety in the deterministic method should depend on the variability of the data. If the data are unreliable with a large variance, a higher factor of safety should be required.

3. If the factor of safety can be expressed in closed-form formulas, the mean factor of safety can be obtained by simply substituting the mean value of each input parameter into the formulas. The variance of the factor of safety can be computed from Taylor's expansion, as indicated by Eq. (10-19).

4. If the factor of safety is determined by complex computer programs and cannot be expressed in closed form, then the mean-value first order second moment method (MFOSM) can be used. For each random variable, the factor of safety must be computed twice, one with a value at one standard deviation above the mean and the other at one standard deviation below the mean, whereas the mean values are used for all other variables. If there are n variables, the factor of safety must be computed $2n$ times. The mean factor of safety is the average of these $2n$ factors of safety, and the variance of the factor of safety can be computed by Eq. (10-22).

5. By assuming the factor of safety as a normal or lognormal distribution with a mean and a variance, the reliability of the design can be evaluated. The reliability based on a lognormal distribution is usually greater than that based on a normal distribution, so the use of normal distribution is on the safe side.

Problems

10.1 A very useful equation to determine variance from the mean is $V[x] = E[x^2] - (E[x])^2$ Prove this equation.

10.2 Given the following eight values of total unit weight in pcf: 124.6, 126.2, 121.6, 117.3, 115.8, 135.6, 130.7, and 125.3, calculate the mean, variance, standard deviation, and coefficient of variation.
[Answer: 124.6 pcf, 42.999, 6.56 pcf, 5.26%]

10.3 Prove that $\rho(x, y) = 1$ when $y = a + bx$ and $\rho(x, y) = -1$ when $y = a - bx$, where a and b are constants. [Hint: Use the equation in Problem 10.1 in the final step of derivation.]

10.4 The following pairs of effective cohesion in psf and effective friction angle in deg, (c', ϕ'), were obtained from piezocone penetration tests: (120, 26.8°), (260, 34.5°), (280, 32.6°), (230, 33.7°), (180, 29.3°), (200, 28.2°), and (150,30.6°). Calculate the mean, the coefficient of variation of both c' and ϕ', and the correlation coefficient between c' and ϕ'.
[Answer: 202.9 psf, 28.6%, 30.81°, 9.4%, 0.689]

10.5 Fig. P10-5 shows an infinite slope with $\beta = 15°$, $d = 30$ ft, and $d_w = 10$ ft. The soil is sand with an effective friction angle of 30° and a coefficient of variation of 0.15, and the coefficient of variation of d_w is 0.25. Determine the mean and variance of factor of safety by Taylor's expansion and then compare them with the MFOSM method.

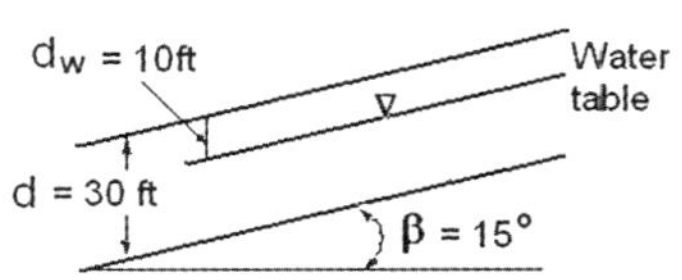

Fig. P10-5.

[Answer: 1.437, 0.07600, 1.443, 0.07683]

10.6 Same as Problem 6.3 with the cross section reproduced and shown in Fig. P10-6. The mean and coefficient of variation of the four random variables are as follows:

Variable	Cohesion	Friction Angle	Total Unit Weight	Pore Pressure ratio
Mean	160 psf	24°	125 pcf	0.05
Coef. of variation	0.4	0.15	0.1	0.2

Determine the factor of safety and its variance by Taylor's expansion.

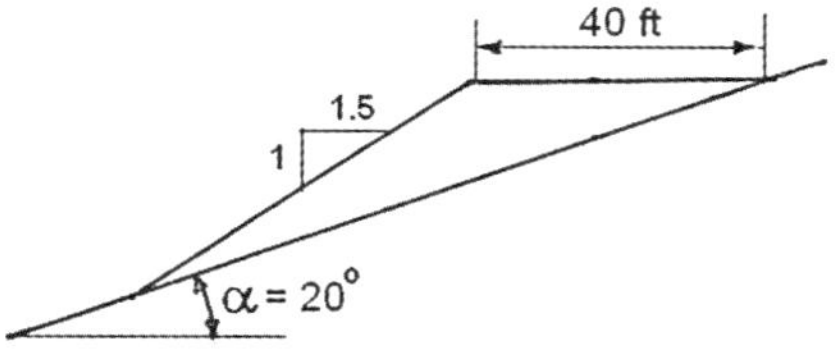

Fig. P10-6.

[Answer: 1.708, 0.08963]

10.7 Same as Problem 10.5, but determine the factor of safety and the reliability based on both normal and lognormal distributions by the MFOSM method.
[Answer: 1.711, 91.97%, 99.89%]

10.8 Same as Problem 6.5 with the cross section reproduced and shown in Fig. P10-8. The mean and coefficient of variation of the three random variables and the correlation coefficient between c and ϕ are as follows:

Variable	Cohesion	Friction Angle	Total Unit Weight	Correlation Coef.
Mean	400 psf	24°	125 pcf	–0.5
Coef. of variation	0.4	0.4	0.1	(between c and ϕ)

Determine the factor of safety and the reliability based on both normal and lognormal distribution by the MFOSM method.

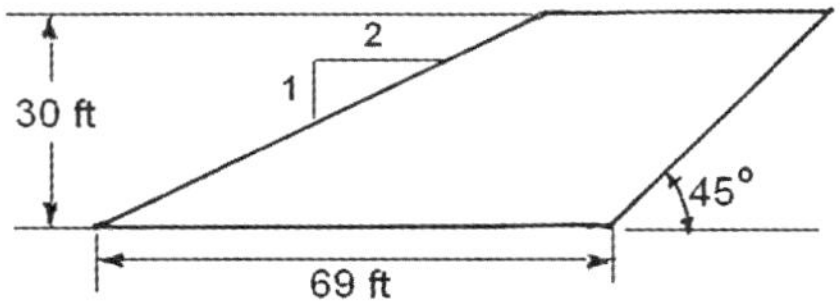

Fig. P10-8.

[Answer: 2.803, 99.18%, 99.99%]

Appendix

Preview of LEAME Computer Software

Thus far, this book has focused on the fundamental principles and methods for analyzing slope stability using the limit equilibrium method. The computer software—known as LEAME, or Limit Equilibrium Analysis of Multilayered Earthworks—which was developed specifically for this purpose is available as a companion product. *LEAME Software and User's Manual: Analyzing Slope Stability by the Limit Equilibrium Method* can be purchased through the ASCE online bookstore or the ASCE Library at http://dx.doi.org/10.1061/9780784477991. The LEAME software can be installed on any computer with a Windows operating system of Windows 95 or later, including the latest Windows 8. The *User's Manual* provides detailed instructions for installing and operating the software to solve a variety of two-dimensional (2D) and three-dimensional (3D) slope stability analyses. The *User's Manual* also contains a chapter demonstrating the use of LEAME software for surface mining operations.

This appendix offers a sampling of typical problems that can be solved using the LEAME software.

LEAME for Two-Dimensional Analysis

To illustrate the capability of LEAME software for 2D analysis, 10 examples are presented here. Detailed solutions using the LEAME software are available, with commentary, in the *User's Manual*.

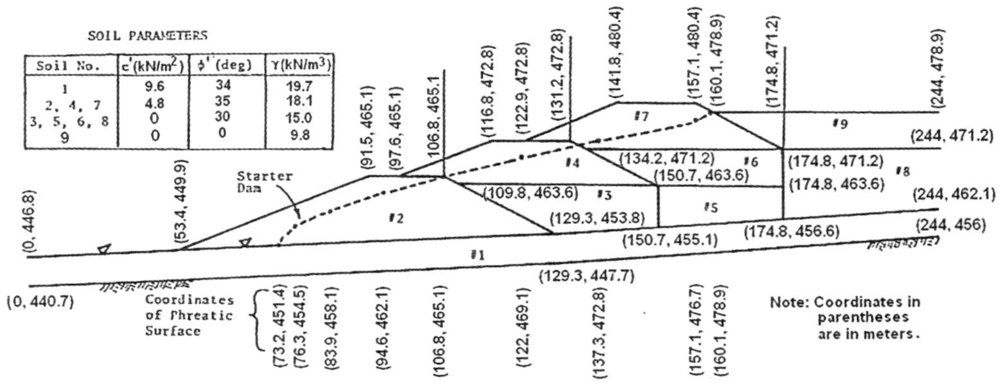

Fig. A-1. *Refuse dam constructed by upstream method*
Note: 1 m = 3.28 ft; 1 kN/m² = 20.9 psf; 1 kN/m³ = 6.36 pcf

2D Example 1: Refuse Dam Constructed by the Upstream Method

This example illustrates the stability analysis of a refuse dam constructed by the upstream method. This type of analysis has widespread applications for analyzing short-term stability during or immediately after construction when the excess pore water pressure in some soils, due to the placement of an overburden, has not been completely dissipated.

Fig. A-1 shows the upstream method of refuse disposal, which is very popular in rugged terrain. First, a starter dam is built by coarse refuse or other earthen materials and the fine refuse in the form of slurry is pumped into the back of the dam. Then the dam is extended upstream in stages, with part of the dam being placed on the settled fine refuse. The dam has a downstream slope of 2.5:1 and an upstream slope of 2:1. The construction is divided into three stages. The first stage involves the construction of the starter dam, the second stage of the lower refuse dam, and the third stage of the upper refuse dam. Both the short-term and long-term stability analyses can be made at the end of each stage. Only the most critical case of short-term stability at the end of stage 3 will be considered. The long-term stability can be obtained by simply assigning the excess pore pressure ratios to 0.

2D Example 2: Steep Slope Reinforced by Geogrids

This example illustrates the use of geogrids to stabilize a steep slope. This type of construction is useful in urban or other built-up areas where space is so limited that a flatter slope just cannot be used.

Fig. A-2 shows a fill slope reinforced by geogrids and placed directly on a rock surface. The fill has a height of 14.4 m (47.2 ft) and a slope of 1:1. A surcharge load of 15 kN/m² (310 psf) is applied on top, as simulated by 0.3 m (1 ft) of soil with a cohesion and friction angle of zero and a total unit weight of 50 kN/m³ (320 pcf). The soil in the fill has a cohesion of zero, a friction angle of 35°, and a

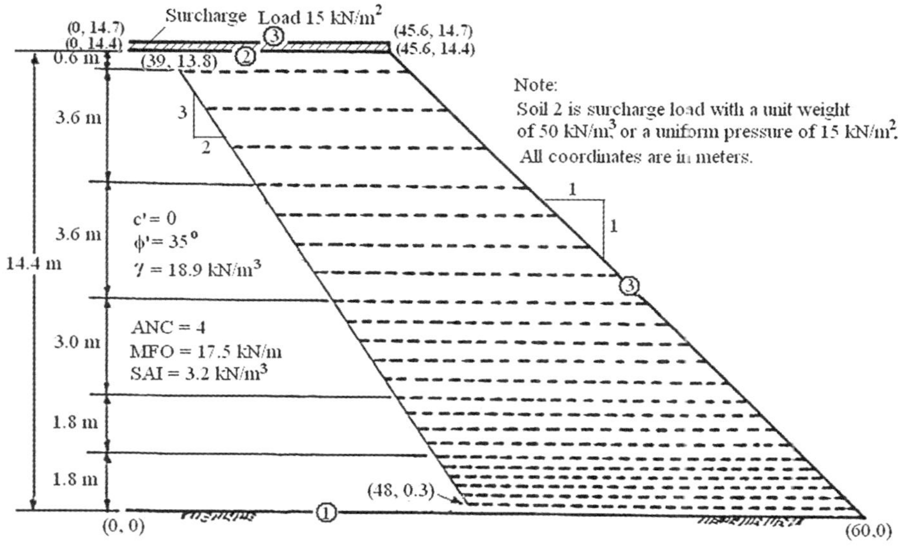

Fig. A-2. *Slope reinforced by geogrids*
Note: 1 m = 3.28 ft; 1 kN/m = 68.5 lb/ft; 1 kN/m² = 20.9 psf; 1 kN/m³ = 6.36 pcf

total unit weight of 18.9 kN/m³ (120 pcf), and there is no seepage. The location of the geogrids is shown in the figure. The left end of the geogrids is the actual end point. Because the resistance of geogrids depends on the overburden pressure, it is assumed ANC (type of forces) = 4, MFO (magnitude of each force) = 17.5 kN/m (1,200 lb/ft), and SAI (soil-anchor interaction) = 3.2 kN/m³ (20.3 pcf). Determine the factor of safety.

2D Example 3: Soil Nails for a Shotcrete Wall

This example illustrates the use of soil nails to stabilize a vertical wall. Because these nails have the same length and capacity and their resistance does not depend on the depth of the overburden, it is more reasonable to assume that ANC = 2.

Fig. A-3 shows a shotcrete wall and the location of the nails. First, a 3.1-m (10.2-ft) vertical cut is made, then the soil nails are installed, and finally a surfacing consisting of steel fiber-reinforced shotcrete is placed on the surface. The soil has an effective cohesion of 9.6 kN/m² (200 psf), an effective friction angle of 25°, and a total unit weight of 18.9 kN/m³ (120 pcf). The applied internal force (MFO) on each row of nails is 65.7 kN/m (4,500 lb/ft). Determine the factor of safety.

2D Example 4: Composite Failure Surfaces

When there is a thin layer of weak material within a slope, part of the failure surfaces most probably will follow the bottom of the weak layer. One of the most

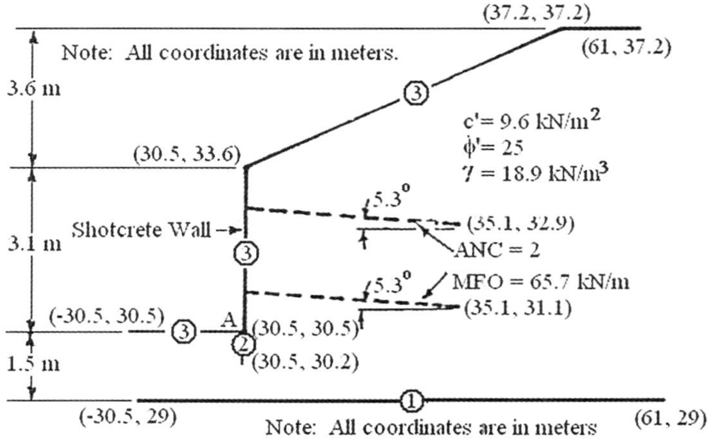

Fig. A-3. Soil nails for shotcrete wall
Note: 1 m = 3.28 ft; 1 kN/m = 68.5 lb/ft; 1 kN/m² = 20.9 psf; 1 kN/m³ = 6.36 pcf

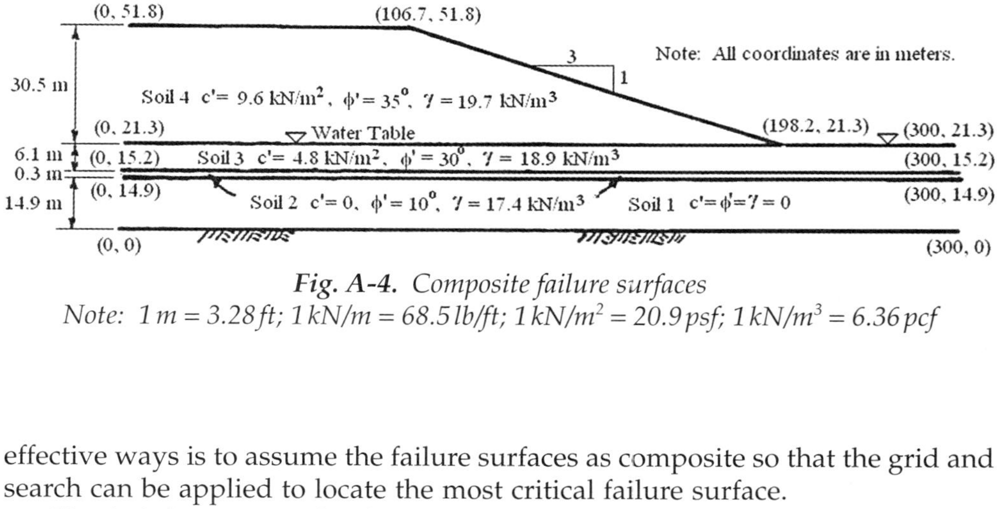

Fig. A-4. Composite failure surfaces
Note: 1 m = 3.28 ft; 1 kN/m = 68.5 lb/ft; 1 kN/m² = 20.9 psf; 1 kN/m³ = 6.36 pcf

effective ways is to assume the failure surfaces as composite so that the grid and search can be applied to locate the most critical failure surface.

Fig. A-4 shows an embankment, 30.5 m (100 ft) high, with a side slope of 3:1. The soil in the embankment has an effective cohesion of 9.6 kN/m^2 (200 psf), an effective friction angle of 35°, and a total unit weight of 19.7 kN/m^3 (125 pcf). The embankment is placed on a foundation soil 6.1 m (20 ft) thick, with an effective cohesion of 4.8 kN/m^2 (100 psf), an effective friction angle of 30°, and a total unit weight of 18.9 kN/m^3 (120 pcf). The water table is on the top of the foundation soil. Below the foundation soil is a thin layer of very weak soil, only 0.3 m (1 ft) in thickness, with a cohesion of zero, an effective friction angle of 10°, and a total unit weight of 17.4 kN/m^3 (110 pcf). Due to the presence of the weak soil, the failure surfaces will follow the bottom of the weak layer instead of cutting into soil 1. Determine the factor of safety.

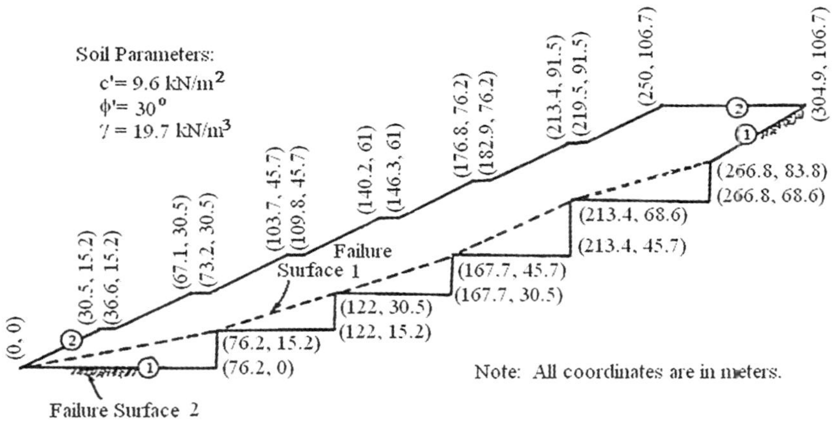

Fig. A-5. Noncircular failure surfaces
Note: 1 m = 3.28 ft; 1 kN/m² = 20.9 psf; 1 kN/m³ = 6.36 pcf

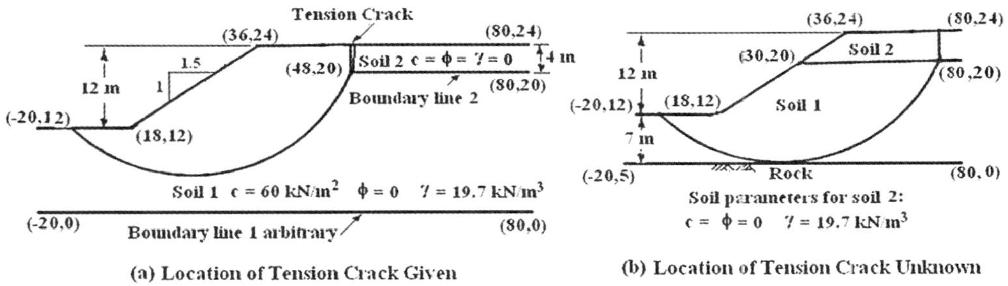

Fig. A-6. Cut slope with tension crack
Note: 1 m = 3.28 ft; 1 kN/m² = 20.9 psf; 1 kN/m³ = 6.36 pcf

2D Example 5: Noncircular Failure Surfaces

Fig. A-5 shows a fill placed on a series of rock benches. The fill material has an effective cohesion of 9.6 kN/m² (200 psf), an effective friction angle of 30°, and a total unit weight of 19.7 kN/m³ (125 pcf). Two noncircular failure surfaces are assumed. The much shorter and smoother failure surface 1, as indicated by the dashed line, is believed to be more critical than failure surface 2, which zigzags along the surface of the benches by following boundary line 1. Compute the static and seismic (seismic coefficient = 0.1) factors of safety for both failure surfaces and determine which surface is more critical.

2D Example 6: Cut Slope with a Tension Crack

This example consists of two different cases: (1) Fig. A-6(a) shows an existing cut slope 12 m (40 ft) high with the depth and location of a tension crack as given. The soil has a cohesion of 60 kN/m² (1,250 psf), a friction angle of 0°, and a total

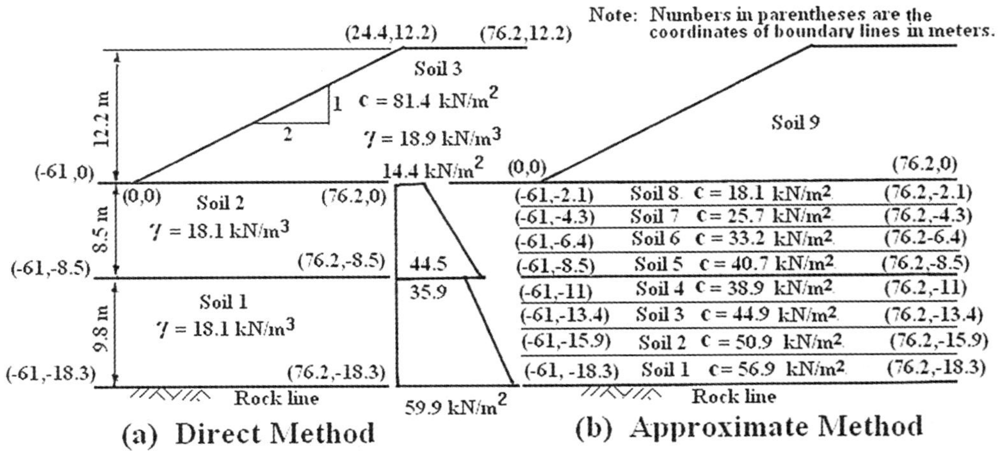

Fig. A-7. Undrained shear strength increasing linearly with depth
Note: 1 m = 3.28 ft; 1 kN/m² = 20.9 psf; 1 kN/m³ = 6.36 pcf

unit weight of 19.7 kN/m^3 (125 pcf). If the circular failure surface passes through the bottom of the tension crack, determine the factor of safety when the tension crack is dry and also when the tension crack is filled with water; (2) Fig. A-6(b) shows a proposed cut slope with the same soil and outside configuration as those shown in Fig. A-6(a) but the rock is located at 7 m (23 ft) below the toe. The predicted depth of the tension crack is 4 m (13 ft), but the slope has not been constructed, and the location of the tension crack is unknown. Determine the factor of safety and the location of the tension crack.

2D Example 7: Undrained Strength Increasing Linearly with Depth

An embankment is placed on a foundation consisting of two layers of clay. The dimensions of the cross section, together with the undrained shear strength and the unit weight of the soils, are shown in Fig. A-7(a). The results of Dutch cone tests indicate that the undrained shear strength of each clay layer varies linearly with depth, as shown by the trapezoidal distribution in the figure. The undrained shear strengths for soil 1 are 35.9 kN/m^2 (750 psf) at the top and 59.9 kN/m^2 (1250 psf) at the bottom, and those for soil 2 are 14.4 kN/m^2 (300 psf) at the top and 44.5 kN/m^2 (930 psf) at the bottom. Determine the factor of safety (1) using the direct method by considering the foundation as two layers, and (2) using the approximate method by dividing each clay layer into four sublayers, as shown in Fig. A-7(b).

2D Example 8: Embankment with Cohesionless Granular Materials

Fig. A-8(a) shows an embankment placed directly on a rock foundation. The soil in the embankment is cohesionless with φ_o = 39°, $\Delta\varphi$ = 7°, and γ = 22.1 kN/m^3

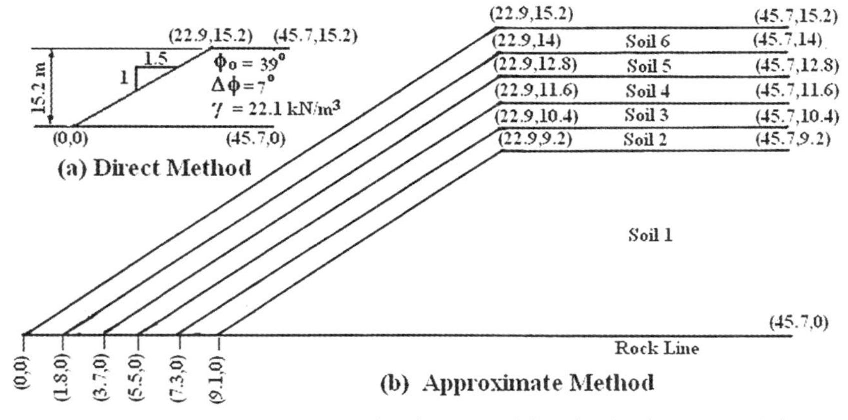

Fig. A-8. Embankment with cohesionless materials
Note: 1 m = 3.28 ft; 1 kN/m² = 20.9 psf; 1 kN/m³ = 6.36 pcf

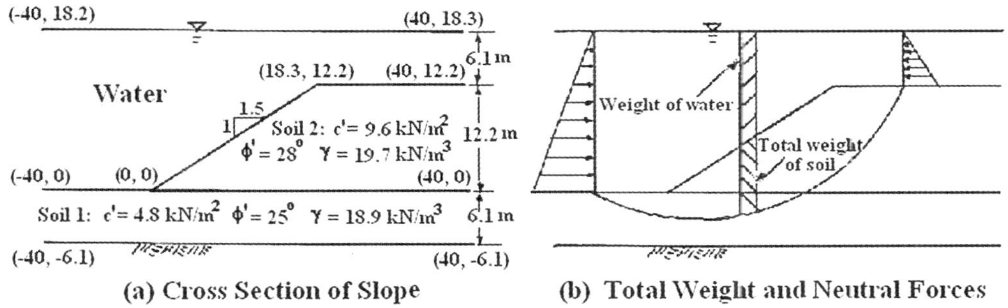

Fig. A-9. Analysis of submerged slope
Note: 1 m = 3.28 ft; 1 kN/m² = 20.9 psf; 1 kN/m³ = 6.36 pcf

(140 pcf). The dimensions of the embankment are shown in the figure. Determine the factor of safety (1) using the direct method by considering the entire embankment as one soil, and (2) using the approximate method by dividing the soil near to the slope surface into a number of sublayers, as shown in Fig. A-8(b).

2D Example 9: Analysis of Submerged Slope

If a slope is submerged under water, as in the case of underwater excavations, a general practice is to ignore the water table and use the submerged weight. Another method, as used in LEAME to solve seepage problems, is to consider the water table as a phreatic surface and use the total weight. The purpose of this example is to determine the factors of safety in both cases. If both cases check closely, the correctness of LEAME in analyzing seepage is further validated.

Fig. A-9(a) shows the cross section of a submerged embankment with a height of 12.2 m (40 ft) and a slope of 1.5:1. The top of the embankment is 6.1 m

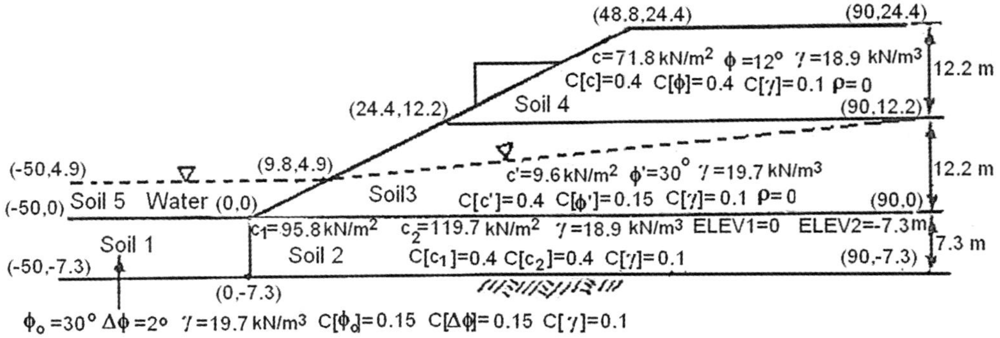

Fig. A-10. *Example on reliability analysis*
Note: 1 m = 3.28 ft; 1 kN/m² = 20.9 psf; 1 kN/m³ = 6.36 pcf

(20 ft) below the water table, and the soil parameters are shown in the figure. Determine the factor of safety for the following two cases: (1) using the submerged weight and ignoring the water table, and (2) using the total weight and boundary neutral forces.

2D Example 10: Example on Reliability Analysis

Although reliability can be determined in all the previous examples by simply specifying PROB as 1, 2, or 3 (1 for high variability, 2 for medium variability, and 3 for low variability), this capability has not been utilized so far. Because of the lack of experience and dependable data, it is anticipated that the users of LEAME will be more interested in the factor of safety than in reliability. This example is the only one in which reliability will be discussed.

Fig. A-10 shows a slope with a phreatic surface and five different soils. Soil 1 is a granular soil with a curved envelope; soil 2 is a clayey soil with an undrained shear strength increasing linearly with depth; soils 3 and 4 are conventional soils with a straight-line envelope; and soil 5 is water with neither cohesion nor internal friction. The parameters for soils 1 and 3 are in terms of effective stress and those for soils 2 and 4 in terms of total stress. The inclusion of all possible types of soil in the same site may not be realistic but is helpful for illustrative purposes. All soil parameters, including the coefficients of variation, are shown in the figure. Determine the factor of safety and the reliability of the design.

LEAME for Three-Dimensional Analysis

To illustrate the capability of LEAME software for 3D analysis, seven examples are presented here. Detailed solutions using the LEAME software are available, with commentary, in the *User's Manual*.

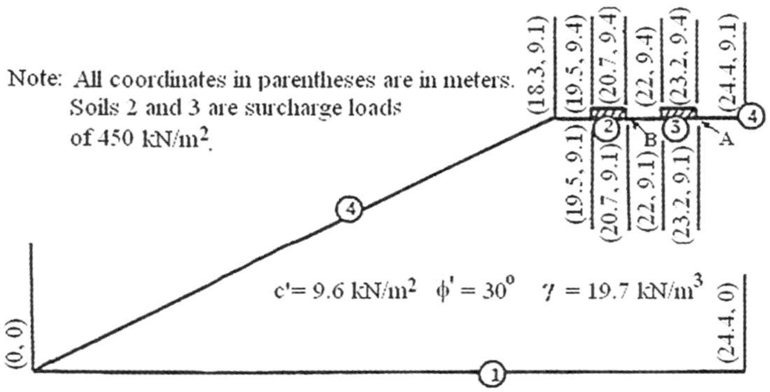

Fig. A-11. *Heavy surcharge loads of limited length*
Note: 1 m = 3.28 ft; 1 kN/m² = 20.9 psf; 1 kN/m³ = 6.36 pcf

3D Example 1: Heavy Surcharge Loads of Limited Length

This example illustrates the application of 3D analysis with ellipsoidal ends to a slope subjected to a heavy load over a limited area. A case in view is the safety to pass extraordinary heavy equipment over an embankment. In 2D analysis, it is assumed that the load and the failure surfaces are infinitely long in the longitudinal direction. This assumption is very conservative and may result in an unsatisfactory factor of safety. If the embankment fails under the load, the failure mass must be spoon-shaped with a limited length. Therefore, the use of 3D analysis is more realistic.

Fig. A-11 shows an embankment subjected to two heavy surcharge loads, each 1.2 m (4 ft) wide with a length of 3 m (9.8 ft) and an intensity of 480 kN/m² (10,000 psf). Soil 1 for the fill has an effective cohesion of 9.6 kN/m² (200 psf), an effective friction angle of 30°, and a total unit weight of 19.7 kN/m³ (125 pcf). Soils 2 and 3 for the surcharge loads are assumed to have a thickness of 0.3 m (1 ft) and a unit weight of 1,600 kN/m³ (10,000 pcf), which is equivalent to a surcharge load of 480 kN/m² (10,000 psf). If the embankment is 18 m (59 ft) long, by the use of LEAME, determine the factors of safety for both 2D and 3D analyses in the same run.

The results of the analysis based on the simplified Bishop method show that the factor of safety is 0.949 for 2D analysis and 1.121 for 3D analysis.

3D Example 2: Failure Surfaces with Planar Ends

If failures occur in a high embankment across a narrow valley with parallel rock banks, two possible types of 3D failures may take place, depending on the interfacial shear strength between the embankment and the rock bank. If the interfacial strength is low, failures will occur along the interface, as well as within the embankment, so the case of 3D analysis with planar ends applies. If the interfacial strength is high, failures will not occur along the interface but will be confined within the embankment only, so the case of 3D analysis with ellipsoidal ends prevails. The former case is illustrated in this example, whereas the latter is in the next example.

Fig. A-12 shows the cross section of an embankment across a narrow valley with steep rock banks. Below the embankment is a thin layer of soft soil, or soil 2. To generate composite failure surfaces, an imaginary rock line, as indicated by the dashed line, is assumed arbitrarily. Because no circles are allowed to pass below the weak layer, all the soil parameters for soil 1 can be assumed 0. However, to avoid the accidental cutting through soil 1 due to improper centers, LEAME will change the cohesion of soil 1 to a large value, say 10,000. This change has no effect on the computed results but will give those circles cutting through soil 1 a very large factor of safety and thus eliminate them from further consideration. The soil parameters are shown in the figure.

The input parameters for end planes are HTW (half length of top width) = 210 m (689 ft), PNTW (point number on ground line for defining top width) = 4, ANEP (angle of end plane) = 0, and SLEP (slope of end plane) = 3. Note that SLEP is the ratio between vertical and horizontal (not between horizontal and vertical), and a value of 3 is a steep slope and should not be confused with the much flatter slope of 3:1. Because the factor of safety depends strongly on the shear strength at the end plane, two different values of SSEP (Shear Strength at End Plane) are assumed: a rough soil-rock interface with SSEP = 1 and a smooth soil-rock interface with SSEP = 0.67. The use of SSEP = 1 indicates that the shear strength of the soils in direct contact with the end plane is considered as the shear strength at the end plane, whereas the use of SSEP = 0.67 implies that only two-thirds of the shear strength of the soils above the end plane is used, as in the design of retaining walls by assuming the wall friction equal to two-thirds of the soil friction. Determine the factors of safety for SSEP of 1 and 0.67.

The factors of safety based on the original Spencer method with SSEP of 1 and 0.67 are 1.640 and 1.569, respectively. The factor of safety for 2D analysis based on the original Spencer method is 1.384. It can be seen that the factor of safety for 3D analysis is much greater than the 2D analysis.

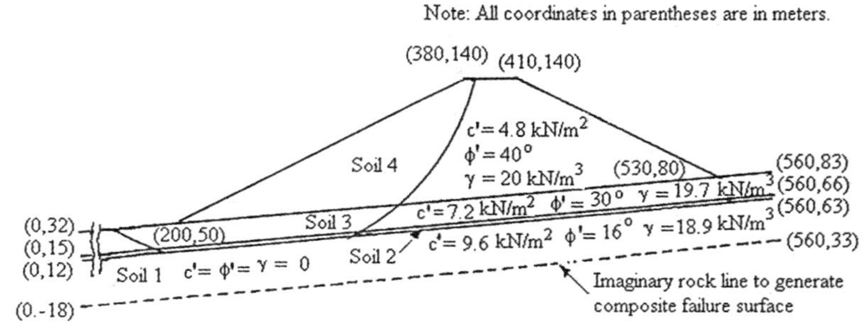

Fig. A-12. *Cross section of an embankment across a narrow valley*
Note: $1\,m = 3.28\,ft$; $1\,kN/m^2 = 20.9\,psf$; $1\,kN/m^3 = 6.36\,pcf$

3D Example 3: Failure Surfaces with Ellipsoidal Ends

The previous example assumes that the failure surface occurs on the end planes. A question immediately arises: Is the factor of safety based on planar ends lower than that based on ellipsoidal ends? In other words, will the failure occur on the end plane rather than inside the embankment? This example will shed some light on this question.

The cross section and soil parameters used for this example are the same as those in the previous example. If the failure surface is cylindrical with ellipsoidal ends, determine the factor of safety. The result of analysis shows that the factor of safety based on the original Spencer method is 1.430, which is smaller than in the previous example.

3D Example 4: Landfill with Geotextiles and Noncircular Failure Surface

This example illustrates the application of 3D analysis with planar ends to a landfill having a weak layer at the bottom. The use of geosynthetic materials, such as geotextiles, geomembranes, or geosynthetic clay liners at the bottom has posed new problems to the stability of landfills. These materials have a very low friction angle and can cause failures to occur through these weaker materials. When such fills are placed in a hollow with steep slopes on three sides, the factor of safety based on 3D analysis may be smaller than that based on the conventional 2D analysis using the most critical cross section at the center. The ability to analyze landfills in three dimensions is an outstanding and important feature of LEAME.

Fig. A-13 shows a landfill with three different materials. The rock toe is constructed of granular materials with an effective cohesion of 4.8 kN/m^2 (100 psf), an effective friction angle of 32°, and a total unit weight of 18.9 kN/m^3 (120 pcf). Layers of geotextiles are placed at the bottom of the fill above a clay liner to facilitate construction and provide drainage. To simulate the very small friction angle between geotextiles, a thin layer of material, say 0.1 m (4 in.) thick with a friction angle of only 9° and a unit weight of 17.4 kN/m^3 (110 pcf), is placed above the clay liner. Because a weak layer exists at the bottom of the fill, all failure surfaces will lie along the bottom of the weak layer, so all the materials below the weak layer, including the clay liner, are immaterial and need not be considered in the stability analysis. The waste material, or soil 3, above the geotextiles has a cohesion of 9.6 kN/m^2 (200 psf), an effective friction angle of 22°, and a total unit weight of 17.4 kN/m^3 (110 pcf). The end plane is defined by the following parameters: HTW = 61 m (200 ft), PNTW = 4, SLEP = 0.5, and ANEP = 20°.

Two potential failure surfaces are assumed. The first failure surface assumes that the failure is along the bottom of the fill, starting from (22.9, 38.1) and ending at (201.2, 59.5), as shown in Fig. A-13. The coordinates of the failure surface are the same as those of boundary line 2, so only the shear strength of soil 2 with a friction angle of 9° and the total unit weight of 17.4 kN/m^3 (110 pcf) for soils 2 and 3 are used in the analysis.

To be sure that the failure surface does not cut horizontally through the rock toe, a second failure surface is assumed, starting from (7.6, 30.5), passing through (30.5, 30.5), and following boundary line 2 to (201.2, 59.5). Both failure surfaces can be analyzed by LEAME at the same time. The results of the analysis based on the original Spencer method show that failure surface 1 has a safety factor of 1.376, which is more critical than the 1.455 for failure surface 2. The factor of safety for 2D analysis is 1.541, which is much greater than the 3D analysis.

3D Example 5: Landfill with Geotextiles and Composite Failure Surfaces

In the previous example, it is quite possible that the most critical failure surface is a composite surface consisting of a noncircular surface near the toe and a circular surface in the interior, rather than a noncircular surface throughout the entire fill. The factor of safety obtained by the composite failure surface in this example will compare with the noncircular failure surface in the previous example to see which is more critical.

To generate a large number of composite surfaces, an imaginary boundary line 1 is added, and the ground line is extended, as shown by the dashed lines in Fig. A-14. The analysis by LEAME reveals that the minimum factor of safety based on the original Spencer method is 1.373 for composite failure surfaces, which is only slightly smaller than the 1.376 for the noncircular failure surfaces.

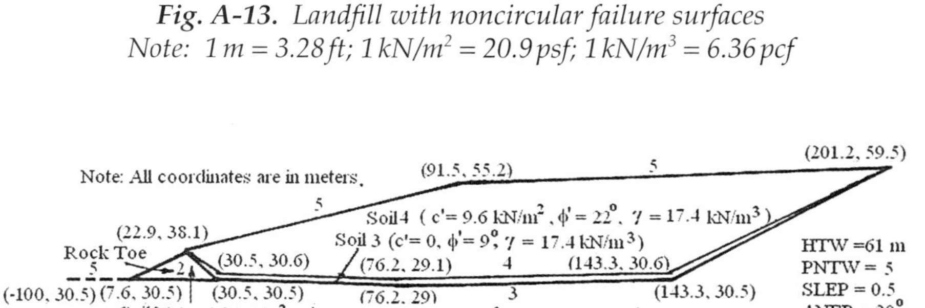

Fig. A-13. Landfill with noncircular failure surfaces
Note: 1 m = 3.28 ft; 1 kN/m² = 20.9 psf; 1 kN/m³ = 6.36 pcf

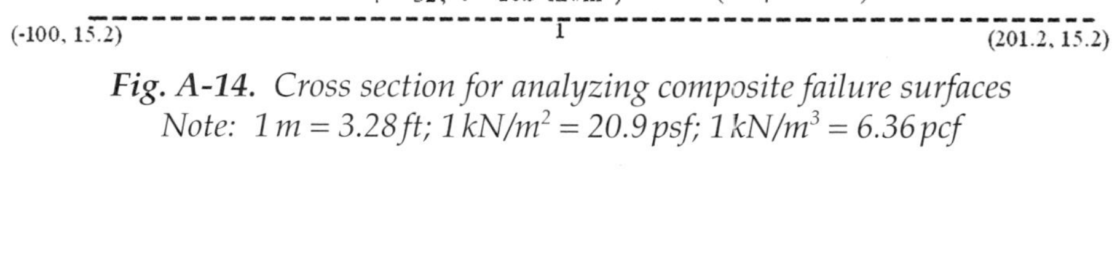

Fig. A-14. Cross section for analyzing composite failure surfaces
Note: 1 m = 3.28 ft; 1 kN/m² = 20.9 psf; 1 kN/m³ = 6.36 pcf

3D Example 6: Effect of Embankment Length on Factor of Safety

When an embankment is placed across a deep valley, such as an earth dam or a valley fill, the most probable failure surface will be spoon-shaped with a cylinder at the center and a half ellipsoid at each end. This type of failure surface can be analyzed by LEAME with ellipsoidal ends, as illustrated in 3D Example 3. The current practice of using the conventional 2D analysis to solve this 3D problem is unrealistic. In the 2D analysis, two assumptions are made. First, the most critical cross section at the center of the valley is used to represent all the cross sections along the length of the embankment. This assumption may not be true, because the height of the embankment usually decreases as the cross section becomes closer to the bank. The consideration of this variation in cross sections requires the provision of different cross sections along the length of the embankment and is too cumbersome and difficult to perform. This assumption, even though unreasonable, is still used in LEAME for 3D analysis, because it will give a more conservative factor of safety.

The second assumption is that the embankment and the failure surface are infinitely long, so a unit length of the embankment can be used to represent the entire embankment. In other words, a unit length of the embankment can slip out freely with no reaction or resistance from the adjacent fill. This complete ignorance of the end effect may be quite significant if the embankment is relatively short. The LEAME program with ellipsoidal ends is designed to consider this end effect. The purpose of this example is to illustrate how the length of the embankment affects the factor of safety obtained.

Fig. A-15 is the cross section of an earth embankment showing the coordinates of the boundary lines, location of the phreatic surface, and soil parameters used for analysis. The minimum factor of safety based on 2D analysis using the simplified Bishop method is 1.470, which is lower than the 1.5 required. If the total length of the embankment is 60 m (196.8 ft), determine the minimum factor of safety.

By varying the length of embankment, it was found that the factor of safety decreased as the length of embankment increased. Therefore, the assumption that the failure surface is most critical when the length of the failure mass is equal to the length of the embankment is valid.

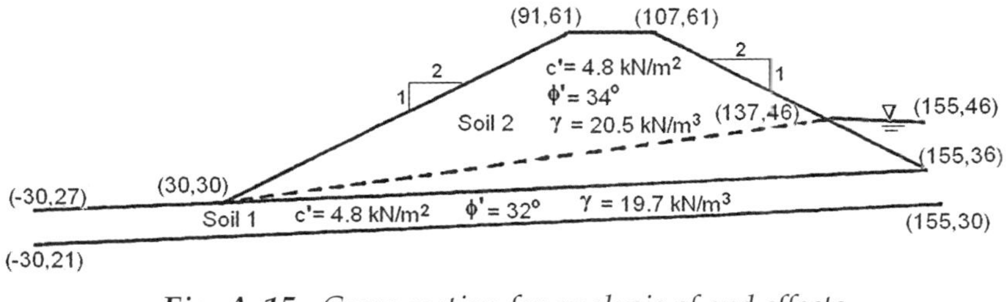

Fig. A-15. *Cross section for analysis of end effects*
Note: 1 m = 3.28 ft; 1 kN/m² = 20.9 psf; 1 kN/m³ = 6.36 pcf

3D Example 7: Effect of Bench Length on Factor of Safety

In 2D analysis, not only the failure surface but also the loading must be infinitely long. If a load is applied over a limited area, the 3D analysis with ellipsoidal ends can be used, as illustrated in 3D Example 1. The same principle can be applied to a short section of steep slope changing gradually to a much flatter slope. The section of steep slope can be considered as a heavy load with a HCL (half cylindrical length) equal to the half length of the steeper section. However, there are two major differences between 3D Example 1 and this case: (1) The half length of the failure mass (HLFM) is defined clearly as the half length of the embankment in 3D Example 1 but not in this case. Because the flatter slope is very stable with a high factor of safety, HLFM should be confined within the transitional and steep sections and not extended into the flatter section; and (2) In 3D Example 1, the cross section is the same throughout the embankment, whereas in this example the half ellipsoid is located in the transitional section with a gradually decreasing slope, so the assumption of the same steep slope for the transitional section is on the safe side with a lower factor of safety.

This method can be applied to the stability analysis of bench fills. In surface mining, a cut is made on a hillside to expose the coal seam, resulting in a bench and a highwall. The Surface Mining Control and Reclamation Act of 1977 requires the return of disturbed land to its original contours. Therefore, the bench created by surface mining must be backfilled to the original slope. If the original slope is quite steep, it may be difficult to achieve the required factor of safety. The factor of safety can be increased by 3D analysis.

Fig. A-16 shows the cross section of a bench fill together with the coordinates of boundary lines and soil parameters. The highwall and bench are mostly of rocky materials, so any failure, if ever present, will be limited within the backfill. The area is well drained, and there is no seepage. If the bench with the steepest section is only 30 m (98 ft) long, determine the factor of safety based on both 2D and 3D analyses.

The results of the analysis based on the simplified Bishop method show that the factor of safety is 1.474 for 2D analysis and 1.504 for 3D analysis.

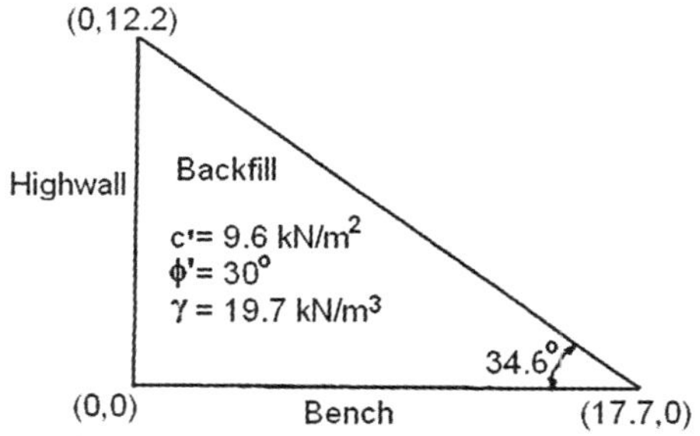

Fig. A-16. *Cross section of a bench fill*
Note: $1\,kN/m^2 = 20.9\,psf$; $1\,kN/m^3 = 6.36\,pcf$

Applications for Surface Mining

As mentioned in the Preface to this book, the original REAME software (Rotational Equilibrium Analysis of Multilayered Embankments) was developed in response to the Surface Mining Control and Reclamation Act of 1977, which requires the stability analysis of spoil banks, hollow fills, and refuse dams created by surface mining. In Chapter 4 of the LEAME *User's Manual*, 10 cases involving various methods of spoil and waste disposal from surface mining are presented to illustrate the practical applications of LEAME. Data files for these cases are included with the software and can be used to run LEAME and obtain the printed results. These examples are real cases that were analyzed by REAME and submitted to the regulatory agencies for the application of mining permits. Because the LEAME presented in this book is quite different from the original REAME, the stability analyses reported herein are not exactly the same as those in the original reports. However, the general procedures and conclusions are about the same.

References

Algermissen, S. T. (1969). *Seismic risk studies in the United States*, U.S. Coast & Geodetic Survey, National Oceanic and Atmospheric Admin., Silver Spring, MD.

Applied Technology Council. (1978). *Tentative provisions for the development of seismic regulations for buildings: A cooperative effort with the design profession, building code interests and the research community*, U.S. Government Printing Office, Washington, DC.

ASTM International. (2010). *Annual book of ASTM standards*, West Conshohocken, PA.

Baecher, G. B., and Christian, J. T. (2003). *Reliability and statistics in geotechnical engineering*, John Wiley & Sons, New York.

Bailey, W. A., and Christian, J. T. (1969). "ICES-LEASE-I: A problem-oriented language for slope stability analysis." *MIT Soil Mechanics Publ. No. 235*, Massachusetts Inst. of Technology, Cambridge, MA.

Beles, A. A., and Stanculescu, I. I. (1958). "Thermal treatment as a means of improving the stability of earth masses." *Geotechnique*, 8(4), 158–165.

Berg, R. R., Anderson, R. P., Rose, R. J., and Chouery-Curtis, V. E. (1990). "Reinforced soil highway slopes." *Transp. Res. Rec. 1288*, Transportation Research Board, Washington, DC, 107, Fig. 10.

Bishop, A. W. (1955). "The use of failure circle in the stability analysis of slopes." *Geotechnique*, 5(1), 7–17.

Bishop, A. W., and Henkel, D. J. (1957). *The measurement of soil properties in the triaxial test*, Edward Arnold, London.

Bishop A. W., and Morgenstern, N. (1960). "Stability coefficients for earth slopes." *Geotechnique*, 10(4), 129–150.

Bjerrum, L. (1972). "Embankments on soft ground." *Proc., ASCE Specialty Conf. on Performance of Earth and Earth-Support Structures*, Purdue University, ASCE, Reston, VA, II, 1–54.

Bjerrum, L., and Simons, N. E. (1960). "Comparison of shear strength characteristics of normally consolidated clays." *Proc., ASCE Specialty Conf. on Shear Strength of Cohesive Soils*, ASCE, Reston, VA, 711–726.

Bowles, J. E. (1984). *Physical and geotechnical properties of soils*, 2nd Ed., McGraw-Hill, New York.

Bray, J. D., Zekkos, D., Kavazanjian, E., Athanasopoulos, G. A., and Reimer, M. F. (2009). "Shear strength of municipal waste." *J. Geotech. Geoenv. Eng.*, 135(6), 709–722.

Broms, B. B., and Flodin, N. (1988). "History of soil penetration testing." *Proc., 1st Int. Symp. on Penetration Testing*, Orlando, FL, 1, 157–220.

Casagrande, A. (1937). "Seepage through dams." *Contributions to soil mechanics*, BSCE (1925–1940.) Paper first published in *J. New Engl. Water Works Assoc.*, June 1937, 295–336.

Casagrande L., Loughney, R. W., and Matich, M. A. J. (1961). "Electro-osmotic stabilization of a high slope in loose saturated silt." *Proc., 5th Int. Conf. on Soil Mechanics and Foundation Engineering*, International Society of Soil Mechanics and Geotechnical Engineering (ISSMGE), London, 2, 555–561.

Casagrande, L., Wade, N., Wakely, M., and Loughney, R. (1981). "Electro-osmosis projects, British Columbia, Canada." *Proc., 10th Int. Conf. on Soil Mechanics and Foundation Engineering*, International Society of Soil Mechanics and Geotechnical Engineering (ISSMGE), London, 3, 607–610.

Cedergren, H. R. (1977). *Seepage, drainage, and flow nets*, 3rd Ed., John Wiley & Sons, New York.

Chowdhury, R. N. (1980). "Landslides as natural hazards: Mechanisms and uncertainties." *Geotech. Eng., SE Asian Soc. of Soil Eng.*, 11(2), 135–180.

Christian, J. T., and Baecher, G. B. (2011). "Unresolved problems in geotechnical risk and reliability." *Geotechnical Risk Assessment & Management*, GeoRisk 2011 Conference, ASCE, Reston, VA, 45–63.

Christian, J. T., Ladd, C. C., and Baecher, G. B. (1994). "Reliability applied to slope stability analysis." *J. Geotech. Eng.*, 120(12), 2180–2207.

Chugh, A. K. (1981). *User information manual, slope stability analysis, Program SSTAB2*, U.S. Dept. of Interior, Bureau of Reclamation, Washington, DC.

Cornell, C. A. (1971). "First-order uncertainty analysis of soils deformation and stability." *Proc., 1st Int. Conf. on Applications of Probability and Statistics in Soil and Structural Engineering*, International Society of Soil Mechanics and Geotechnical Engineering (ISSMGE), London, 129–144.

Cruden, D. M., and Varnes, D. J. (1996). "Landslide types and processes." *Landslides: Investigations and mitigations, Special Rep. 247*, Transportation Research Board, Washington, DC, 36–75.

D'Appolonia Consulting Engineers, Inc. (1975). *Engineering and design manual: Coal refuse disposal facilities*, Mining Enforcement and Safety Administration, U.S. Department of Interior, Washington, DC.

Dennis, T. H., and Allan, R. J. (1941). "Slide problem: Storms do costly damages on state highways yearly." *California Highways and Public Works*, 20, 1–3.

Drnevich, V. P., Gorman, T., and Hopkins, T. C. (1974). "Shear strength of cohesive soils and friction sleeve resistance." *Proc., European Symp. on Penetration Testing*, National Swedish Institute for Building Research, Stockholm, 2:2, 129–132.

Duncan, J. M. (1996). "State of the art: Limit equilibrium and finite-element analysis of slopes." *J. Geotech. and Geoenv. Eng.*, 122(7), 577–596.

Duncan, J. M., and Wright, S. G. (2005). *Soil strength and slope stability*, John Wiley & Sons, New York.

Eid, H. T., Evans, W. D., and Sherry, P. E. (2000). "Municipal solid waste slope failure: Waste and foundation soil properties." *J. Geotech. and Geoenv. Eng.*, 126(5), 391–407.

Federal Highway Administration (FHWA). (2000). "Mechanically stabilized earth walls and reinforced soil slope: Design and construction guidelines." *Rep. FHWA-NHI-00–043*, Washington, DC.

Federal Register. (1977). "Mineral resources." Part II, Title 30, Chapter VII, Part 715, Office of Surface Mining Reclamation and Enforcement, U.S. Dept. of Interior, Washington, DC.

Fellenius, W. (1936). "Calculation of the stability of earth dams." *Trans., 2nd Congress on Large Dams*, U.S. Government Printing Office, Washington, DC, 4, 445–462.

Fredlund, D. G., and Dahlman, A. E. (1972). "Statistical geotechnical properties of glacial Lake Edmonton sediments." *Proc., Int. Conf. on Statistics and Probability in Civil Engineering*, Hong Kong Univ. Press, Hong Kong (also distributed by Oxford Univ. Press, London).

Gedney, D., and Weber, W. (1978). "Design and construction of soil slopes." Chapter 8, *Landslides: Analysis and control, Special Rep. 176*, R. L. Schuster and R. J. Krizek, eds., Transportation Research Board, Washington, DC.

Gilboy, G. (1933). "Hydraulic-fill dams." *Proc., Int. Comm. on Large Dams*, World Power Conference, Stockholm.

Gray, D. H., and Leiser, A. T. (1982). *Biotechnical slope protection and erosion control*, Van Nostrand Reinhold, New York.

Gray, D. H., and Leiser, A. T. (1992). "Biotechnical stabilization of cut and fill slopes." *Proc., Stability and Performance of Slopes and Embankments: II*, Berkeley, CA, *ASCE Geotech. Spec. Publ. No. 31*, Reston, VA, 1395–1410.

Gray, D. H., and Leiser, A. T. (1995). "Biotechnical stabilization of steep slopes." *Transp. Res. Rec.1474*, Transportation Research Board, Washington, DC.

Handy, R. L., and Williams, W. W. (1967). "Chemical stabilization on an active landslide." *Civil Eng.*, 37(8), 62–65.

Hansen, W. R. (1965). "Effect of the earthquake of March 27, 1964 at Anchorage, Alaska." *USGS Professional Paper 542-A*, U.S. Geological Survey, Reston, VA.

Harr, M. E. (1962). *Groundwater and seepage*, McGraw-Hill, New York.

Harr, M. E. (1977). *Mechanics of particulate media*, McGraw-Hill, New York.

Haugen, J. J., and DiMillio, A. F. (1974). "A history of recent shale problems in Indiana." *Highway Focus*, 6(3), 15–21.

Henkel, D. J. (1967). "Local geology and the stability of natural slopes." *J. Soil Mech. and Found. Div. ASCE*, 93(SM4), 437–446.

Hill, R. A. (1934). "Clay stratum dried out to prevent landslides." *Civil Eng.*, 4(8), 403–407.

Hopkins, T. C., Allen, D. L., and Deen, R. C. (1975). "Effect of water on slope stability." *Res. Rep. 435*, Div. of Research, Kentucky Dept. of Transportation, Frankfort, KY.

Hovland, H. J. (1977). "Three-dimensional slope stability analysis method." *J. Geotech. Eng. Div. ASCE*, 103(GT9), 971–986.

Huang, Y. H. (1975). "Stability charts for earth embankments." Fig. 1, p. 4 and Fig. 4, p. 7, *Transp. Res. Rec. 548*, Transportation Research Board, Washington, DC.

Huang, Y. H. (1977a). "Stability of mine spoil banks and hollow fills." *Proc., Conf. on Geotechnical Practice for Disposal of Solid Waste Materials*, Ann Arbor, MI, Specialty Conference of the Geotechnical Engineering Division, ASCE, Reston, VA, 407–427.

Huang, Y. H. (1977b). "Stability coefficients for sidehill benches." *J. Geotech. Eng. Div. ASCE*, 103(GT5), 467–481.

Huang, Y. H. (1978a). "Stability charts for sidehill fills." *J. Geotech. Eng. Div. ASCE*, 104(GT5), 659–663.

Huang, Y. H. (1978b). "Stability of spoil banks and hollow fills created by surface mining." *Res. Rep. IMMR34-RRR1-78*, Inst. for Mining and Mineral Research, Univ. of Kentucky, Lexington, KY.

Huang, Y. H. (1979). "Stability charts for refuse dams." *Proc., 5th Kentucky Coal Refuse Disposal and Utilization Seminar and Stability Analysis of Refuse Dam Workshop*, Univ. of Kentucky, Lexington, KY, 57–65.

Huang, Y. H. (1980). "Stability charts for effective stress analysis of nonhomogeneous embankments." *Transp. Res. Rec. 749*, Transportation Research Board, Washington, DC, 72–74.

Huang, Y. H. (1981). "Line of seepage in earth dams on inclined ledge." *J. Geotech. Eng. Div. ASCE*, 107(GT5), 662–667.

Huang, Y. H. (1983). *Stability Analysis of Earth Slopes*, Van Nostrand Reinhold Company, New York.

Huang, Y. H. (1986). "Unsteady-state phreatic surface in earth dams." *J. Geotech. Eng. Div. ASCE*, 112(1), 93–98.

Huang, Y. H., and Avery, M. C. (1976). "Stability of slopes by the logarithmic-spiral method." *J. Geotech. Eng. Div. ASCE*, 102(GT1), 41–49.

Hungr, O., Salgado, F. M., and Byrne, P. M. (1989). "Evaluation of a three-dimensional method of stability analysis." *Canadian Geotech. J.*, 26, 679–686.

Hunter, J. H., and Schuster, R. L. (1971). "Chart solutions for stability analysis of earth slopes." *Highway Res. Rec. 345*, Highway Research Board, Washington, DC, 77–89.

Hutchinson, J. N. (1988). "General report: Morphological and geotechnical parameters of landslides in relation to geology and geohydrology." *Proc., 5th Int. Symp. on Landslides*, Rotterdam, International Society of Soil Mechanics and Geotechnical Engineering (ISSMGE), London, 3–35.

International Business Machines Corp. (IBM). (1970). *System/360 scientific subroutine package, version III, programmer's manual*, Armonk, NY.

Janbu, N. (1954). "Application of composite failure surface for stability analysis." *Proc., European Conf. on Stability of Earth Slopes*, International Society of Soil Mechanics and Geotechnical Engineering (ISSMGE), London.

Janbu, N. (1973). "Slope stability computation." *Embankment-dam engineering, Casagrande volume*, R. C. Hirschfeld and S. J. Poulos, eds., John Wiley & Sons, New York, 47–86.

Janbu, N., Bjerrum, L., and Kjaernsli, B. (1956). "Soil mechanics applied to some engineering problems." *Publ. 16*, Norwegian Geotechnical Inst., Oslo.

Karlsson, R., and Viberg, L. (1967). "Ratio c/p' in relation to liquid limit and plasticity index, with special reference to Swedish clays." *Proc.,*

Geotechnical Conf. International Society of Soil Mechanics and Geotechnical Engineering (ISSMGE), London, 1, 43–47.

Kavazanjian, E., Jr. (2001). "Mechanical properties of municipal solid waste." *Proc., 8th Int. Waste Management and Land Filling Symp.*, International Society of Soil Mechanics and Geotechnical Engineering (ISSMGE), London, 415–424.

Kavazanjian, E., Jr., Matasovic, N., Bonaparte, R., and Schmertmann, G. R. (1995). "Evaluation of MSW properties for seismic analysis." *Geotech. Special Publ. 46*, ASCE, Reston, VA, 1126–1141.

Kenney, T. C. (1959). "Discussion." *J. Soil Mech. and Found. Div. ASCE*, 85(SM3), 67–79.

Kulhawy, F. H., and Mayne, P. W. (1990). "Manual on estimating soil properties for foundation design." *Rep. EPRI EL-6800*, Electric Power Research Inst., Palo Alto, CA.

Ladd, C. C., and Foott, R. (1974). "New design procedure for stability of soft clays." *J. Geotech. Eng. Div. ASCE*, 100(GT7), 763–786.

Ladd, C. C., Moh, Z. C., and Gifford, D. G. (1972). "Statistical analysis of undrained strength of soft Bangkok clay." *Proc., Int. Conf. on Statistics and Probability in Civil Eng.*, Hong Kong Univ. Press, Hong Kong (also distributed by Oxford Univ. Press, London).

Lambe, T. W., and Whitman, R. V. (1969). *Soil mechanics*, John Wiley & Sons, New York.

Lumb, P. (1966). "Variability of Natural Soils." *Canadian Geotech. J.*, 3(2), May, 1966.

Lumb, P. (1972). "Precision and accuracy of soil tests." *Proc., Int. Conf. on Statistics and Probability in Civil Engineering*, Hong Kong Univ. Press, Hong Kong (also distributed by Oxford Univ. Press, London).

Leshchinski, D., and San, K. (1994). "Pseudostatic seismic stability of slopes: Design charts." *J. Geotech. Eng. Div. ASCE*, 120(9), 1514–1532.

Lowe, J., and Karafiath, L. (1959). "Stability of earth dams upon drawdown." *Proc., 1st Pan American Conf. on Soil Mechanics and Foundation Engineering*, International Society of Soil Mechanics and Geotechnical Engineering (ISSMGE), London, 2, 537–552.

Makdisi, F. I., and Seed, H. B. (1978). "A simplified procedure for estimating dam and embankment earthquake-induced deformations." *J. Geotech. Eng. Div. ASCE*, 107(7), 849–867.

Marachi, N. D., Chan, C. K., Seed, H. B., and Duncan, J. M. (1969). "Strength and deformation characteristics of rockfill materials." *Rep. TE 69-5*, Dept. of Civil Engineering, Univ. of California, Berkeley, CA.

Matsuo, M., and Kuroda, K. (1974). "Probabilistic approach to design of embankments." *Soils and Foundations*, Japanese Soc. of Soil Mech. and Foundations Eng., 14(2).

Mayne, P. W., and Campanella, R. G. (2005). "Versatile site characterization by seismic piezocone." *Proc., 16th Int. Conf. on Soil Mechanics and Geotechnical Engineering*, International Society of Soil Mechanics and Geotechnical Engineering (ISSMGE), London, 2, 721–724.

Mendez, C. (1971). "Computerized slope stability: The sliding block problem." *Tech. Rep. No. 21*, Purdue Univ. Water Resources Research Center, Lafayette, IN.

Meyerhof G. G. (1956). "Penetration tests and bearing capacity of piles." *J. Soil Mech. and Found. Div. ASCE*, 82(1), paper 886.

Mines Branch, Canada. (1972). *Tentative design guide for mine waste embankments in Canada*. Natural Resources Canada (formerly Dept. of Energy, Mines, and Resources), Ottawa, Ontario, Canada.

Morgenstern, N. R. (1963). "Stability charts for earth slopes during rapid drawdown." *Geotechnique*, 13(2), 121–131.

Morgenstern, N., and Price, V. E. (1965). "The analysis of the stability of general failure surfaces." *Geotechnique*, 15(1), 79–93.

Morse, R. K. (1972). "The importance of Proper Soil Units for Statistical Analysis," *Int. Conf. on Statistics and Probability in Civ. Eng.*, Hong Kong University Press, Hong Kong, also distributed by Oxford University Press, London.

National Coal Board. (1970). *Spoil heaps and lagoons: Technical handbook*, London.

Neumann, F. (1954). *Earthquake intensity and related ground motion*, Univ. of Washington Press, Seattle, WA.

O'Colman, E., and Trigo, R. J. (1970). "Design and construction of tie-back sheet pile wall." *Highway Focus*, 2(5), 63–71.

Oka, Y., and Wu, T. (1990). "System reliability of slope stability." *J. Geotech. Eng. Div. ASCE*, 116(8), 1185–1189.

Peck, R. B. (1967). "Stability of natural slopes." *J. Soil Mech. and Found. Div. ASCE*, 93(SM4), 403–417.

Peck, R. B., and Ireland, H. O. (1953). "Investigation of stability problems." *Bull. 507*, American Railway Engineering Assn., Chicago, 1116–1128.

Petersen, M. D., et al. (2008). *2008 United States national seismic hazard maps*, U.S. Geological Survey Fact Sheet, U.S. Geological Survey, Reston, VA.

Robertson, P. K., and Campanella, R. G. (1983). "Interpretation of cone penetration tests. 1: Sands." *Canadian Geotech. J.*, 20(4), 718–733.

Root, A. W. (1958). "Prevention of landslides." Fig. 8.12, p. 179, *Landslides and engineering practice, Special Rep. 29*, E. B. Eckel, ed., Highway Research Board, Washington, DC.

Royster, D. L. (1974). "Construction of a reinforced earth fill along I-40 in Tennessee." *Proc., 25 Highway Geology Symp.*, International Society of Soil Mechanics and Geotechnical Engineering (ISSMGE), London, 76–93.

Schmertmann, J. H. (1975). "Measurement of in-situ shear strength: State of the art review." *Proc., ASCE Specialty Conf. on In-Situ Measurement of Soil Properties*, North Carolina State Univ., ASCE, Reston, VA, 2, 57–138.

Schultze, E. (1972). "Frequency distributions and correlations of soil properties." *Proc., Int. Conf. on Statistics and Probability in Civil Engineering*, Hong Kong Univ. Press, Hong Kong (also distributed by Oxford Univ. Press, London).

Schultze, E. (1975). "The general significance of statistics for the civil engineer." *Proc., 2nd Int. Conf. on Application of Statistics and Probability to Soil and*

Structural Engineering, International Society of Soil Mechanics and Geotechnical Engineering (ISSMGE), London.

Schuster, R. L., and Krizek, R. J. (1978). "Landslides analysis and control." *Special Rep. 176*, Transportation Research Board, Washington, DC.

Seed, H. B., Idriss, I. M., Lee, K. L., and Makdisi, F. I. (1975a). "Dynamic analysis of the slide in the lower San Fernando dam during the earthquake of February 9, 1971." *J. Geotech. Eng. Div. ASCE*, 101(GT9), 889–911.

Seed, H. B., Lee, K. L., Idriss, I. M., and Makdisi, F. I. (1975b). "The slides in the San Fernando dams during the earthquake of February 9, 1971." *J. Geotech. Eng. Div. ASCE*, 101(GT7), 651–688.

Senneset K., Sandven, R., and Janbu, N. (1989). "Evaluation of soil parameters from piezocone tests." *Transp. Res. Rec. 1235*, Table 2, p. 28; Fig. 8, p. 30; and Fig. 9, p. 31, Transportation Research Board, Washington, DC.

Shannon and Wilson, Inc. (1968). *Slope stability investigation vicinity of Prospect Park, Trunk Highway No. 94, Contract No. 12605.* Report prepared for State of Minnesota Department of Highways, Project No. W-1311-01, Shannon and Wilson, Inc., Seattle, WA.

Shen, C. K., Bang, S., Herrmann, L. R., and Romstad, K. M. (1978). "A reinforced lateral earth support system." *Proc., Symp. on Earth Reinforcement, Pittsburgh*, ASCE, Reston, VA, 764–793.

Shen, C. K., Bang, S., Romstad, K. M., Kulchin, L., and Denatale, J. S. (1981a). "Field measurements of an earth support system." *J. Geotech. Eng. Div. ASCE*, 107(GT12), 1625–1642.

Shen, C. K., Herrmann, L. R., Romstad, Bang, S., Kim, Y. S., and Denatale, J. S. (1981b). "In situ earth reinforcement lateral support system." *Rep. No. 81-03*, Dept. of Civil Engineering, Univ. of California, Davis, CA.

Singh, A. (1972). "How reliable is the factor of safety in foundation engineering?" *Proc., Int. Conf. on Statistics and Probability in Civil Engineering*, Hong Kong Univ. Press, Hong Kong (also distributed by Oxford Univ. Press, London).

Skempton, A. W. (1964). "Long-term stability of clay slopes." *Geotechnique*, 14(2), 77–101.

Smith T. W., and Forsyth, R. A. (1971). "Potrero Hill slide and correction." *J. Soil Mech. and Found. Div. ASCE*, 97(SM3), 541–564.

Smith, T. W., and Cedergren, H. R. (1962). "Cut slope design in field testing of soils and landslides." *ASTM Special Tech. Publ. 322, Field Testing of Soils*, ASTM International, West Conshohocken, PA, 135–158.

Sowers, G. F. (1979). *Introductory soil mechanics and foundations: Geotechnical engineering*, 4th Ed., Macmillan, New York.

Spencer, E. (1967). "A method of analysis of the stability of embankments assuming parallel inter-slice forces." *Geotechnique*, 17(1), 11–26.

Spencer, E. (1973). "Thrust line criterion in embankment stability analysis." *Geotechnique*, 23(1), 85–100.

Stark, T., and Eid, H. (1994). "Drained residual strength of cohesive soils." *J. Geotech. Eng. Div. ASCE*, 120(5), 856–871.

Taylor, D. W. (1937). "Stability of Earth Slopes." *J. Boston Society Civ. Eng.*, 24, 197–246.

Taylor, D. W. (1948). *Fundamentals of soil mechanics*, John Wiley & Sons, New York.
Terzaghi, K., and Peck, R. B. (1967). *Soil mechanics in engineering practice*, John Wiley & Sons, New York.
Turner, A. K., and Schuster, R. L., eds. (1996). "Landslides: Investigation and mitigation." *Special Rep. 247*, Transportation Research Board, Washington, DC.
U.S. Army Corps of Engineers (USACE). (1970). "Engineering and design: Stability analysis of earth and rock-filled dams." *Engineer Manual EM 1110-2-1902*, Washington, DC.
USACE. (1986). "Seepage analysis and control for dams." *Engineering Manual EM 1110-2-1901*, Washington, DC.
U.S. Bureau of Reclamation (USBR). (1973). *Design of small dams*, 2nd Ed., U.S. Government Printing Office, Washington, DC.
USBR. (1974). *Earth manual*, 2nd Ed., U.S. Dept. of Interior, Denver, CO.
U.S. Navy. (1971). "Design manual, soil mechanics, foundations, and earth structures." *NAVFAC DM-7*, Naval Facilities Engineering Command, Washington, DC.
Varnes, D. J. (1978). "Slope movement types and processes." Fig. 2.2, p. 11, *Landslides: Analysis and control, Special Rep. 176R*, L. Schuster and R. J. Krizek, eds., Transportation Research Board, Washington, DC.
Whitman, R. V., and Moore, P. J. (1963). "Thoughts concerning the mechanics of slope stability analysis." *Proc., 2nd Pan American Conf. on Soil Mechanics and Foundation Engineering*, Associacao Brasileira de Mecanica dos Solos, Sao Paulo, Brazil, 1, 391–411.
Wong, K. S., and Duncan, J. M. (1974). "Hyperbolic stress-strain parameters for nonlinear finite element analysis of stresses and movements in soil masses." *Geotech. Eng. Rep. TE 74-3*, Dept. of Civil Engineering, Univ. of California, Berkeley, CA.
Wu, T., Tang, W., Sangrey, D., and Baecher, G. (1989). "Reliability of offshore foundation: State of the art." *J. Geotech. Eng. Div. ASCE*, 115(2), 157–178.
Zaruba, Q., and Mencl, V. (1969). *Landslides and their control*, Elsevier, New York.

Index

Page numbers followed by *e*, *f*, and *t* indicate equations, figures, and tables, respectively.

M

R